T0184606

Lecture Notes in Mathematics 1706

Springer

Berlin
Heidelberg
New York
Barcelona
Hong Kong
London
Milan
Paris
Singapore
Tokyo

Sergei Yu. Pilyugin

Shadowing
in Dynamical Systems

 Springer

Author

Sergei Yu. Pilyugin
Faculty of Mathematics and Mechanics
St. Petersburg State University
Bibliotechnaya pl., 2, Petrodvorets
198904 St. Petersburg, Russia

E-mail: sp@spil.usr.pu.ru

Cataloging-in-Publication Data applied for

Die Deutsche Bibliothek - CIP-Einheitsaufnahme

Piljugin, Sergej Ju.:
Shadowing in dynamical systems / Sergei Yu. Pilyugin. - Berlin ;
Heidelberg ; New York ; Barcelona ; Hong Kong ; London ; Milan ;
Paris ; Singapore ; Tokyo : Springer, 1999
   (Lecture notes in mathematics ; 1706)
   ISBN 3-540-66299-5

Mathematics Subject Classification (1991): 58Fxx, 34Cxx, 65Lxx, 65Mxx

ISSN 0075-8434
ISBN 3-540-66299-5 Springer-Verlag Berlin Heidelberg New York

Typesetting: Camera-ready T$_E$X output by the author
SPIN: 10650213   41/3143-543210 - Printed on acid-free paper

To my sons Sergei and Kirill

Preface

Let (X, r) be a metric space and let ϕ be a homeomorphism mapping X onto itself.

A d-pseudotrajectory of the dynamical system ϕ is a sequence of points $\xi = \{x_k \in X : k \in \mathbb{Z}\}$ or $\xi = \{x_k \in X : k \in \mathbb{Z}_+\}$ such that

$$r(\phi(x_k), x_{k+1}) < d.$$

Usually, a pseudotrajectory is considered as a result of application of a numerical method to our dynamical system ϕ. In this case, the value d measures one-step errors of the method and round-off errors.

The notion of a pseudotrajectory plays an important role in the general qualitative theory of dynamical systems. It is used to define some types of invariant sets (such as the chain-recurrent set [Con] or chain prolongations [Pi2]).

We say that a point x (ϵ, ϕ)-shadows a pseudotrajectory $\xi = \{x_k\}$ if the inequalities

$$r(\phi^k(x), x_k) < \epsilon$$

hold. Thus, the existence of a shadowing point for a pseudotrajectory ξ means that ξ is close to a real trajectory of ϕ.

The mostly studied shadowing property of dynamical systems is the POTP (the pseudoorbit tracing property). A system ϕ is said to have the POTP if given $\epsilon > 0$ there exists $d > 0$ such that for any d-pseudotrajectory ξ there is a point x that (ϵ, ϕ)-shadows ξ.

From the numerical point of view, if ϕ has the POTP, then numerically obtained trajectories (on arbitrarily long time intervals) reflect the real behavior of trajectories of ϕ.

If ψ is a dynamical system C^0-close to ϕ, then obviously any trajectory of ψ is a d-pseudotrajectory of ϕ with small d. Thus, if ϕ has the POTP, then any trajectory of the "perturbed" system ψ is close to a trajectory of ϕ. Hence, we may consider the POTP as a weak form of stability of ϕ with respect to C^0-small perturbations.

Theory of shadowing was developed intensively in recent years and became a significant part of the qualitative theory of dynamical systems containing a lot of interesting and deep results. This book is an introduction to the main methods of shadowing.

The book is addressed to the following three main groups of readers.

The main expected group of readers are specialists in the qualitative theory of dynamical systems and its applications. For them, the author tried to describe a unified approach based on shadowing results for sequences of mappings of Banach spaces. It is shown that this approach can be applied to establish the classical shadowing property and limit shadowing properties in a neighborhood of a hyperbolic set, shadowing properties of structurally stable dynamical systems (both diffeomorphisms and flows), and some other classes of shadowing properties. In addition, we present a systematic treatment of connections between the shadowing theory and the classical fields of the global qualitative theory of dynamical systems (such as the theories of topological stability and of structural stability).

Next, some parts of the book (Sects. 1.1, 1.2.1, 1.2.2, 1.2.3, 1.3.1, 1.3.2, 3.1, 3.2, and 4.1) can be included into courses or used for the first acquaintance with the theory of shadowing by advanced students with basic training in dynamical systems. For this purpose, main definitions and results are illustrated by a lot of simple (maybe, too simple for specialists) examples.

Proofs of basic results of the theory and description of some important general constructions contained in the sections mentioned above are given with all details and with necessary background from functional analysis.

Finally, the book is addressed to specialists in numerical methods for dynamical systems. Some recent conferences (for example, the Conference on Dynamical Numerical Analysis, Georgia Tech, December 1995) showed that the idea of shadowing plays now an important role in this field and that "numerical dynamics" specialists need a detailed survey of results and methods of the shadowing theory.

It was an intention of the author to describe two "numerically oriented" shadowing approaches. The first one is based on methods for verification of numerically obtained data. These methods allow to establish the existence of a real trajectory near a computed one and to give the corresponding error bounds (see [Cho2, Cho3, Coo1-Coo6, Gr, Ham, Sau2] and others). The second approach establishes shadowing properties of dynamical systems generated by numerical methods (for example, discretizations of a parabolic PDE are realized as finite-dimensional diffeomorphisms [Ei1], see Sect. 4.4). These results allow us to study the influence of errors in application of numerical methods on unbounded time intervals.

The book consists of 4 chapters. Chapter 1 is devoted to "local shadowing", i.e., shadowing in a neighborhood of an invariant set. We introduce the main shadowing properties in Sect. 1.1 and discuss relations between these properties.

Section 1.2 is devoted to the classical shadowing result – the Shadowing Lemma by Anosov [Ano2] and Bowen [Bo2]. This result states that a diffeomorphism has the POTP in a neighborhood of its hyperbolic set. It is shown that this shadowing property is Lipschitz, i.e., if Λ is a hyperbolic set of a diffeomorphism ϕ, then there exist constants $d_0, L > 0$ and a neighborhood U of Λ such that for any sequence $\{x_k\} \subset U$ with

$$r(\phi(x_k), x_{k+1}) \le d \le d_0 \tag{0.1}$$

there exists a point x with the property

$$r(\phi^k(x), x_k) \le Ld. \tag{0.2}$$

In Subsect. 1.2.1, we describe main properties of hyperbolic sets. It is shown that there exists a Lyapunov metric in a neighborhood of a hyperbolic set. In Subsect. 1.2.2, we give a detailed proof of the Shadowing Lemma applying the approach of Anosov [Ano2].

Let Λ be a hyperbolic set of a diffeomorphism $\phi : \mathbb{R}^n \to \mathbb{R}^n$. Assume that, for a sequence $\{x_k\}$ belonging to a small neighborhood of Λ, inequalities (0.1) hold. Following Anosov, we study a functional equation to find a sequence $\{v_k \in \mathbb{R}^n\}$ such that

$$\phi(x_k + v_k) = x_{k+1} + v_{k+1} \qquad (0.3)$$

(i.e., the sequence $\{x_k + v_k\}$ is a trajectory of ϕ), and the point $x = x_0 + v_0$ satisfies inequalities (0.2).

Subsection 1.2.3 is devoted to the so-called "theorem on a family of ϵ-trajectories" [Ano2] and some of its applications. This statement says that if Λ is a hyperbolic set for a diffemorphism ϕ, and if a neighborhood of Λ contains a family of approximate trajectories of a diffeomorphism ψ, C^1-close to ϕ, then it is possible to shadow all this family by trajectories of ψ.

In Subsect. 1.2.4, Bowen's proof of the Shadowing Lemma is given. We also describe an abstract construction modelling the method of Bowen (the so-called Smale space) introduced by Ruelle [Ru].

In Sect. 1.3, we introduce the main technical tool applied in this book – "abstract" shadowing results for sequences of mappings of Banach spaces [Pi4]. In the case of a hyperbolic set of a diffeomorphism ϕ, we can introduce "local" mappings

$$\phi_k(v) = \phi(x_k + v) - x_{k+1},$$

then Eqs. (0.3) have the form

$$\phi_k(v_k) = v_{k+1}. \qquad (0.4)$$

Generalizing this approach, we consider sequences $\{H_k\}$ of Banach spaces and mappings $\phi_k : H_k \to H_k$, and look for sequences $\{v_k \in H_k\}$ satisfying (0.4). Obviously, inequalities (0.1) are equivalent to the estimates

$$|\phi_k(0)| \le d \le d_0, \qquad (0.5)$$

and inequalities (0.2) are equivalent to the estimates

$$|v_k| \le Ld. \qquad (0.6)$$

In Subsect. 1.3.1, we study sequences $\{\phi_k\}$ with (possibly) nonivertible linear parts. Theorem 1.3.1 (Subsect. 1.3.1) gives sufficient conditions under which there exist constants d_0 and L such that inequalities (0.5) imply the existence of a sequence $\{v_k\}$ satisfying (0.6). Such a sequence is considered as a local shadowing trajectory for the sequence $\{\phi_k\}$.

In Subsect. 1.3.2, we obtain conditions for the uniqueness of a shadowing trajectory in this case. Subsection 1.3.3 contains a general scheme of application of Theorem 1.3.1. This scheme is applied to establish an analog of the Shadowing Lemma for one-sided sequences. Subsection 1.3.4 is devoted to two theorems on shadowing for sequences of mappings of Banach spaces proved by Chow, Lin, and Palmer [Cho1] and by Steinlein and Walther [Ste1].

Pliss [Pli2, Pli3] obtained necessary and sufficient conditions under which a family of systems of linear differential equations has uniformly bounded solutions. We "translate" these results into the shadowing language in Subsect. 1.3.5 and apply them to "affine" mappings

$$\phi_k(v) = A_k v + w_{k+1}, \ v \in \mathbb{R}^n.$$

It is shown that the developed method can be also applied to nonlinear mappings ϕ_k (Theorem 1.3.7).

In the statement of the shadowing problem above, the values

$$d_k = r(x_{k+1}, \phi(x_k))$$

are assumed to be uniformly small. One can impose another condition on these values,

$$d_k \to 0 \text{ as } k \to \infty,$$

and look for a point x such that the values

$$h_k = r(\phi^k(x), x_k) \to 0 \text{ as } k \to \infty.$$

The corresponding shadowing property (called the limit shadowing property [Ei2]) is studied in Sect. 1.4. It is shown in Subsect. 1.4.1 that in a neighborhood of its hyperbolic set a diffeomorphism has this property.

In Subsect. 1.4.2, we investigate the rate of convergence of the values h_k in terms of d_k. It is shown that if a sequence $\{x_k : k \geq 0\}$ belongs to a neighborhood U of a hyperbolic set of a diffeomorphism ϕ, and, for some $p \geq 1$, the $\mathcal{L}_p$ norm

$$\|\{d_k\}\|_p = \left(\sum_{k \geq 0} d_k^p \right)^{1/p}$$

is small, then there is a point x such that

$$\|\{h_k\}\|_p \leq L \|\{d_k\}\|_p$$

with a constant L depending only on the neighborhood U.

In Subsect. 1.4.3, we pass from the spaces $\mathcal{L}_p$ to their weighted analogs, the spaces $\mathcal{L}_{\bar{r},p}$ with norms

$$\|\{d_k\}\|_{\bar{r},p} = \left(\sum_{k \geq 0} r_k d_k^p \right)^{1/p}, \ \bar{r} = \{r_k\}.$$

We show that it is possible to establish the corresponding "$\mathcal{L}_{\bar{r},p}$-shadowing property" in a neighborhood of a compact invariant set, not necessarily hyperbolic, under appropriate conditions on the weight sequence $\bar{r}$. These conditions are formulated in terms of the so-called Sacker-Sell spectrum [Sac].

Another possibility to establish the "$\mathcal{L}_{\bar{r},p}$"-shadowing is to assume that the weight sequence $\bar{r}$ grows "fast enough" (Theorems 1.4.6 and 1.4.8).

Hirsch studied in [Hirs4] asymptotic pseudotrajectories, i.e., sequences $\{x_k\}$ such that

$$\overline{\lim}_{k\to\infty} d_k^{1/k} \leq \lambda, \text{ where } \lambda \in (0,1).$$

The main result of Hirsch [Hirs4] on shadowing of asymptotic pseudotrajectories is described in Subsect. 1.4.4.

Section 1.5 is devoted to shadowing in flows generated by autonomous systems of ordinary differential equations. Technically, the shadowing problem for flows is significantly more complicated than the one for discrete dynamical systems.

We prove in this section that if Λ is a hyperbolic set of a flow containing no rest points, then in a neighborhood of Λ pseudotrajectories are shadowed by real trajectories, and the shadowing is Lipschitz with respect to the "errors".

In Chap. 2, we study connections between shadowing properties and the classical "global" properties of dynamical systems, such as topological stability and structural stability. We consider dynamical systems on a closed smooth manifold.

Walters [Wa2] and Morimoto [Morim1] showed that a topologically stable homeomorphism has the POTP. We prove this statement in Sect. 2.1. It is also shown in this section that an expansive homeomorphism having the POTP is topologically stable.

Section 2.2, the main part of Chap. 2, is devoted to shadowing properties of structurally stable dynamical systems. In Subsect. 2.2.1, we prove that a structurally stable flow has a Lipschitz shadowing property [Pi4]. We begin to work with a flow since this case is technically more difficult than the case of a diffeomorphism (mostly due to the possible coexistence of rest points and of nonwandering trajectories that are not rest points). The main statement (Theorem 2.2.3) is reduced to shadowing results for sequences of mappings of Banach spaces with noninvertible "linear parts" (see Sect. 1.3). It was an intention of the author to make the presentation of Theorem 2.2.3 maximally "self-contained". Due to this reason, we give a detailed proof of the existence of Robinson's "compatible extensions of stable and unstable bundles" [Robi1] (see Lemma 2.2.9).

Shadowing for structurally stable diffeomorphisms is studied in Subsect. 2.2.2. It is shown that a structurally stable diffeomorphism has the Lipschitz shadowing property. Sakai noted that the POTP is "uniform" in a C^1-neighborhood of a structurally stable diffeomorphism [Sak1] and that the C^1 interior of the set of diffeomorphisms with the POTP consists of structurally stable diffeomorphisms [Sak2]. We prove that the Lipschitz shadowing property is also uniform [Beg] and that a diffeomorphism in the C^1 interior of diffeomorphisms with this property is structurally stable.

Without additional assumptions on the dynamical system, no necessary conditions for the POTP are known. In Sect. 2.3, we study necessary and sufficient conditions of shadowing for two-dimensional diffeomorphisms satisfying Axiom A. We show that in this case the POTP is equivalent to the so-called C^0

transversality condition [Sak3], and the Lipschitz shadowing property is equivalent to the strong transversality condition (and hence to structural stability).

In the same section, we describe results of Plamenevskaya [Pla2] on weak shadowing for two-dimensional Axiom A diffeomorphisms. A diffeomorphism ϕ is said to have the weak shadowing property if given $\epsilon > 0$ there exists $d > 0$ such that for any d-pseudotrajectory ξ there is a trajectory $O(x)$ of ϕ with the property

$$\xi \subset N_\epsilon(O(x))$$

(here $N_\epsilon(O(x))$ is the ϵ-neighborhood of $O(x)$). An example of an Axiom A diffeomorphism of the two-torus T^2 with finite nonwandering set shows that necessary and sufficient conditions for weak shadowing have complicated structure, they are connected with arithmetical properties of eigenvalues of the derivative $D\phi$ at saddle fixed points.

Section 2.4 is devoted to the long-standing problem of genericity of the shadowing property when the dimension of the manifold is arbitrary. We formulate (without a proof) a theorem from [Pi5] stating that a C^0-generic homeomorphism has the POTP.

In Chap. 3, we study the shadowing problem for some special classes of dynamical systems. Section 3.1 is devoted to one-dimensional systems. We prove two theorems of Plamenevskaya [Pla1]. The first one gives necessary and sufficient conditions under which a homeomorphism of the circle has the POTP. In the second theorem, sufficient conditions for the limit shadowing property are obtained. As a corollary, it is shown that if a homeomorphism of the circle has the POTP, then it has the limit shadowing property.

In Sect. 3.2, we consider linear and linearly induced systems. Following Morimoto [Morim3], we show that a linear mapping $\phi(x) = Ax$ has the POTP if and only if the matrix A is hyperbolic. Conditions of the POTP are known for a wide class of linearly induced systems, we treat in detail the so-called spherical linear transformations of the unit sphere $S^n \subset \mathbb{R}^{n+1}$ defined by the formula

$$\phi(x) = \frac{Ax}{|Ax|}.$$

We prove the following theorem of Sasaki [Sas]: ϕ has the POTP if and only if the eigenvalues of the matrix A have different absolute values.

The second part of Chap. 3 is devoted to two special classes of infinite-dimensional dynamical systems. In both cases, the shadowing problem is reduced to the same problem for auxiliary finite-dimensional systems.

The first class, lattice systems, is studied in Sect. 3.3. Usually, the following three types of solutions for lattice systems are investigated: steady-state solutions, travelling waves, and spatially-homogeneous solutions. Under appropriate conditions, these solutions are governed by finite-dimensional diffeomorphisms. We describe conditions [Af3] under which an approximately static (approximately travelling or approximately spatially-homogeneous) pseudotrajectory is shadowed by a steady-state solution (correspondingly, by a travelling wave or by a spatially-homogeneous solution).

Section 3.4 is devoted to shadowing in nonlinear evolution systems on Hilbert spaces. It is assumed that the evolution system $\mathcal{S}$ generated by a parabolic PDE

$$u_t = u_{xx} + f(u) \tag{0.7}$$

has Morse-Smale structure on its global attractor $\mathcal{A}$. We show that $\mathcal{S}$ has a type of Lipschitz shadowing property in a neighborhood of the global attractor $\mathcal{A}$ [Lar4].

In the last chapter, we describe some applications of the shadowing theory to numerical investigation of dynamical systems. Section 4.1 is devoted to methods of verification of numerically obtained data. We prove two theorems of Chow and Palmer [Cho1, Cho2] on finite shadowing in one-dimensional and multidimensional systems.

Coomes, Koçak, and Palmer [Coo1-Coo6] developed a theory of "practical" shadowing for ordinary differential equations. Their methods allow to establish the existence of a real trajectory near a computed one. Section 4.2 is devoted to one of their methods, the method of periodic shadowing [Coo2]. This method gives computable error bounds for the distance between a computed closed trajectory and a real one.

In Sect. 4.3, we consider connections between pseudotrajectories of a dynamical system and its "spectral" characteristics such as Lyapunov exponents and the Morse spectrum. In Subsect. 4.3.1, we study the problem of approximate evaluation of Lyapunov exponents. It is shown that in the evaluation of the upper Lyapunov exponent on a hyperbolic set, the resulting errors are Lipschitz with respect to the errors of the method and to round-off errors [Cor1].

We show in Subsect. 4.3.2 that symbolic images of a dynamical system generated by partitions of its phase space [Os1] can be applied to approximate its Morse spectrum [Os2].

In Sect. 4.4, we investigate qualitative properties of semi-implicit discretizations of (0.7). In Subsect. 4.4.1, we study finite-dimensional diffeomorphisms generated by discretizations and the global attractors for these diffeomorphisms [Ei1]. It is shown that, for a generic nonlinearity $f(u)$, these global attractors have Morse-Smale structure, hence we can apply the theory of shadowing for structurally stable systems (Chap. 2).

In Subsect. 4.4.2, we apply shadowing results obtained in Sect. 3.4 to give explicit estimates (in terms of time and space steps) for the differences between approximate and exact solutions on unbounded time intervals [Lar4].

Cooperation with colleagues was very important during the preparation of this book. Special thanks are to V. Afraimovich, R. Corless, T. Eirola, S. Larsson, O. Nevanlinna, G. Osipenko, V. Pliss, O. Plamenevskaya, and G. Sell.

The author is grateful to S.-N. Chow, M. Hurley, A. Katok, H. Koçak, K. Odani, C. Robinson, and K. Sakai for the attention they paid to the manuscript of this book during its preparation.

The author's research was partially supported by INTAS grant 96–1158.

The manuscript of the book was completed during the author's visit to the University of Alberta, Edmonton, Canada, supported by the Distinguished

Visitors Fund of the University of Alberta and by the Natural Science and Engineering Research Council of Canada. The author expresses his deep gratitude to Applied Mathematics Institute and Department of Mathematical Sciences of the University of Alberta and especially to Professor H. Freedman.

Contents

Chapter 1. Shadowing Near an Invariant Set 1
  1.1 Basic Definitions 1
  1.2 Shadowing Near a Hyperbolic Set for a Diffeomorphism 9
    1.2.1 Hyperbolic Sets 10
    1.2.2 The Classical Shadowing Lemma 14
    1.2.3 Shadowing for a Family of Approximate Trajectories 22
    1.2.4 The Method of Bowen 27
  1.3 Shadowing for Mappings of Banach Spaces 34
    1.3.1 Shadowing for a Sequence of Mappings 35
    1.3.2 Conditions of Uniqueness 40
    1.3.3 Application to the Classical Shadowing Lemma 43
    1.3.4 Theorems of Chow-Lin-Palmer and Steinlein-Walther 45
    1.3.5 Finite-Dimensional Case 53
  1.4 Limit Shadowing 63
    1.4.1 Limit Shadowing Property 64
    1.4.2 $\mathcal{L}_p$-Shadowing 68
    1.4.3 The Sacker-Sell Spectrum and Weighted Shadowing 71
    1.4.4 Asymptotic Pseudotrajectories 84
  1.5 Shadowing for Flows 88

Chapter 2. Topologically Stable, Structurally Stable, and Generic Systems 103
  2.1 Shadowing and Topological Stability 103
  2.2 Shadowing in Structurally Stable Systems 109
    2.2.1 The Case of a Flow 109
    2.2.2 The Case of a Diffeomorphism 145
  2.3 Shadowing in Two-Dimensional Diffeomorphisms 158
  2.4 C^0-Genericity of Shadowing for Homeomorphisms 172

Chapter 3. Systems with Special Structure 173
  3.1 One-Dimensional Systems 173
  3.2 Linear and Linearly Induced Systems 182
  3.3 Lattice Systems 189
  3.4 Global Attractors for Evolution Systems 202

Chapter 4. Numerical Applications of Shadowing 219
  4.1 Finite Shadowing 219
  4.2 Periodic Shadowing for Flows 223
  4.3 Approximation of Spectral Characteristics 233
    4.3.1 Evaluation of Upper Lyapunov Exponents 233

4.3.2 Approximation of the Morse Spectrum 238
4.4 Discretizations of PDEs 244
4.4.1 Shadowing in Discretizations 245
4.4.2 Discretization Errors on Unbounded Time Intervals 255

References 259

Index 269

List of Main Symbols

$\mathbb{R}$ – the set of real numbers;
$\mathbb{R}^n$ – the Euclidean n-space;
$\mathbb{Z}$ – the set of integer numbers;
$\mathbb{Z}_+ = \{k \in \mathbb{Z} : k \geq 0\}$;
$\mathbb{N}$ – the set of natural numbers;
$\mathbb{C}$ – the set of complex numbers;
$GL(n, \mathbb{R})$ $(GL(n, \mathbb{C}))$ – the group of invertible linear transformations of $\mathbb{R}^n$ (respectively, of $\mathbb{C}^n$);

I is the identity operator (or the unit matrix);

For a set A in a topological space, $\overline{A}$ is the closure of A, $\mathrm{Int}\,A$ is the interior of A, and ∂A is the boundary of A;

For sets A, B in a metric space (X, r), $N_a(A)$ is the a-heighborhood of A,

$$\mathrm{diam}\,A = \sup_{x,y \in A} r(x, y)$$

and

$$\mathrm{dist}(A, B) = \inf_{x \in A, y \in B} r(x, y);$$

For a linear mapping A,

$$\|A\| = \sup_{|v|=1} |Av|$$

is the operator norm of A;

For a Banach space $\mathcal{B}$, $\mathcal{B}(r)$ is the closed ball of radius r centered at 0;

For a smooth manifold M, $T_x M$ is the tangent space of M at x and TM is the tangent bundle of M;

For a smooth mapping f, $Df(x)$ is the derivative of f at x;

$:=$ means "equal by definition".

1. Shadowing Near an Invariant Set

1.1 Basic Definitions

Let (X, r) be a metric space. A homeomorphism ϕ mapping X onto itself generates a dynamical system

$$\Phi : \mathbb{Z} \times X \to X$$

by the formula

$$\Phi(m, x) = \phi^m(x), \ m \in \mathbb{Z}, \ x \in X.$$

The trajectory $O(x)$ of a point $x \in X$ in the dynamical system Φ is the set

$$O(x) = \{\Phi(m, x) : m \in \mathbb{Z}\}.$$

Usually, we identify the homeomorphism ϕ with the dynamical system Φ it generates (and call ϕ a dynamical system).

A continuous mapping $f : X \to X$ generates a semi-dynamical system by the formula

$$\Psi(m, x) = f^m(x), \ m \in \mathbb{Z}_+, \ x \in X.$$

The trajectory $O(x)$ of a point $x \in X$ in the semi-dynamical system Ψ is the set

$$O^+(x) = \{\Psi(m, x) : m \in \mathbb{Z}_+\}.$$

We also identify f and Ψ.

The main objects of investigation in this book are dynamical systems and their approximate trajectories (pseudotrajectories) defined below.

Let either $K = \mathbb{Z}$ or $K = \mathbb{Z}_+$. Fix $d > 0$.

Definition 1.1 *We say that a sequence $\xi = \{x_k \in X : k \in K\}$ is a "d-pseudotrajectory" (or a "d-pseudoorbit") of a dynamical system ϕ on K if the inequalities*

$$r(\phi(x_k), x_{k+1}) < d, \ k \in K,$$

hold.

Fix $\epsilon > 0$.

Definition 1.2 *We say that a point $x \in X$ "(ϵ, ϕ)-shadows" (or "(ϵ, ϕ)-traces") a d-pseudotrajectory $\xi = \{x_k\}$ on K if the inequalities*

$$r(\phi^k(x), x_k) < \epsilon, \ k \in K, \tag{1.1}$$

hold.

Below, if only one dynamical system ϕ is considered, we will usually write simply "x (ϵ)-shadows ξ". This will lead to no confusion.

In this book, we mostly study the shadowing properties of dynamical systems defined below.

Let Y be a subset of X.

Definition 1.3 *We say that the dynamical system ϕ has the "POTP" (the "pseudoorbit tracing property") on Y if given $\epsilon > 0$ there exists $d > 0$ such that for any d-pseudotrajectory ξ on $\mathbb{Z}$ with $\xi \subset Y$ there is a point x that (ϵ, ϕ)-shadows ξ on $\mathbb{Z}$. If this property holds with $Y = X$, we say that ϕ has the POTP.*

Remark. [1] Our usual term in this book is "shadowing" (not "tracing"), but we preserve the term "POTP", since it became standard.

Sometimes a d-pseudotrajectory of a dynamical system ϕ is considered as a result of a small random perturbation of ϕ. In this case, the POTP means that near trajectories of a "randomly perturbed" ϕ there are real trajectories of ϕ. Due to this reason, some authors use the term *stochastic stability* instead of the POTP [Morim1]. Note that shadowing properties of random dynamical systems were studied, for example, in [Bl1, Cho2]. We do not consider random dynamical systems in this book.

Remark. [2] An analogous property connected with pseudotrajectories on $\mathbb{Z}_+$ is called the POTP$_+$.

It is easy to show that there exist systems which do not have the POTP.

Example 1.4 Consider the circle S^1 with coordinate $x \in [0, 1)$ and a homeomorphism ϕ of S^1 generated by the mapping $f(x) \equiv x$. Fix $d > 0$ and take a sequence of points $\xi = \{x_k : k \in \mathbb{Z}\} \subset S^1$ such that $x_0 = 0$ and

$$x_{k+1} = x_k + \frac{d}{2}(\mathrm{mod}\ 1), \ k \in \mathbb{Z}.$$

Obviously, ξ is a d-pseudotrajectory for ϕ. Any trajectory of ϕ is a fixed point. It is easy to see that, for $d < 2/3$ and for any trajectory p of ϕ, the set ξ is not contained in the $(1/3)$-neighborhood of p. This means that for $\epsilon = 1/3$ there exists no d with the property described in Definition 1.3.

Let us describe some simple relations between the introduced notions.

Lemma 1.1.1.

   (a) *Assume that X is compact and that ϕ has the following "finite shadowing property" on $Y \subset X$: given $\epsilon > 0$ there exists $d > 0$ such that if for a set $\{x_0, \ldots, x_m\} \subset Y$ the inequalities $r(x_{k+1}, \phi(x_k)) < d$ hold for $0 \le k \le m - 1$, then there is a point $x \in X$ with $r(\phi^k(x), x_k) < \epsilon$ for $0 \le k \le m - 1$. Then ϕ has the POTP on Y.*

   (b) *Assume that a set $Y \subset X$ is negatively invariant, i.e., $\phi^{-k}(x) \in Y$ for $x \in Y$ and $k \ge 0$. If ϕ has the POTP on Y, then ϕ has the POTP$_+$ on Y.*

Proof. (a) Let $\epsilon > 0$ be given. Find a corresponding $d > 0$ given by the "finite shadowing property". Let $\xi = \{x_k : k \in \mathbb{Z}\} \subset Y$ be a d-pseudotrajectory for ϕ. Fix $m > 0$ and set $x'_k = x_{k-m}$. By our assumption, there is a point $y_m \in X$ such that

$$r(\phi^k(y_m), x'_k) < \epsilon, \ 0 \le k \le 2m.$$

Set $w_m = \phi^m(y_m)$. Then

$$r(\phi^k(w_m), x_k) < \epsilon, \ |k| \le m.$$

Let w be a limit point of the sequence w_m (for simplicity we assume that w_m converges to w). Passing to the limit as $m \to \infty$ in the last inequality, we see that

$$r(\phi^k(w), x_k) \le \epsilon, \ k \in \mathbb{Z},$$

hence ξ is (2ϵ)-shadowed by w.

   (b) Let $\epsilon > 0$ be fixed. Find a corresponding d given by Definition 1.3. Let $\xi = \{x_k : k \ge 0\} \subset Y$ be a d-pseudotrajectory of ϕ on $\mathbb{Z}_+$. Consider the sequence $\xi' = \{x'_k : k \in \mathbb{Z}\}$ such that $x'_k = x_k$ for $k \ge 0$ and $x'_k = \phi^k(x_0)$ for $k < 0$. Obviously, ξ' is a d-pseudotrajectory of ϕ belonging to Y. By the choice of d, there is a point x that (ϵ)-shadows ξ' on $\mathbb{Z}$. It follows immediately that x (ϵ)-shadows ξ on $\mathbb{Z}_+$. $\qquad \square$

   This lemma shows that if X is compact, then ϕ has the POTP if and only if ϕ has the POTP$_+$.

Definition 1.5 *We say that the dynamical system ϕ has the "LpSP" (the "Lipschitz shadowing property") on Y if there exist positive constants L, d_0 such that for any sequence $\{x_k \in Y : k \in \mathbb{Z}\}$ with*

$$r(\phi(x_k), x_{k+1}) \le d \le d_0, \ k \in \mathbb{Z},$$

there is a point x such that the inequalities

$$r(\phi^k(x), x_k) \le Ld, \ k \in \mathbb{Z}, \tag{1.2}$$

hold. If this property holds with $Y = X$, we say that ϕ has the LpSP.

Remark. An analogous property connected with pseudotrajectories on $\mathbb{Z}_+$ is called the LpSP$_+$.

Of course, a statement analogous to Lemma 1.1.1 is true for the LpSP and LpSP$_+$. Let us give a simple example of a dynamical system that has the POTP on some set but does not have the LpSP on this set.

Example 1.6 Consider a diffeomorphism $\phi : \mathbb{R} \to \mathbb{R}$ such that $\phi(x) = x + x^2 \text{sgn}(x)$, where $\text{sgn}(x)$ is the sign of x. Set

$$W = \{x \in \mathbb{R} : |x| \le 1\}.$$

Let us show that ϕ has the POTP on W. Fix arbitrary $\epsilon > 0$. Denote $b = \epsilon^2/2$.

Take a d-pseudotrajectory $\xi = \{x_k : k \in \mathbb{Z}\} \subset W$ with $d \le b$. We claim that

$$|x_k| < \epsilon \text{ for } k \in \mathbb{Z}. \tag{1.3}$$

To obtain a contradiction, assume that there exists $|x_k| \ge \epsilon$. If $x_k \ge \epsilon$, we obtain the inequality

$$x_{k+1} > \phi(x_k) - d \ge x_k + b,$$

and similar inequalities

$$x_{k+m} > x_k + mb \text{ for } m \ge 0$$

which imply that the sequence x_m leaves W as m grows. The case $x_k \le -\epsilon$ is considered similarly.

It follows from (1.3) that any d-pseudotrajectory ξ is (ϵ)-shadowed by the point $x = 0$.

Let us show that ϕ does not have the LpSP in W. Take a small $d > 0$ and consider $x_k = \sqrt{d}, k \in \mathbb{Z}$. Then we have

$$|\phi(x_k) - x_{k+1}| = x_k^2 = d.$$

Assume that ϕ has the LpSP in W with constants L, d_0. For small d, the inequalities

$$|\phi^k(x) - x_k| \le Ld, \ k \in \mathbb{Z},$$

are possible only if $x = 0$, since the trajectories of points $x \ne 0$ leave W. Hence, our assumption leads to the inequality $\sqrt{d} \le Ld$ which is contradictory for small d.

Of course, the main reason for the absence of the LpSP in a neighborhood of the fixed point $x = 0$ in our example is the equality $D\phi(0) = 1$. This equality means that $x = 0$ is a nonhyperbolic fixed point.

One of the main goals of this book is to show that "hyperbolicity implies the LpSP" for a wide class of dynamical systems (we hope that the exact meaning of this statement will be clear to the reader after reading the book).

Other shadowing properties are also considered. We define some of them below. In addition, let us mention the rotational shadowing property [Bar], the asymptotically shadowing property [Che], the bi-shadowing property [AlN]. We refer the reader to the original publications for the details.

Now let us pay some attention to the problem of uniqueness of the shadowing trajectory.

We give the following natural definition. Let again ϕ be a dynamical system on a metric space (X, r).

Definition 1.7 *We say that ϕ has the "SUP" (the "shadowing uniqueness property") on a set $Y \subset X$ if there exists a constant $\epsilon > 0$ such that any d-pseudotrajectory $\xi = \{x_k : k \in \mathbb{Z}\} \subset Y$ is (ϵ)-shadowed by not more than one point x. If $Y = X$, we say that ϕ has the SUP.*

The following statement (Lemma 1.1.2) shows that this property is almost equivalent to the well-known expansivity property. First let us define this property.

Definition 1.8 *The system ϕ is called "expansive" on a set $Y \subset X$ if there exists $\Delta > 0$ such that if for two points x, y the inclusions*

$$\phi^k(x), \phi^k(y) \in Y, \ k \in \mathbb{Z},$$

and the inequalities

$$r(\phi^k(x), \phi^k(y)) \leq \Delta, \ k \in \mathbb{Z},$$

hold, then $x = y$. If $Y = X$, the system ϕ is called expansive. We say that Δ is an "expansivity constant".

Lemma 1.1.2.

(a) Assume that ϕ has the SUP on a set $Y \subset X$ with constant ϵ. Then ϕ is expansive on Y, and any $\Delta \in (0, \epsilon)$ is an expansivity constant of ϕ on Y.

(b) Assume that for a set Y there exists a number $\Delta > 0$ and a set Y_1 such that ϕ is expansive on Y with expansivity constant Δ, and the Δ-neighborhood of Y_1 is a subset of Y. Then ϕ has the SUP on Y_1 with $\epsilon = \Delta/2$.

Proof. (a) Assume that for x, y we have

$$\phi^k(x), \phi^k(y) \in Y, \ k \in \mathbb{Z},$$

and

$$r(\phi^k(x), \phi^k(y)) \leq \Delta, \ k \in \mathbb{Z}.$$

Since $\xi = \{\phi^k(x)\}$ is a trajectory of ϕ, it is a d-pseudotrajectory for ϕ with any $d > 0$. Obviously, it is (ϵ)-shadowed by x. It follows from the relations above that $\xi \subset Y$ and that it is (ϵ)-shadowed by y. This implies that $x = y$.

(b) Now assume that for a d-pseudotrajectory $\xi = \{x_k : k \in \mathbb{Z}\} \subset Y_1$ there exist points x, y such that

$$r(\phi^k(x), x_k) < \epsilon \text{ and } r(\phi^k(y), x_k) < \epsilon, \ k \in \mathbb{Z}.$$

It follows that $\{\phi^k(x)\}, \{\phi^k(y)\} \subset Y$, and

$$r(\phi^k(x), \phi^k(y)) < 2\epsilon = \Delta, \ k \in \mathbb{Z}.$$

This implies that $x = y$.   □

Remark. Thus, if a homeomorphism ϕ has the SUP, then ϕ is expansive. This means that the class of homeomorphisms having both the POTP and SUP coincides with the class of expansive homeomorphisms with the POTP. This last class was studied by many authors (see a detailed survey [Ao1] and the books [Ak, Ao2]). An expansive homeomorphism having the POTP is often called *topologically Anosov* [Ao2]. It was shown by Hiraide [Hira1, Hira2] that if a manifold X admits a topologically Anosov homeomorphism, then strong restrictions are imposed on the structure of X. For example, the only compact surface with the mentioned property is a 2-torus.

We devote the last part of this introductory section to the proof of a simple technical result.

Sometimes it is easier to establish the POTP (or the LpSP) not for the given dynamical system ϕ but for ϕ^ν with some $\nu \in \mathbb{N}$. It is easy to show that ϕ has the POTP (or it is Lipschitz and has the LpSP) if and only if ϕ^ν has the same property (with other constants). This approach was applied by Newhouse [Gu2] and others. We prove here a slightly more general statement (for sequences of mappings of Banach spaces and their finite superpositions) concerning Lipschitz shadowing.

Let $H_k, k \in \mathbb{Z}$, be a sequence of Banach spaces, we denote by $|.|$ norms in H_k. Consider a sequence of mappings

$$\psi_k : H_k \to H_{k+1}.$$

For a natural number ν and $k \in \mathbb{Z}$ we denote

$$\Psi_{k,\nu} = \psi_{k+\nu-1} \circ \ldots \circ \psi_k.$$

Take $z_m \in H_{m\nu}, m \in \mathbb{Z}$, and construct a sequence $x_k \in H_k$ so that $x_{m\nu} = z_m$ and

$$x_k = \psi_{k-1} \circ \ldots \circ \psi_{m\nu+1} \circ \psi_{m\nu}(z_m) \text{ for } k = m\nu + m_1, \ 0 \le m_1 \le \nu - 1. \quad (1.4)$$

Lemma 1.1.3. *Assume that there exist sets* $V_k \subset V_k' \subset H_k$ *and a number* $\mathcal{L} > 0$ *such that*

$$\psi_{k+i} \circ \psi_{k+i-1} \circ \ldots \circ \psi_k(V_k) \subset V_{k+i+1}'$$

for $k \in \mathbb{Z}, 1 \le i \le \nu - 1$, *and any* ψ_k *is Lipschitz on* V_k' *with Lipschitz constant* $\mathcal{L}$.

Assume, in addition, that, for sequences $y_k \in V_k, \xi_m = y_{m\nu} \in V_{m\nu}$, and $z_m \in V_{m\nu}$, the inequalities

$$|\psi_k(y_k) - y_{k+1}| \leq d, \ k \in \mathbb{Z}, \tag{1.5}$$

$$|z_m - \xi_m| \leq \epsilon, \ m \in \mathbb{Z},$$

hold with some $d, \epsilon > 0$. Then
  (1) for $m \in \mathbb{Z}$ we have

$$|\Psi_{m\nu,\nu}(\xi_m) - \xi_{m+1}| \leq L_1 d, \ m \in \mathbb{Z}, \tag{1.6}$$

where

$$L_1 = 1 + \mathcal{L} + \ldots + \ldots \mathcal{L}^{\nu-1};$$

(2) for the sequence $x_k, k \in \mathbb{Z}$, constructed according to formula (1.4), we have the inequalities

$$|x_k - y_k| \leq \mathcal{L}^{\nu-1}\epsilon + L_1 d, \ k \in \mathbb{Z}.$$

Proof. The proof is straightforward. Take $m \in \mathbb{Z}$. By our conditions, $\xi_m = y_{m\nu}$. It follows from (1.5) that

$$|y_{m\nu+1} - \psi_{m\nu}(\xi_m)| \leq d.$$

Since $\psi_{m\nu+1}$ is Lipschitz on $V'_{m\nu+1}$, we see that

$$|y_{m\nu+2} - \psi_{m\nu+1} \circ \psi_{m\nu}(\xi_m)| \leq$$

$$\leq |y_{m\nu+2} - \psi_{m\nu+1}(y_{m\nu+1})| + |\psi_{m\nu+1}(y_{m\nu+1}) - \psi_{m\nu+1} \circ \psi_{m\nu}(\xi_m)| \leq d + \mathcal{L}d.$$

Repeating this process, we obtain the estimate

$$|\Psi_{m\nu,\nu}(\xi_m) - \xi_{m+1}| = |\Psi_{m\nu,\nu}(y_{m\nu}) - \xi_{m+1}| \leq$$

$$\leq d(1 + \mathcal{L} + \ldots + \mathcal{L}^{\nu-1}) = L_1 d.$$

This proves (1.6).
  Now we take arbitrary $k \in \mathbb{Z}$ and represent it in the form

$$k = m\nu + m_1, \ 0 \leq m_1 \leq \nu - 1.$$

By the conditions of our lemma,

$$|x_{m\nu} - y_{m\nu}| = |z_m - \xi_m| \leq \epsilon.$$

Let us estimate

$$|x_{m\nu+1} - y_{m\nu+1}| \leq |\psi_{m\nu}(z_m) - \psi_{m\nu}(\xi_m)| +$$

$$+ |\psi_{m\nu}(y_{m\nu}) - y_{m\nu+1}| \leq \mathcal{L}\epsilon + d.$$

Similarly we estimate

$$|x_{m\nu+2} - y_{m\nu+2}| \leq |\psi_{m\nu+1} \circ \psi_{m\nu}(z_m) - \psi_{m\nu+1}(y_{m\nu+1})| +$$

$$+ |\psi_{m\nu+1}(y_{m\nu+1}) - y_{m\nu+2}| \leq \mathcal{L}(\mathcal{L}\epsilon + d) + d = \mathcal{L}^2\epsilon + (1 + \mathcal{L})d.$$

Continuing this process, we see that

$$|x_{m\nu+\nu-1} - y_{m\nu+\nu-1}| \leq \mathcal{L}^{\nu-1}\epsilon + (1 + \mathcal{L} + \ldots + \mathcal{L}^{\nu-2})d < \mathcal{L}^{\nu-1}\epsilon + L_1 d.$$

This completes the proof. $\square$

Let us apply this lemma to show that a diffeomorphism ϕ of $\mathbb{R}^n$ has the LpSP on a neighborhood of a compact invariant set Λ if and only if $\Phi = \phi^\nu$ with some $\nu \in \mathbb{N}$ has the LpSP on a neighborhood of Λ.

First we assume that ϕ has the LpSP with constants L, d_0 on a bounded neighborhood W of Λ. Let $\mathcal{L}$ be a Lipschitz constant for ϕ on W.

Find a neighborhood $W_0 \subset W$ of Λ such that

$$\phi^i(W_0) \subset W, \ 1 \leq i \leq \nu - 1.$$

Take a sequence $\{z_m : m \in \mathbb{Z}\} \subset W_0$ such that

$$|\Phi(z_m) - z_{m+1}| \leq d \leq d_0.$$

Construct the corresponding sequence $\{x_k : k \in \mathbb{Z}\}$ by formula (1.4) with $\psi_k \equiv \phi$. It follows from the choice of W_0 that $\{x_k\} \subset W$. Since $\phi(x_k) = x_{k+1}$ for $k \neq m\nu - 1, m \in \mathbb{Z}$, and

$$|\phi(x_{m\nu-1}) - x_{m\nu}| = |\Phi(z_{m-1}) - z_m| \leq d,$$

there exists a point x such that

$$|\phi^k(x) - x_k| \leq Ld, \ k \in \mathbb{Z},$$

then

$$|\Phi^m(x) - z_m| = |\phi^{m\nu}(x) - x_{m\nu}| \leq Ld, \ m \in \mathbb{Z}.$$

Now we assume that the diffeomorphism $\Phi = \phi^\nu$ has the LpSP on a bounded neighborhood W of Λ with constants L, d_0. Let again $\mathcal{L}$ be a Lipschitz constant for ϕ on W. Find a neighborhood W_0 of Λ with the same property as above.

Take a neighborhood W_1 of Λ and a positive number Δ such that W_2, the Δ-neighborhood of W_1, is a subset of W_0.

Now we set $H_k = \mathbb{R}^n, \psi_k = \phi, V_k = W_2, V_k' = W$ for $k \in \mathbb{Z}$. Take a sequence $\{y_k : k \in \mathbb{Z}\} \subset W_1$ such that

$$|\phi(y_k) - y_{k+1}| \leq d \leq d_1,$$

where

$$d_1 = \min\left(\frac{d_0}{L_1}, \frac{\Delta}{LL_1}\right). \tag{1.7}$$

It follows from the first statement of Lemma 1.1.3 that the sequence $\{\xi_m\}$, where $\xi_m = y_{m\nu}$, belongs to W and satisfies the inequalities

$$|\Phi(\xi_m) - \xi_{m+1}| \leq L_1 d.$$

Hence, there exists a point z_0 such that for $z_m = \Phi^m(z_0)$ we have

$$|z_m - \xi_m| \leq LL_1 d, \ m \in \mathbb{Z}.$$

By (1.7), the inequalities

$$|z_m - \xi_m| \leq \Delta$$

hold. Since $\xi_m \in W_1$, we see that $z_m \in W_2 \subset W_0$. Hence, we can apply the second statement of Lemma 1.1.3 to show that, for the sequence $\{x_k\}$ constructed according to (1.4), the inequalities

$$|x_k - y_k| \leq L_2 d, \ k \in \mathbb{Z},$$

hold, where $L_2 = \mathcal{L}^{\nu-1} LL_1 + L_1$. Note that by construction the sequence $\{x_k\}$ is a trajectory of ϕ. This shows that ϕ has the LpSP on W_1.

1.2 Shadowing Near a Hyperbolic Set for a Diffeomorphism

In this section, we describe two classical proofs of the so-called Shadowing Lemma, the main result about shadowing near a hyperbolic set of a diffeomorphism. These proofs were given by Anosov [Ano2] and Bowen [Bo2].

Different proofs of similar statements were given later by Conley [Con], Robinson [Robi3], Newhouse [Gu2], Ekeland [Ek], Lanford [Lanf], Bronshtein [Bro], Meyer and Sell [Me], Shub [Shu2], Palmer [Palm2], Blank [Bl1], Kruger and Trubetzkoy [Kr], Katok and Hasselblatt [Katok2], Reinfelds [Re], and other authors. Below we mention some methods to establish shadowing which are of interest for the practice.

The method of Anosov reduces the shadowing problem to a functional equation in a proper Banach space. We show in this book that the approach based on solving a functional equation is applicable to a very wide class of shadowing problems. That is why we describe its origin, the classical proof based on the ideas of Anosov, in Subsect. 1.2.2 with all the details (we hope that reading the proof of Theorem 1.2.3 will help the reader to understand its generalizations).

In [Ano2], Anosov established the Shadowing Lemma as a particular case of a more general statement, the so-called "theorem on a family of ϵ-trajectories". This statement and some of its applications are described in Subsect. 1.2.3.

The method of Bowen is "geometric", and it uses more detailed information about the behavior of the investigated system near its hyperbolic set. Subsection 1.2.4 is devoted to the method of Bowen and to an abstract construction modelling this method (the so-called Smale space) introduced by Ruelle [Ru].

To avoid unessential technical difficulties, we consider in this section diffeomorphisms of $\mathbb{R}^n$ (instead of the general case of a smooth manifold). The generalization to the case of a manifold (with application of exponential mappings) is described in detail in Chap. 2.

1.2.1 Hyperbolic Sets

This subsection is preliminary, we devote it to the study of some basic properties of hyperbolic sets.

Let ϕ be a diffeomorphism of class C^1 of $\mathbb{R}^n$. We denote by $|.|$ the usual norm of $\mathbb{R}^n$.

Definition 1.9 *We say that a set Λ is "hyperbolic" for a diffeomorphism ϕ if*
(a) Λ is compact and ϕ-invariant;
(b) there exist constants $C > 0, \lambda_0 \in (0,1)$, and families of linear subspaces $S(p), U(p)$ of $\mathbb{R}^n$, $p \in \Lambda$, such that
(b.1) $S(p) \oplus U(p) = \mathbb{R}^n$;
(b.2) $D\phi(p)T(p) = T(\phi(p))$, $p \in \Lambda$, $T = S, U$;
(b.3)

$$|D\phi^m(p)v| \leq C\lambda_0^m |v| \text{ for } v \in S(p), \ m \geq 0; \tag{1.8}$$

$$|D\phi^{-m}(p)v| \leq C\lambda_0^m |v| \text{ for } v \in U(p), \ m \geq 0. \tag{1.9}$$

Remark. It is easy to show (see [Pil], for example) that the families of subspaces $S(p), U(p)$ (usually called the *hyperbolic structure* on Λ) are continuous on Λ. Below we call the numbers C, λ_0 the *hyperbolicity constants* of Λ.

In many proofs below, we apply the so-called *Lyapunov* (or *adapted*) norm [Ano1] on a neighborhood of a hyperbolic set. With respect to this norm, estimates (1.8),(1.9) hold with $C = 1$ and $\lambda \in (\lambda_0, 1)$ instead of λ_0. Let us prove its existence.

Lemma 1.2.1. *Let Λ be a hyperbolic set of ϕ with hyperbolicity constants C, λ_0. Given $\epsilon > 0, \lambda \in (\lambda_0, 1)$ there exists a neighborghood $W = W(\epsilon, \lambda)$ of Λ with the following properties. There exist positive constants N', δ, a C^∞ norm $|.|_x$ for $x \in W$, and continuous (not necessarily $D\phi$-invariant) extensions S', U' of S, U to W such that*
(1) $S'(x) \oplus U'(x) = \mathbb{R}^n$, $x \in W$;
(2) for $x, y \in W$ with $|y - \phi(x)| < \delta$, the mapping $\Pi_y^s D\phi(x)$ $(\Pi_y^u D\phi(x))$ is an isomorphism between $S'(x)$ and $S'(y)$ (respectively, between $U'(x)$ and $U'(y)$), and the inequalities

$$|\Pi_y^s D\phi(x)v|_y \leq \lambda |v|_x, \ |\Pi_y^u D\phi(x)v|_y \leq \epsilon |v|_x, \ v \in S'(x); \tag{1.10}$$

$$|\Pi_y^u D\phi(x)v|_y \geq \frac{1}{\lambda}|v|_x, \ |\Pi_y^s D\phi(x)v|_y \leq \epsilon |v|_x, \ v \in U'(x), \tag{1.11}$$

hold, where Π_x^s is the projection onto $S'(x)$ parallel to $U'(x)$, and $\Pi_x^u = I - \Pi_x^s$;
(3)

$$\frac{1}{N'}|v|_x \le |v| \le N'|v|_x \text{ for } x \in W, \ v \in \mathbb{R}^n. \tag{1.12}$$

Remark. Take $x \in \Lambda, y = \phi(x)$ in Lemma 1.2.1 to see that $|.|_x$ is a Lyapunov norm on W.

Proof. First we construct a continuous norm $|.|_x$ with the desired properties. Fix $\mu \in (\lambda_0, \lambda)$ and find a natural ν such that

$$C\left(\frac{\lambda_0}{\mu}\right)^{\nu+1} < 1. \tag{1.13}$$

Take a point $p \in \Lambda$ and a vector $v \in \mathbb{R}^n$, represent $v = v^s + v^u \in S(p) \oplus U(p)$, and set

$$|v|_p^2 = (|v^s|_p^2 + |v^u|_p^2)^{1/2},$$

where

$$|v^s|_p = \sum_{j=0}^{\nu} \mu^{-j}|D\phi^j(p)v^s|, \quad |v^u|_p = \sum_{j=0}^{\nu} \mu^{-j}|D\phi^{-j}(p)v^u|.$$

For v^s we obtain the estimate

$$|D\phi(p)v^s|_{\phi(p)} = \sum_{j=0}^{\nu} \mu^{-j}|D\phi^j(\phi(p))D\phi(p)v^s| =$$

$$= \mu\left(\sum_{j=1}^{\nu} \mu^{-j}|D\phi^j(p)v^s| + \mu^{-\nu-1}|D\phi^{\nu+1}(p)v^s|\right) \le$$

$$\le \mu\left(\sum_{j=1}^{\nu} \mu^{-j}|D\phi^j(p)v^s| + \mu^{-\nu-1}C\lambda_0^{\nu+1}|v^s|\right) \le \mu|v^s|_p$$

(we applied (1.13) in the last inequality).
   Similarly, for v^u we have

$$|D\phi(p)v^u|_{\phi(p)} = \sum_{j=0}^{\nu} \mu^{-j}|D\phi^{-j}(\phi(p))D\phi(p)v^u| =$$

$$= \mu^{-1}\left(\sum_{j=0}^{\nu-1} \mu^{-j}|D\phi^{-j}(p)v^u| + \mu|D\phi(p)v^u|\right).$$

Since $w := D\phi^{-\nu}(p)v^u = D\phi^{-\nu-1}(\phi(p))D\phi(p)v^u$, it follows from (1.9) that

$$|w| \le C\lambda_0^{\nu+1}|D\phi(p)v^u|,$$

hence we arrive to the inequality

$$|D\phi(p)v^u|_{\phi(p)} \geq$$

$$\geq \mu^{-1}\left(\sum_{j=0}^{\nu-1}\mu^{-j}|D\phi^{-j}(p)v^u| + \mu^{-\nu}\frac{\mu^{\nu+1}}{C\lambda_0^{\nu+1}}|D\phi^{-\nu}(p)v^u|\right) \geq \frac{1}{\mu}|v^u|_p.$$

By construction, the obtained norm $|.|_p$ is continuous on the set Λ (we recall that the families S, U are continuous). Extend S, U to continuous (but not $D\phi$-invariant) families S', U' on a closed neighborhood (i.e., the closure of a neighborhood) W_0 of Λ so that statement (1) of our lemma holds. Now we extend $|.|_x$ to W_0 (decreasing W_0, if necessary) and fix a constant N' with property (1.12) for $x \in W_0$.

For points $x \in \Lambda, y = \phi(x)$, the mapping $\Pi_y^s D\phi(x)$ $(\Pi_y^u D\phi(x))$ is an isomorphism between $S(x)$ and $S(y)$ (respectively, between $U(x)$ and $U(y)$), and the relations

$$\|\Pi_y^s D\phi(x)|_{S(x)}\| \leq \mu, \ \Pi_y^u D\phi(x)|_{S(x)} = 0,$$

$$\|\Pi_y^u D\phi(x)|_{U(x)}\| \geq 1/\mu, \ \Pi_y^s D\phi(x)|_{U(x)} = 0$$

hold (the operator norms are taken with respect to $|.|_x$). Since $\Pi_x^s, \Pi_x^u, \phi(x)$, and $D\phi(x)$ are uniformly continuous, given arbitrary $\epsilon > 0$ we obviously can find a neighborhood $W(\epsilon, \lambda) \subset W_0$ and a number δ with the desired properties. To complete the proof of our lemma, it remains to approximate $|.|_x$ by a C^∞ norm (and to decrease the neighborhood W, if necessary) so that all the estimates remain true.   □

Note that by our construction the neighborhood W is bounded.

Now we take a bounded neighborhood V of the hyperbolic set Λ for our diffeomorphism ϕ, a diffeomorphism ψ of class C^1 in $\mathbb{R}^n$, and define the number

$$\rho_{1,V}(\phi,\psi) = \sup_{x \in V}|\phi(x) - \psi(x)| + \sup_{x \in V}\|D\phi(x) - D\psi(x)\|.$$

The proof of Lemma 1.2.1 shows that the following statement holds.

Lemma 1.2.2. *The neighborhood W, the norm $|.|_x$, and the extensions S', U' of the hyperbolic structure on Λ have the following property. There exists $\delta_1 > 0$ such that if for a diffeomorphism ψ the inequality*

$$\rho_{1,W}(\phi,\psi) < \delta_1 \tag{1.14}$$

holds, then for $x, y \in W$ with $|y - \psi(x)| < \delta_1$ we have

$$|\Pi_y^s D\psi(x)v|_y \leq \lambda|v|_x, \ |\Pi_y^u D\psi(x)v|_y \leq \epsilon|v|_x, \ v \in S'(x); \tag{1.15}$$

$$|\Pi_y^u D\psi(x)v|_y \geq \frac{1}{\lambda}|v|_x, \ |\Pi_y^s D\psi(x)v|_y \leq \epsilon|v|_x, \ v \in U'(x). \tag{1.16}$$

Now we describe the main geometric structures generated by a hyperbolic set, the so-called *local stable and unstable manifolds* $W_\Lambda^s(x)$ and $W_\Lambda^u(x)$.

Let Λ be a hyperbolic set for a C^1 diffeomorphism ϕ of $\mathbb{R}^n$, we assume that the Euclidean norm is Lyapunov in a neighborhood U of Λ, so that inequalities (1.10), (1.11) hold for $x, y \in U$. A proof of the following statement (the generalized stable manifold theorem) can be obtained from Theorem 2.1 of Chap. 1 and Theorems 1.2,1.3 of Chap. 4 in the book [Pli1].

To formulate the theorem, we need the following construction. Take a point $p \in \mathbb{R}^n$ and a number $\Delta > 0$. Assume that $\mathbb{R}^n$ is represented as $\mathbb{R}^k \times \mathbb{R}^{n-k}$ with coordinates y in $\mathbb{R}^k$ and z in $\mathbb{R}^{n-k}$ so that $p = (0,0)$. We say that a set $D \subset \mathbb{R}^n$ is a C^1 Δ-disk centered at p if there is a C^1-mapping f from the set

$$\{y \in \mathbb{R}^k : |y| \le \Delta\}$$

into $\mathbb{R}^{n-k}$ such that $f(0) = 0$ and

$$D = \{(y, f(y)) : y \in \mathbb{R}^k, |y| \le \Delta\}.$$

Theorem 1.2.1. *There exists a neighborhood U of Λ and numbers $\Delta, \Delta_1, \mathcal{K} > 0, \nu \in (0,1)$ with the following properties.*

For any point $x \in U$ such that $\phi^k(x) \in U$ for $k_1 < k < k_2$ with $k_1 < 0, k_2 > 0$ (infinite values of k_1, k_2 are admissible), there exist C^1 Δ-disks $W_\Delta^s(x)$ and $W_\Delta^u(x)$ centered at x and such that
   (1)

$$\phi(W_\Delta^s(x)) \subset W_\Delta^s(\phi(x)) \text{ if } k_2 > 1, \ \phi^{-1}(W_\Delta^u(x)) \subset W_\Delta^u(\phi^{-1}(x)) \text{ if } k_1 < -1;$$

   (2) if $y_1, y_2 \in W_\Delta^s(x)$, then

$$|\phi^l(y_1) - \phi^l(y_2)| \le \nu^l |y_1 - y_2| \text{ for } l \ge 0 \text{ while } l < k_2;$$

if $y_1, y_2 \in W_\Delta^u(x)$, then

$$|\phi^l(y_1) - \phi^l(y_2)| \ge \nu^{-l} |y_1 - y_2| \text{ for } l \le 0 \text{ while } l > k_1 \text{ and } \nu^{-l} |y_1 - y_2| < \Delta;$$

   (3) if $k_2 = \infty$, and for a point y the inequalities

$$|\phi^k(x) - \phi^k(y)| \le \Delta$$

hold for $k \ge 0$, then $y \in W_\Delta^s(x)$; if $k_1 = -\infty$, and for a point y the inequalities

$$|\phi^k(x) - \phi^k(y)| \le \Delta$$

hold for $k \le 0$, then $y \in W_\Delta^u(x)$;
   (4) if for $x, y \in U$ we have $|x - y| \le \Delta_1$, then the disks $W_\Delta^s(x), W_\Delta^u(y)$ have a unique point z of intersection, and

$$|z - x|, |z - y| \le \mathcal{K}|x - y|.$$

1.2.2 The Classical Shadowing Lemma

The following statement is usually called the Shadowing Lemma.

Theorem 1.2.2. *If Λ is a hyperbolic set for a diffeomorphism ϕ, then there exists a neighborhood W of Λ such that ϕ has the POTP on W. In addition, we can find a neighborhood W_1 of Λ such that ϕ has the SUP on W_1.*

We apply here the method of Anosov to prove that in a neighborhood of a hyperbolic set a diffeomorphism has a stronger shadowing property (the LpSP instead of the POTP), and then reduce Theorem 1.2.2 to Theorem 1.2.3. It should be noted that both classical proofs of the Shadowing Lemma by Anosov and by Bowen really provide Lipschitz dependence of ϵ on d in (1.1).

Theorem 1.2.3. *If Λ is a hyperbolic set for a diffeomorphism ϕ, then there exists a neighborhood W of Λ such that ϕ has the LpSP on W. In addition, we can take the number d_0 such that, for $d \leq d_0$, the point x with property (1.2) is unique.*

Remark. A similar statement for the LpSP$_+$ (without the uniqueness of a shadowing trajectory) is also true (see Theorem 1.3.3). Note that since a small neigborhood of a hyperbolic set is not always negatively invariant, we cannot refer to an analog of Lemma 1.1.1 for the LpSP and LpSP$_+$.

One of the main technical problems in the proofs of shadowing statements below is to establish that some operators in Banach spaces are invertible. We will often apply the following statement.

Lemma 1.2.3. *(1) Let B be a Banach space. Consider a linear operator $A : B \to B$ such that $\|A\| = \lambda < 1$. Then the operator $I - A$ is invertible, and*

$$\|(I - A)^{-1}\| \leq \frac{1}{1 - \lambda}.$$

(2) Let B be a Banach space represented as

$$B = B^s \oplus B^u. \tag{1.17}$$

Assume that for a linear operator $A : B \to B$ represented as

$$\begin{bmatrix} A^{ss} & A^{su} \\ A^{us} & A^{uu} \end{bmatrix}$$

according to (1.17), we have

$$\|A^{ss}\|, \|(A^{uu})^{-1}\| \leq \lambda \tag{1.18}$$

for some $\lambda \in (0, 1)$.

If for $\nu > 0$ the inequalities

$$\nu_1 = \nu \frac{1}{1 - \lambda} < 1$$

and

$$\|A^{su}\|, \|A^{us}\| \leq \nu$$

hold, then the operator $I - A$ is invertible, and

$$\|(I - A)^{-1}\| \leq R(\lambda, \nu) = \frac{1}{(1 - \lambda)(1 - \nu_1)}.$$

Proof. To prove statement (1), consider the operator C defined by

$$C = \sum_{k=0}^{\infty} A^k.$$

Obviously, the series on the right converges, and the estimate

$$\|C\| \leq \sum_{k=0}^{\infty} \|A\|^k = \sum_{k=0}^{\infty} \lambda^k = \frac{1}{1 - \lambda}$$

holds.

Let us write

$$(I - A)C = \sum_{k=0}^{\infty} A^k - \sum_{k=1}^{\infty} A^k = I.$$

Similarly one shows that $C(I - A) = I$. This proves (1).

To prove (2), we first consider the operator A_0 given by

$$\begin{bmatrix} A^{ss} & 0 \\ 0 & A^{uu} \end{bmatrix}.$$

Let us show that $I - A_0$ is invertible and that

$$(I - A_0)^{-1} = C_0, \tag{1.19}$$

where C_0 is represented as

$$\begin{bmatrix} C_0^s & 0 \\ 0 & C_0^u \end{bmatrix}$$

with

$$C_0^s = \sum_{k=0}^{\infty} (A^{ss})^k, \quad C_0^u = -\sum_{k=1}^{\infty} (A^{uu})^{-k}.$$

Obviously, it follows from (1.18) that the series defining C_0^s, C_0^u converge, and that

$$\|C_0^s\| \leq \frac{1}{1 - \lambda}, \quad \|C_0^u\| \leq \frac{\lambda}{1 - \lambda},$$

hence,

$$||C_0|| \leq \frac{1}{1-\lambda}. \tag{1.20}$$

Analogously to (1), we obtain the equalities

$$(I - A^{ss})C_0^s = \sum_{k=0}^{\infty}(A^{ss})^k - \sum_{k=1}^{\infty}(A^{ss})^k = I$$

and

$$(I - A^{uu})C_0^u = -\sum_{k=1}^{\infty}(A^{uu})^{-k} + \sum_{k=0}^{\infty}(A^{uu})^{-k} = I.$$

Similarly one shows that $C_0^s(I - A^{ss}) = I$ and $C_0^u(I - A^{uu}) = I$, hence (1.19) holds.

Set $A' = A_0 - A$, $A'' = I - A_0$. It follows from our conditions and from (1.20) that

$$||A'|| \leq \nu, \ ||A'C_0|| \leq \nu_1 < 1. \tag{1.21}$$

Define the operator

$$C = C_0 \sum_{k=0}^{\infty}(-1)^k(A'C_0)^k.$$

By the second inequality in (1.21), the series in the definition of C converges, and

$$||C|| \leq ||C_0||\frac{1}{1-\nu_1} \leq R(\lambda, \nu).$$

Let us show that $(I - A)C = I$. Indeed, since $A''C_0 = I$, we have

$$(I - A)C = (A'' + A')C_0 \sum_{k=0}^{\infty}(-1)^k(A'C_0)^k =$$

$$= \sum_{k=0}^{\infty}(-1)^k(A'C_0)^k + \sum_{k=0}^{\infty}(-1)^k(A'C_0)^{k+1} = I.$$

It is easy to see that if $(I - A)C = I$ and C maps B onto B, then $C = (I - A)^{-1}$. Indeed, in this case for any $x \in B$ we can find $y \in B$ such that $x = Cy$. Since $(I-A)x = (I-A)Cy = y$, we see that $C(I-A)x = Cy = x$, hence $C(I-A) = I$.

Thus, to complete the proof, we have to show that C maps B onto B.

Represent $C = C_0C_1$ and set $C_2 = A'C_0$. By construction, C_0 maps B onto B. Since $||C_2|| \leq \nu_1$, the same reasons as in the proof of (1) show that

$$C_1 = \sum_{k=0}^{\infty}(-1)^kC_2^k = (I + C_2)^{-1},$$

hence C_1 also maps B onto B.   □

Now we prove Theorem 1.2.3.

Proof. Let Λ be a hyperbolic set with hyperbolicity constants C, λ_0. Take numbers $\lambda \in (\lambda_0, 1)$ and $\epsilon > 0$ such that the inequality

$$\frac{\epsilon}{1 - \lambda} < 1$$

holds. Set $R = R(\lambda, \epsilon)$ (given by Lemma 1.2.3).

For the fixed λ, ϵ, we can find, by Lemma 1.2.1, a neighborhood $W = W(\lambda, \epsilon)$, a norm $|.|_x$, and families of subspaces S', U' such that inequalities (1.10), (1.11) hold. Let us show that ϕ has the LpSP on W.

Take a sequence $\xi = \{x_k\}$ in W such that

$$|\phi(x_k) - x_{k+1}| \le d.$$

By (1.12), the inequalities

$$|\phi(x_{k-1}) - x_k|_{x_k} \le N'|\phi(x_{k-1}) - x_k| \le N'd$$

hold. Assume that we can find δ_0, L_0 such that for any sequence $\xi = \{x_k\}$ satisfying the inequalities above with $N'd \le \delta_0$, there is a point x with the property

$$|\phi^k(x) - x_k|_{x_k} \le L_0 N'd, \ k \in \mathbb{Z}.$$

Then

$$|\phi^k(x) - x_k| \le L_0 (N')^2 d, \ k \in \mathbb{Z},$$

hence ϕ has the LpSP on W with $d_0 = \delta_0/N', L = L_0(N')^2$. Thus, without loss of generality, we may assume that the usual Euclidean norm $|.|$ is Lyapunov in W. To simplify notation, we write S, U instead of S', U'.

Fix a sequence $\xi = \{x_k\}$ such that

$$|\phi(x_k) - x_{k+1}| \le d,$$

take a sequence $\eta = \{y_k \in \mathbb{R}^n : k \in \mathbb{Z}\}$, and set $y_k = x_k + v_k$. Then η is a trajectory of ϕ if and only if

$$v_{k+1} = \phi(x_k + v_k) - x_{k+1}, \ k \in \mathbb{Z}. \tag{1.22}$$

Obviously, we have

$$\phi(x_k + v_k) - x_{k+1} = [\phi(x_k) - x_{k+1}] + D\phi(x_k)v_k + h_k(v_k), \tag{1.23}$$

where $h_k(0) = 0$ and $Dh_k(0) = 0$.

Let $\mathcal{B}$ be the Banach space of sequences $\bar{v} = \{v_k \in \mathbb{R}^n : k \in \mathbb{Z}\}$ with the norm

$$||\bar{v}|| = \sup_{k \in \mathbb{Z}} |v_k|.$$

Recall that for $r > 0$ we denote

$$\mathcal{B}(r) = \{\bar{v} \in \mathcal{B} : ||\bar{v}|| \le r\}.$$

Since our neighborhood W is bounded, there exist constants $C_1, d_1 > 0$, and a function $b(s) \to 0$ for $s \to 0$ such that

$$\|D\phi(x)\| \le C_1 \text{ for } x \in W$$

and

$$|h_k(v_k) - h_k(v'_k)| \le b(\max(\|\bar{v}\|, \|\bar{v}'\|))|v_k - v'_k| \text{ for } \|\bar{v}\|, \|\bar{v}'\| \le d_1.$$

For $\bar{v} \in \mathcal{B}$ we set $F(\bar{v}) = \bar{v}'$, where

$$v'_k = \phi(x_{k-1} + v_{k-1}) - x_k.$$

Since $|\phi(x_k) - x_{k+1}| \le d$, it follows from (1.23) and from the estimates above that for $\bar{v} \in \mathcal{B}(d_1)$ we have

$$|\phi(x_k + v_k) - x_{k+1}| \le d + (C_1 + b(d_1))|v_k|,$$

hence F is an operator from $\mathcal{B}(d_1)$ to $\mathcal{B}$. By (1.22), η is a trajectory of ϕ if and only if $\bar{v}$ is a fixed point of F.

Now we represent

$$\mathcal{B} = \mathcal{B}^s \oplus \mathcal{B}^u, \tag{1.24}$$

where

$$\mathcal{B}^s = \{\bar{v} \in \mathcal{B} : v_k \in S(x_k), k \in \mathbb{Z}\}, \quad \mathcal{B}^u = \{\bar{v} \in \mathcal{B} : v_k \in U(x_k), k \in \mathbb{Z}\}$$

(recall that we denote by $S(x), U(x)$ the extended families on W).

It follows from (1.23) that F is differentiable at 0 with $(DF(0)\bar{v})_{k+1} = D\phi(x_k)v_k$. Take

$$v_k = v_k^s + v_k^u, \quad v_k^s \in S(x_k), \quad v_k^u \in U(x_k),$$

and represent $w_{k+1} = D\phi(x_k)v_k$ as

$$w_{k+1} = w_{k+1}^s + w_{k+1}^u, \quad w_{k+1}^s \in S(x_{k+1}), \quad w_{k+1}^u \in U(x_{k+1}).$$

Denote $z = x_{k+1}$. Then we have

$$w_{k+1}^s = w_{k+1}^{ss} + w_{k+1}^{su},$$

where

$$w_{k+1}^{ss} = \Pi_z^s D\phi(x_k)v_k^s, \quad w_{k+1}^{su} = \Pi_z^s D\phi(x_k)v_k^u.$$

Similarly,

$$w_{k+1}^u = w_{k+1}^{us} + w_{k+1}^{uu},$$

where

$$w_{k+1}^{us} = \Pi_z^u D\phi(x_k)v_k^s, \quad w_{k+1}^{uu} = \Pi_z^u D\phi(x_k)v_k^u.$$

Take $d < \delta$ (where δ is given by Lemma 1.2.1 for the fixed λ, ϵ), so that the inequality $|z - \phi(x_k)| < \delta$ holds. Then it follows from Lemma 1.2.1 that

$$|w_{k+1}^{ss}| \le \lambda|v_k^s|, \ |w_{k+1}^{su}| \le \epsilon|v_k^u|, \ |w_{k+1}^{us}| \le \epsilon|v_k^s|, \ |w_{k+1}^{uu}| \ge \frac{1}{\lambda}|v_k^u|.$$

If $\bar{v}^s = \{v_k^s\}, \bar{w}^{ss} = \{w_k^{ss}\}$, then obviously $\|\bar{w}^{ss}\| \le \lambda\|\bar{v}^s\|$ and so on. Hence, $DF(0)$ is represented with respect to decomposition (1.24) in the form

$$\begin{bmatrix} A^{ss} & A^{su} \\ A^{us} & A^{uu} \end{bmatrix},$$

where

$$\|A^{ss}\|, \|(A^{uu})^{-1}\| \le \lambda, \ \|A^{su}\|, \|A^{us}\| \le \epsilon.$$

Now (by our choice of λ, ϵ) Lemma 1.2.3 implies that the operator $I - DF(0)$ in $\mathcal{B}$ is invertible, and

$$\|(I - DF(0))^{-1}\| \le R. \tag{1.25}$$

Set

$$G(\bar{v}) = (I - DF(0))^{-1}(F(\bar{v}) - DF(0)\bar{v}).$$

Clearly, the equation $F(\bar{v}) = \bar{v}$ is equivalent to $G(\bar{v}) = \bar{v}$.

Set $H(\bar{v}) = F(\bar{v}) - DF(0)\bar{v}$, i.e., $H(\bar{v})_k = [\phi(x_{k-1}) - x_k] + h_{k-1}(v_{k-1})$. Then

$$\|H(0)\| \le d. \tag{1.26}$$

It follows from properties of h_k that for $\bar{v}, \bar{v}' \in \mathcal{B}(d_1)$ we have

$$\|H(\bar{v}) - H(\bar{v}')\| \le b(\max(\|\bar{v}\|, \|\bar{v}'\|))\|\bar{v} - \bar{v}'\|. \tag{1.27}$$

We deduce from (1.25)-(1.27) that if $d < \delta$, then

$$\|G(0)\| \le Rd \text{ and } \|G(\bar{v}) - G(\bar{v}')\| \le Rb(\max(\|\bar{v}\|, \|\bar{v}'\|))\|\bar{v} - \bar{v}'\|$$

for $\bar{v}, \bar{v}' \in \mathcal{B}(d_1)$. Take $d_2 \in (0, \min(\delta, d_1))$ such that $Rb(d_2) < 1/2$. It follows that

$$\|G(\bar{v}) - G(\bar{v}')\| \le \frac{1}{2}\|\bar{v} - \bar{v}'\|$$

for $\bar{v}, \bar{v}' \in \mathcal{B}(d_2)$, hence G contracts on the ball $\mathcal{B}(d_2)$.

Set

$$d_0 = \frac{d_2}{2R}, \ L = 2R.$$

Take $d \le d_0$ (note that $R > 1$, hence $d_0 < d_2$). For $\bar{v} \in \mathcal{B}(Ld)$ we have

$$\|G(\bar{v})\| \le \|G(0)\| + \|G(\bar{v}) - G(0)\| \le Rd + \frac{1}{2}2Rd = Ld.$$

We see that G maps the ball $\mathcal{B}(Ld)$ into itself and contracts on it, hence there is a unique fixed point $\bar{v}^*$ of G (and also of F) in this ball. Obviously, the point $x = x_0 + v_0^*$ satisfies inequalities (1.2). The uniqueness of the fixed point implies the uniqueness of the shadowing trajectory. □

Remark. It is easy to see that, for $d \le d_0$, the mapping G has a unique fixed point not only in the ball $\mathcal{B}(Ld)$ but also in the larger ball $\mathcal{B}(Ld_0)$. Take

$\epsilon > 0$ and a neighborhood W_0 of Λ such that $2\epsilon < Ld_0$ and $N_\epsilon(W_0) \subset W$. We claim that ϕ has the SUP on W_0 with constant ϵ. Indeed, assume that for a d-pseudotrajectory $\xi = \{x_k\} \subset W_0$ there exist points x, y such that

$$|\phi^k(x) - x_k| < \epsilon, \ |\phi^k(y) - x_k| < \epsilon.$$

Then we have $|\phi^k(x) - \phi^k(y)| < Ld_0$. The sequence $\xi' = \{\phi^k(x)\}$ is a d-pseudotrajectory with any $d > 0$. We have $\xi' \subset W$. This d-pseudotrajectory is (Ld_0)-shadowed by both points x and y. The uniqueness of a fixed point of G in $\mathcal{B}(Ld_0)$ implies that $x = y$. This proves our claim.

Now we reduce Theorem 1.2.2 to Theorem 1.2.3. Obviously, its first statement follows from the first statement of Theorem 1.2.3. Formally, the second statement (uniqueness) is not a corollary of Theorem 1.2.2, but it is easily proved using Lemma 1.1.2. Indeed, by the remark above our system ϕ has the SUP on W_0 with a constant ϵ. It follows from part (a) of Lemma 1.1.2 that ϕ is expansive on W_0 with expansivity constant $\Delta \in (0, \epsilon)$. Obviously, we can find a neighborhood W_1 of Λ and a number $\delta \in (0, \Delta/2)$ such that $N_{2\delta}(W_1) \subset W_0$. By part (b) of Lemma 1.1.2, ϕ has the SUP on W_1 with constant δ.

Let us describe an example of application of Theorem 1.2.2. Assume that p is a hyperbolic fixed point of a diffeomorphism ϕ of $\mathbb{R}^n$ (i.e., the invariant set $\Lambda = \{p\}$ satisfies Definition 1.9). It is easy to see that this condition is equivalent to the following one: all eigenvalues λ_i of the matrix $D\phi(p)$ satisfy the inequalities $|\lambda_i| \neq 1$.

We also assume that there exists a point $q \neq p$ such that

$$\phi^k(q) \to p \text{ as } |k| \to \infty. \tag{1.28}$$

A point q with property (1.28) is called a *homoclinic point* of the fixed point p. It follows from Theorem 1.2.1 that for fixed $\Delta > 0$ there exists a natural number m such that

$$q \in \phi^m(W^u_\Delta(p)) \cap \phi^{-m}(W^s_\Delta(p)).$$

The sets $W_1 = \phi^m(W^u_\Delta(p))$ and $W_2 = \phi^{-m}(W^s_\Delta(p))$ are diffeomorphic to C^1 disks, we say that q is a *transverse homoclinic point* if the point q is a point of transverse intersection of the disks W_1 and W_2.

We introduce the set

$$\Gamma = \overline{O(q)}.$$

Obviously, Γ is a compact invariant set of ϕ.

It is well known that if q is a transverse homoclinic point, then Γ is a hyperbolic set of ϕ (it is a useful exercise for the reader to prove this statement).

Poincaré was the first who showed that the existence of a homoclinic point implies a very complicated structure of trajectories in a neighborhood of the set Γ. We establish here an important property of neighborhoods of transverse homoclinic trajectories.

Theorem 1.2.4. *If q is a transverse homoclinic point, then any neighborhood of the set Γ contains a countable family of different periodic points of ϕ.*

Proof. Take an arbitrary neighborhood U of the set Γ. We take the neighborhood U bounded and so small that ϕ has the POTP and the SUP on U (this is possible by Theorem 1.2.2). Then ϕ has the POTP and the SUP on Γ. Take $\Delta > 0$ such that any d-pseudotrajectory $\xi \subset \Gamma$ is (Δ)-shadowed by not more than one point.

Find a number $\epsilon \in (0, \Delta)$ such that $N_\epsilon(\Gamma) \subset U$. We assume, in addition, that

$$2\epsilon < |p - q|. \tag{1.29}$$

Fix a number d corresponding to ϵ according to Definition 1.3 (with $Y = \Gamma$). Let $\mathcal{L}$ be a Lipschitz constant of ϕ in U.

It follows from (1.28) that there exist numbers $l < 0, m > 0$ such that for the points $q_0 = \phi^l(q)$ and $q_1 = \phi^m(q)$ the inequalities

$$|q_0 - p| < d, \ |q_1 - p| < \frac{d}{\mathcal{L}}, \tag{1.30}$$

and the inclusion $q_1 \in U$ hold. Set $N = m - l + 2$ and define a sequence $\xi = \{x_k : k \in \mathbb{Z}\}$ as follows. Represent $k \in \mathbb{Z}$ in the form $k = k_0 + k_1 N$ with $0 \leq k < N$, and set $x_k = p$ if $k_0 = 0$, and $x_k = \phi^{k_0-1}(q_0)$ if $k_0 \neq 0$. We obtain a periodic sequence with period N,

$$\ldots, \ x_0 = p, \ x_1 = q_0, \ \ldots, \ x_{N-1} = q_1, \ x_N = p, \ x_{N+1} = q_0, \ \ldots$$

It follows from (1.30) that

$$|\phi(x_0) - x_1| = |p - q_0| < d \text{ and } |\phi(x_{N-1}) - x_N| \leq \mathcal{L}|q_1 - p| < d,$$

hence ξ is a d-pseudotrajectory of ϕ. By the choice of d, there is a point x that (ϵ)-shadows ξ. Set $y = \phi^N(x)$. Since $x_k = x_{k+N}$ for all k, we have

$$|\phi^k(y) - x_k| = |\phi^{k+N}(x) - x_{k+N}| < \epsilon, \ k \in \mathbb{Z},$$

i.e., the point y also (ϵ)-shadows ξ. The inclusion $\xi \subset \Gamma$ and the SUP on Γ (with constant $\Delta > \epsilon$) imply that $x = y$, hence x is a periodic point of ϕ.

There exists k such that $|\phi^k(x) - q| < \epsilon$. It follows from (1.29) that $x \neq p$. Relations (1.28) imply that the set Γ contains no periodic points of ϕ different from p, hence for the finite set $O(x)$ we have $O(x) \cap \Gamma = \emptyset$. Thus, we can find a neighborhood U' of Γ such that $O(x) \cap U' = \emptyset$. The same reasons as above show that U' contains a periodic point $x' \neq p$, and so on. Hence, there is a countable family of different periodic points of ϕ in U. Note that we proved that the trajectories of these points are in U. □

Remark. Palmer [Palm1, Palm2] applied the shadowing approach to describe the structure of the set of all trajectories belonging to a small neighborhood of the trajectory of a transverse homoclinic point for a diffeomorphism. Later, he did the same for flows [Palm3]. Stoffer [Sto] proved a special shadowing lemma to study transverse homoclinic orbits for sequences of mappings of the plane. Steinlein and Walther [Ste1, Ste2] gave a nonstandard definition of a hyperbolic set for a mapping in a Banach space and proved the corresponding shadowing result. This result can be applied to study homoclinic trajectories for infinite-dimensional semi-dynamical system. We prove a variant of this theorem of Steinlein-Walther in Subsect. 1.3.4 (see Theorem 1.3.4).

Let Λ be an invariant set of a dynamical system ϕ. We say that Λ is *locally maximal* (or *isolated*) if there exists a neighborhood V of Λ such that the inclusion $O(p) \subset V$ implies that $p \in \Lambda$. It is easy to understand that if Λ is locally maximal, and ϕ has the POTP in a neighborhood of Λ, then pseudotrajectories belonging to a small neighborhood of Λ are shadowed by points of Λ. The set Γ defined before Theorem 1.2.4 is an example of a hyperbolic set which is not locally maximal (any its neighborhood contains a periodic trajectory not belonging to Γ).

1.2.3 Shadowing for a Family of Approximate Trajectories

It was mentioned that Anosov established the Shadowing Lemma in [Ano2] as a particular case of a more general statement, the so-called "theorem on a family of ϵ-trajectories". This statement says that if Λ is a hyperbolic set for a diffemorphism ϕ, and if a small neighborhood of Λ contains a family of approximate trajectories of a diffeomorphism ψ, C^1-close to ϕ, then it is possible to shadow all this family by trajectories of ψ.

Let us formulate the theorem. We assume that Λ is a hyperbolic set for a C^1-diffeomorphism ϕ of $\mathbb{R}^n$. Consider a topological space X and a homeomorphism τ of X. For a continuous mapping $f : X \to \mathbb{R}^n$ and for a diffeomorphism ψ of $\mathbb{R}^n$, set

$$P(f \circ \tau, \psi \circ f) = \sup_{x \in X} |f \circ \tau(x) - \psi \circ f(x)|.$$

Obviously, the equality $P(f \circ \tau, \psi \circ f) = 0$ means that f maps trajectories of τ to trajectories of ψ. If the number $P(f \circ \tau, \psi \circ f)$ is small, f maps trajectories of τ to "approximate" trajectories of ψ.

Theorem 1.2.5. *There exist neigborhoods W_0, W of the set Λ and positive numbers L, d_0 with the following property. For any diffeomorphism ψ of $\mathbb{R}^n$ with*

$$\rho_{1,W_0}(\phi, \psi) < d_0 \tag{1.31}$$

and for any continuous mapping $f : X \to W$ such that

$$P(f \circ \tau, \psi \circ f) < d \le d_0, \tag{1.32}$$

there exists a continuous mapping $g : X \to \mathbb{R}^n$ such that

(1) $g \circ \tau = \psi \circ g$;

(2) $\sup_{x \in X} |f(x) - g(x)| \leq Ld$.

In addition, if for a continuous mapping $h : X \to \mathbb{R}^n$ analogs of statements (1),(2) (with h instead of g) hold, then $h = g$.

Proof. We repeat the beginning of the proof of Theorem 1.2.3. For a hyperbolic set Λ of a diffeomorphism ϕ with hyperbolicity constants C, λ_0, we fix numbers $\lambda \in (\lambda_0, 1)$ and $\epsilon > 0$ such that the inequality

$$\frac{\epsilon}{1 - \lambda} < 1$$

holds. Set $R = R(\lambda, \epsilon)$ (see Lemma 1.2.3).

For the fixed λ, ϵ we find, by Lemma 1.2.1, a neighborhood $W = W(\lambda, \epsilon)$, a norm $|.|_x$, and families of subspaces S', U' such that inequalities (1.10), (1.11) hold. By the same reasons as in Theorem 1.2.3, we may assume that the usual Euclidean norm $|.|$ is Lyapunov in W. We again write S, U instead of S', U'.

Fix a number $d_1 > 0$ and let W_0 be the d_1-neighborhood of W. Fix δ_1 given by Lemma 1.2.2. We consider below a diffeomorphism ψ such that $\rho_{1,W_0}(\phi, \psi) < \delta_1$.

For $p \in W$ and $w \in \mathbb{R}^n$ denote

$$h_p^{\phi}(w) = \phi(p + w) - \phi(p) - D\phi(p)w.$$

We take $d_2 \in (0, d_1)$ so small that if $p \in W$ and $|w|, |w'| \leq d_2$, then the segment joining the points $p + w$ and $p + w'$ belongs to W_0.

Since the neighborhood W is bounded, $D\phi$ is uniformly continuous on W_0. Hence, it easily follows from the standard formula

$$h_p^{\phi}(w) - h_p^{\phi}(w') = \int_0^1 D\phi(\theta(s))(w - w')ds - D\phi(p)(w - w') \tag{1.33}$$

(where $\theta(s), s \in [0, 1]$, parametrizes the segment joining the points $p + w$ and $p + w'$) that we can find d_2 having the following property. If $p \in W$ and $|w|, |w'| \leq d_2$, then

$$|h_p^{\phi}(w) - h_p^{\phi}(w')| \leq \frac{1}{4R}|w - w'|.$$

Now we take $d_3 \in (0, d_2)$ such that $8Rd_3 < 1$. If

$$\rho_{1,W_0}(\phi, \psi) < d_3,$$

then the formula for $h_p^{\psi}(w)$ similar to (1.33) and the inequality $\|D\phi - D\psi\| < d_3$ (which holds in W_0) imply the inequality

$$|h_p^{\psi}(w) - h_p^{\psi}(w')| \leq \left(\frac{1}{4R} + 2d_3\right)|w - w'| \leq \frac{1}{2R}|w - w'| \tag{1.34}$$

for $p \in W$ and $|w|, |w'| \leq d_2$.

Denote $Y = f(X)$ and let $\mathcal{B}$ be the space of continuous vector fields on Y. It is well known (see Chap. 0 in the book [Ni1]) that $\mathcal{B}$ with the norm

$$||v|| = \sup_{p \in Y} |v(p)|$$

is a Banach space.

Take f such that (1.32) holds (d_0 will be chosen later). We will find the continuous mapping $g : X \to \mathbb{R}^n$ in the form $g(x) = f(x) + v(f(x))$, where $v \in \mathcal{B}$. The equality $g \circ \tau = \psi \circ g$ reduces to the following equation for the vector field v:

$$f(\tau(x)) + v(f(\tau(x))) = \psi(f(x) + v(f(x))). \tag{1.35}$$

We rewrite Eq.(1.35) at $x, z = \tau^{-1}(x)$ instead of $\tau(x), x$,

$$f(x) + v(f(x)) = \psi(f(z) + v(f(z))),$$

or, equvalently,

$$v(f(x)) = \psi(f(z) + v(f(z))) - f(x). \tag{1.36}$$

It follows from condition (1.32) that for any $x \in X$ the inequality

$$|f(\tau(x)) - \psi(f(x))| < d$$

holds. This is equivalent to the condition

$$|f(x) - \psi(f(\tau^{-1}(x)))| < d$$

or

$$|f(x) - \psi(f(z))| < d, \; x \in X. \tag{1.37}$$

Now we define the following operator F: for $v \in \mathcal{B}$ we set $F(v) = \tilde{v}$, where

$$\tilde{v}(f(x)) = \psi(f(z) + v(f(z))) - f(x)$$

(recall that $z = \tau^{-1}(x)$).

It follows from (1.36) that, for the mapping $g(x) = f(x) + v(f(x))$, statement (1) of our theorem holds if and only if v is a fixed point of the operator F.

Let us write

$$\psi(f(z) + v(f(z))) - f(x) =$$
$$= [\psi(f(z)) - f(x)] + D\psi(f(z))v(f(z)) + h^{\psi}_{f(z)}(v(f(z))), \tag{1.38}$$

where

$$h^{\psi}_{f(z)}(0) = 0, \; Dh^{\psi}_{f(z)}(0) = 0.$$

Since the neighborhood W is bounded, there exists a constant C_1 such that

$$||D\phi|| \le C_1 \text{ on } W.$$

The inequality

$$\rho_{1,W_0}(\phi,\psi) < \delta_1$$

implies that

$$\|D\psi\| \leq C_2 = C_1 + \delta_1 \text{ on } W.$$

Let $\mathcal{B}(r)$ again denote the ball $\{v \in \mathcal{B} : \|v\| \leq r\}$ for $r > 0$.
It follows from (1.34), (1.37) that for $v \in \mathcal{B}(d_2)$ we have

$$\|F(v)\| \leq d + \left(C_2 + \frac{1}{2R}\right)\|v\|,$$

hence F is an operator from $\mathcal{B}(d_2)$ to $\mathcal{B}$.
We deduce from (1.38) that F is differentiable at 0, and

$$(DF(0)v)(f(x)) = D\psi(f(z))v(f(z)).$$

Now we represent $\mathcal{B}$ in the form $\mathcal{B}^s \oplus \mathcal{B}^u$, where

$$\mathcal{B}^s = \{v \in \mathcal{B} : v(f(x)) \in S(f(x)), x \in X\},$$

$$\mathcal{B}^u = \{v \in \mathcal{B} : v(f(x)) \in U(f(x)), x \in X\}.$$

Take $x \in X, z = \tau^{-1}(x)$, and denote $y = f(x)$. It follows from Lemma 1.2.2, from our choice of ψ, and from (1.37) (with $d < \delta_1$) that, for $w \in S(f(z))$, the inequalities

$$|\Pi_y^s D\psi(f(z))w| \leq \lambda|w|, \ |\Pi_y^u D\psi(f(z))w| \leq \epsilon|w|$$

hold. For $w \in U(f(z))$ we obtain the inequalities

$$|\Pi_y^s D\psi(f(z))w| \leq \epsilon|w|, \ |\Pi_y^u D\psi(f(z))w| \geq \frac{1}{\lambda}|w|.$$

The same reasons as in the proof of Theorem 1.2.3 show that we can represent $DF(0)$ with respect to the decomposition $\mathcal{B} = \mathcal{B}^s \oplus \mathcal{B}^u$ in the form

$$\begin{bmatrix} A^{ss} & A^{su} \\ A^{us} & A^{uu} \end{bmatrix}$$

with

$$\|A^{ss}\|, \|(A^{uu})^{-1}\| \leq \lambda, \ \|A^{su}\|, \|A^{us}\| \leq \epsilon.$$

Now Lemma 1.2.3 implies that the operator $I - DF(0)$ is invertible, and

$$\|(I - DF(0))^{-1}\| \leq R.$$

We again define

$$G(v) = (I - DF(0))^{-1}(F(v) - DF(0)v)$$

and consider the equation $G(v) = v$, equivalent to $F(v) = v$.
In this case, for $H(v) = F(v) - DF(0)v$ we have the representation

$$H(v)(f(x)) = [\psi(f(z)) - f(x)] + h^\psi_{f(z)}(v(f(z))).$$

By (1.37),
$$\|H(0)\| \le d.$$
By (1.34), for $v, v' \in L(d_2)$ the inequality

$$\|H(v) - H(v')\| \le \frac{1}{2R}\|v - v'\|$$

holds.

Now we set
$$L = 2R, \ d_0 = \frac{\min(d_3, \delta_1)}{2R},$$
and repeat word-to-word the end of the proof of Theorem 1.2.3 (note that $|f(x) - g(x)| = |v(f(x))|$ for $x \in X$). □

To obtain the Shadowing Lemma (Theorem 1.2.2) from Theorem 1.2.5, one can proceed as follows. Take $X = \mathbb{Z}$ (with discrete topology) and set $\tau(k) = k + 1$. Let $\xi = \{x_k\}$ be a sequence in a neighborhood of a hyperbolic set Λ for a diffeomorphism ϕ such that

$$|\phi(x_k) - x_{k+1}| \le d.$$

Define $f : X \to \mathbb{R}^n$ by the equalities $f(k) = x_k$. Then

$$P(f \circ \tau, \phi \circ f) = \sup_{k \in \mathbb{Z}} |x_{k+1} - \phi(x_k)| \le d.$$

Apply Theorem 1.2.5 to find a mapping $g : X \to \mathbb{R}^n$ such that

$$P(g \circ \tau, \phi \circ g) = 0 \text{ and } |f(k) - g(k)| \le Ld.$$

For the points $y_k = g(k)$, the equality above means that $y_{k+1} = \phi(y_k)$, i.e., $\{y_k\}$ is a trajectory of ϕ. The inequality above shows that $|x_k - y_k| \le Ld$.

Following [Katok1], we describe below an important example of application of Theorem 1.2.5 (the reader can find other interesting applications in [Katok1]).

Let us recall the definition of a nonwandering point for a dynamical system ϕ on a topological space Y.

Definition 1.10 *A point p is called "nonwandering" for (Y, ϕ) if for any neighborhood V of p and for any $N_0 > 0$ there is a natural number N such that $N > N_0$ and*
$$\phi^N(V) \cap V \ne \emptyset.$$

Theorem 1.2.6. *Let Λ be a hyperbolic set of a diffeomorphism ϕ. Assume that x is a nonwandering point of the restriction $\phi|_\Lambda$. Then any neighborhood of x contains a periodic point of ϕ.*

Proof. Take an arbitrary neighborhood U of x. Find $a > 0$ such that the ball of radius a centered at x belongs both to U and to the neighborhood W of Λ given by Theorem 1.2.5.

Take $d < d_0$ (from the theorem) such that $Ld < a/2$, and denote by V the intersection of the $d/2$-neighborhood of x with the set Λ. Since V is a neighborhood of x in Λ, we deduce from definition that there exists a natural number N such that

$$\phi^N(V) \cap V \neq \emptyset, \text{ hence } V_0 = \phi^{-N}(V) \cap V \neq \emptyset.$$

Take a point $y \in V_0$, it follows that $\phi^N(y) \in V$. Consider the space $X = \{0, 1, \ldots, N-1\}$ (with discrete topology) and the shift homeomorphism $\tau : X \to X, \tau(k) = k + 1(\mathrm{mod}\ N)$.

Define $f : X \to \mathbb{R}^n$ by $f(k) = \phi^k(y)$ (note that $y \in \Lambda$, hence $f(k) \in \Lambda \subset W$). Then

$$f \circ \tau(k) = \phi^{k+1}(y) = \phi \circ f(k) \text{ for } k = 0, \ldots, N-2,$$

$$f \circ \tau(N-1) = f(0) = y, \ \phi \circ f(N-1) = \phi^N(y),$$

hence

$$P(f \circ \tau, \phi \circ f) = |\phi^N(y) - y| < \frac{d}{2}.$$

By Theorem 1.2.5, there exists a mapping $g : X \to \mathbb{R}^n$ such that $|g(x) - f(x)| \leq Ld < a/2$ and $g \circ \tau = \phi \circ g$. Let $p_k = g(k)$, then the equalities

$$g \circ \tau(k) = \phi \circ g(k), \ k = 0, \ldots, N-1,$$

have the form

$$p_1 = \phi(p_0), \ p_2 = \phi(p_1), \ \ldots, \ p_0 = \phi(p_{N-1}),$$

i.e., p_0 is a periodic point of ϕ. Since

$$|y - x| < \frac{a}{2}, \ |p_0 - y| < \frac{a}{2},$$

we see that $p_0 \in U$. Note that the mapping g is not necessarily one-to-one, so possibly the minimal period of p_0 is less than N. $\square$

1.2.4 The Method of Bowen

We describe here the proof of the Shadowing Lemma given by Bowen [Bo2]. As was mentioned, this approach uses not only the definition of a hyperbolic set, but applies more detailed information about the behavior of a diffeomorphism near its hyperbolic set. Later the ideas of this method were generalized [Ru], we describe this generalization at the end of this subsection.

Bowen's proof of the Shadowing Lemma.

We will prove only the existence part of Theorem 1.2.3, the proof of the uniqueness part is the same. Let Λ be a hyperbolic set for a diffeomorphism ϕ. Fix a neighborhood U and numbers $\mathcal{K}, \Delta, \Delta_1, \nu$ given by Theorem 1.2.1. Find a natural number m such that

$$2\nu^m \mathcal{K} < 1. \tag{1.39}$$

We can take the neighborhood U bounded, hence ϕ is Lipschitz on U. It follows from Lemma 1.1.3 that there is a number L_1 with the following property: if for a sequence $\xi = \{x_k\} \subset U$ we have

$$|\phi(x_k) - x_{k+1}| \leq d,$$

then

$$|x_{k+j} - \phi^j(x_k)| \leq L_1 d \text{ for } k \in \mathbb{Z},\ 0 \leq j \leq m. \tag{1.40}$$

Since Λ is a compact invariant set, there exists $\epsilon > 0$ such that if $y \in \Lambda$ and $|x - y| < 2\epsilon$, then $\phi^m(x) \in U$ for $0 \leq k \leq m$. Let U_0 be the ϵ-neighborhood of Λ, it follows from the choice of ϵ that $U_0 \subset U$.

Find $d_0 > 0$ such that the inequality

$$2L_1 d_0 < \Delta_1 \tag{1.41}$$

is fulfilled, and take $d < d_0$.

Now take a sequence $\xi = \{x_k\} \subset U_0$ such that

$$|\phi(x_k) - x_{k+1}| \leq d,$$

fix a natural number r, and consider the set $\{x_0, \ldots, x_{rm}\}$.

Let us construct points $y_0, \ldots, y_k$ as follows. Set $y_0 = x_0$. We deduce from (1.40) and (1.41) that

$$|\phi^m(y_0) - x_m| = |\phi^m(x_0) - x_m| \leq L_1 d < 2L_1 d_0 < \Delta_1,$$

hence by Theorem 1.2.1 there is a point

$$y_1 \in W_\Delta^u(\phi^m(y_0)) \cap W_\Delta^s(x_m)$$

such that

$$|y_1 - \phi^m(y_0)| \leq \mathcal{K} L_1 d < 2\mathcal{K} L_1 d,$$
$$|y_1 - x_m| \leq \mathcal{K} L_1 d < 2\mathcal{K} L_1 d.$$

We use induction to construct points $y_k, 1 \leq k \leq r$, such that

$$y_k \in W_\Delta^u(\phi^m(y_{k-1})) \cap W_\Delta^s(x_{mk}), \tag{1.42}$$

$$|y_k - \phi^m(y_{k-1})| \leq 2\mathcal{K} L_1 d, \tag{1.43}$$

$$|y_k - x_{km}| \leq 2\mathcal{K} L_1 d. \tag{1.44}$$

Assume that we have constructed points $y_1, \ldots, y_k$ such that (1.42)-(1.44) hold. Since $y_k \in W_\Delta^s(x_{km})$, the inequality

$$|\phi^m(y_k) - \phi^m(x_{km})| < 2\nu^m K L_1 d < L_1 d$$

holds.

Since

$$|\phi^m(x_{km}) - x_{(k+1)m}| \leq L_1 d,$$

we see that

$$|\phi^m(y_k) - x_{(k+1)m}| < 2L_1 d,$$

thus we can find points

$$y_{k+1} \in W_\Delta^u(\phi^m(y_k)) \cap W_\Delta^s(x_{(k+1)m})$$

such that analogs of (1.43), (1.44) are true.

Take $x = \phi^{-rm}(y_r)$, and consider s, i such that $0 \leq s \leq r$, $0 \leq i \leq m$. Since

$$\phi^{sm+i}(x) = \phi^{sm-rm+i}(y_r),$$

we can write

$$|\phi^{sm+i}(x) - \phi^i(y_s)| \leq |\phi^{-m+i}(y_{s+1}) - \phi^i(y_s)| +$$

$$+|\phi^{-2m+i}(y_{s+2}) - \phi^{-m+i}(y_{s+1})| + \ldots + |\phi^{(s-r)m+i}(y_r) - \phi^{(s-r+1)m+i}(y_{r-1})| =$$

$$= \sum_{t=s+1}^{r} |\phi^{(s-t)m+i}(y_t) - \phi^{(s-t+1)m+i}(y_{t-1})| =$$

$$= \sum_{t=s+1}^{r} |\phi^{(s-t)m+i}(y_t) - \phi^{(s-t)m+i}(\phi^m(y_{t-1}))| \leq$$

$$\leq 2K L_1 d \sum_{t=s+1}^{r} \nu^{(t-s)m-i} < 2K L_1 d \nu^{-i} \sum_{j=1}^{\infty} \nu^{jm} \leq 2K L_1 d \frac{1}{1-\nu^m}.$$

Here we take (1.42), (1.43) into account.

Hence,

$$|\phi^{sm+i}(x) - \phi^i(y_s)| < L_2 d,$$

where

$$L_2 = 2K L_1 \frac{1}{1-\nu^m}.$$

It follows from (1.44) and (1.40) that

$$|\phi^i(y_s) - x_{sm+i}| \leq |\phi^i(y_s) - \phi^i(x_{sm})| + |\phi^i(x_{sm}) - x_{sm+i}| \leq$$

$$\leq 2K L_1 d + L_1 d, \ 0 \leq i \leq m$$

(to estimate the first term we take into account that $y_s \in W_\Delta^s(x_{sm})$). Finally we obtain, for $k \in \{0, \ldots, rm\}$, the estimates

$$|\phi^k(x) - x_k| \le Ld,$$

where

$$L = (2K + 1)L_1 + L_2.$$

Consider an arbitrary natural number r and set

$$x'_k = x_{k-rm}, \quad -rm \le k \le rm.$$

It follows from our reasoning above that there exists a point x' such that

$$|\phi^k(x') - x'_k| \le Ld, \; 0 \le k \le 2rm.$$

Hence, for the point $x^r = \phi^{rm}(x')$ the inequalities

$$|\phi^k(x^r) - x_k| \le Ld, \; -rm \le k \le rm,$$

hold. Let x be a limit point of the sequence x^r as $r \to \infty$ (to simplify notation, we assume that $x^r \to x$). Then, passing to the limit for $r \to \infty$, we see that

$$|\phi^k(x) - x_k| \le Ld, \; k \in \mathbb{Z}.$$

This completes the proof.

Smale spaces.

Ruelle developed in [Ru] an abstract theory of the so-called Smale spaces. Let us describe this approach (with slight modifications).

Let (X, r) be a compact metric space. It is assumed that

(SS1) there exists a positive number η and a continuous mapping

$$[,] : \{(x, y) \in X \times X : r(x, y) < \eta\} \to X$$

such that $[x, x] = x$, and

$$[[x, y], z] = [x, z], \; [x, [y, z]] = [x, z],$$

when the two sides of these relations are defined.

Define, for $\delta > 0$ and $x \in X$, the sets

$$V_\delta^s(x) = \{u : u = [u, x] \text{ and } r(x, u) < \delta\},$$

$$V_\delta^u(x) = \{v : v = [x, v] \text{ and } r(x, v) < \delta\}.$$

Now we state the second assumption.

(SS2) There exists a homeomorphism ϕ of X and numbers $\delta > 0, \nu \in (0, 1)$ such that

$$\phi([x, y]) = [\phi(x), \phi(y)] \tag{1.45}$$

when both sides are defined, and

$$r(\phi^n(y), \phi^n(z)) \leq \nu^n r(y, z) \text{ if } y, z \in V_\delta^s(x), \ n > 0, \qquad (1.46)$$

$$r(\phi^{-n}(y), \phi^{-n}(z)) \leq \nu^n r(y, z) \text{ if } y, z \in V_\delta^u(x), \ n > 0. \qquad (1.47)$$

Definition 1.11 *We define a "Smale space" to be a compact metric space* (X, r) *with a map* $[,]$ *and a homeomorphism* ϕ *satisfying (SS1) and (SS2) for suitable* η, δ, ν. *Below* (X, ϕ) *denotes this space.*

The introduced notion is an abstract model of the behavior of a diffeomorphism near its hyperbolic set described in Theorem 1.2.1. A small neighborhood of a hyperbolic set cannot be invariant, thus it is impossible to apply the model directly (the only possible direct application is the case of an *Anosov diffeomorphism*, i.e., of a diffeomorphism ϕ on a smooth closed manifold M such that M is a hyperbolic set for ϕ [Ano1]).

Nevertheless, the model reflects the main properties used in Bowen's proof of the Shadowing Lemma. The sets $V_\delta^s(x)$ and $V_\delta^u(x)$ are analogs of $W_\Delta^s(x)$ and $W_\Delta^u(x)$ in Theorem 1.2.1, and $[x, y]$ corresponds to the unique point of intersection of $W_\Delta^u(x)$ and $W_\Delta^s(y)$. Inequalities (1.46) and (1.47) are analogs of the corresponding properties (see statement (2) of Theorem 1.2.1).

Theorem 1.2.7. *If* (X, ϕ) *is a Smale space, then* ϕ *has the POTP.*

Proof. Let η, δ, ν be the defining constants for (X, ϕ). Fix $\epsilon > 0$. Take $\delta_0 \in (0, \delta)$ and such that

$$\delta_0 \left(1 + \frac{1}{1-\nu}\right) < \frac{\epsilon}{2}. \qquad (1.48)$$

Since the functions $r(x, [x, y])$ and $r(y, [x, y])$ are continuous for x, y with small $r(x, y)$ and vanish for $x = y$, there exists $\delta_1 \in (0, \eta)$ such that

$$r(x, [x, y]), r(y, [x, y]) < \delta_0 \text{ for } r(x, y) < \delta_1.$$

Find a natural number m such that

$$\nu^m \delta_0 < \delta_1.$$

Now we fix a number $d > 0$ such that

$$r(\phi^i(x_{sm}), x_{sm+i}) < \min\left(\delta_1 - \nu^m \delta_0, \frac{\epsilon}{2}\right) \qquad (1.49)$$

for any d-pseudotrajectory $\xi = \{x_k : k \in \mathbb{Z}\}$ of ϕ, any natural s, and any $0 \leq i \leq m$.

The existence of a number d with the formulated property is established similarly to the proof of Lemma 1.1.3 but since ϕ is not necessarily Lipschitz, one has to refer to the uniform continuity of ϕ (we leave the details to the reader).

Let us show that this d has the property described in Definition 1.3. Like in Bowen's proof above, we again fix a natural number r and construct points $y_0, \ldots, y_r$ such that $y_0 = x_0$,

$$y_k \in V_\delta^s(x_{km}) \cap V_\delta^u(\phi^m(y_{k-1})),$$

and

$$r(y_k, \phi(y_{k-1})) < \delta_0, \ r(y_k, x_{km}) < \delta_0 \qquad (1.50)$$

for $1 \le k \le r$. Assume that $y_1, \ldots, y_{k-1}$ are constructed. Since $y_{k-1} \in V_\delta^s(x_{(k-1)m})$, we deduce from the second inequality in (1.50) for $k - 1$ that

$$r(\phi^m(x_{(k-1)m}), \phi^m(y_{k-1})) < \nu^m \delta_0.$$

By the choice of d,

$$r(x_{km}, \phi^m(y_{k-1})) \le r(x_{km}, \phi^m(x_{(k-1)m})) + r(\phi^m(x_{(k-1)m}), \phi^m(y_{k-1})) < \delta_1.$$

Hence, we can set

$$y_k = [\phi^m(y_{k-1}), x_{km}].$$

It follows from the choice of δ_1 that inequalities (1.50) are satisfied.

To show that $y_k \in V_\delta^s(x_{km})$, we apply the definition of the set $V_\delta^s(x_{km})$ and (SS1). We can write

$$[y_k, x_{km}] = [[\phi^m(y_{k-1}), x_{km}], x_{km}] = [\phi^m(y_{k-1}), x_{km}] = y_k.$$

Now the desired inclusion follows from the second inequality in (1.50). Similarly one shows that $y_k \in V_\delta^u(\phi^m(y_{k-1}))$.

We again take $x = \phi^{-rm}(y_r)$. The estimate of

$$r(\phi^{sm+i}(x), \phi^i(y_s)), \ 0 \le s \le r, \ 0 \le i \le m,$$

is obtained similarly to the one in Bowen's proof, and we give here the final expression,

$$r(\phi^{sm+i}(x), \phi^i(y_s)) \le \frac{\delta_0}{1 - \nu}$$

(we take into account that $1 - \nu^m \ge 1 - \nu$).

Finally we see (taking (1.48) into account) that

$$r(\phi^{sm+i}(x), x_{sm+i}) \le r(\phi^{sm+i}(x), \phi^i(y_s)) + r(\phi^i(x_{sm}), \phi^i(y_s)) +$$

$$+ r(\phi^i(x_{sm}), x_{sm+i}) \le \frac{\delta_0}{1 - \nu} + \delta_0 + \frac{\epsilon}{2} < \epsilon.$$

It remains to repeat the passing to the limit for $r \to \infty$ to complete the proof. □

Remark. Since $[,]$ and ϕ are continuous (and not Lipschitz), we can prove only the POTP (instead of the LpSP) for the Smale space.

Let us show that the model of Ruelle has one more important property characteristic for neighborhoods of hyperbolic sets.

Theorem 1.2.8. *If (X, ϕ) is a Smale space, then ϕ is expansive.*

Proof. Find $\Delta \in (0, \eta)$ such that

$$r(x, [x, y]), r(y, [x, y]) < \delta \text{ for } r(x, y) \leq \Delta.$$

We claim that Δ is an expansivity constant for ϕ.

Take $x, y \in X$ with the property

$$r(\phi^n(x), \phi^n(y)) \leq \Delta, \ n \in \mathbb{Z}.$$

It follows from (1.45) that for $n > 0$ we have

$$z := \phi^n([y, x]) = [\phi^n(y), \phi^n(x)].$$

By the choice of Δ,

$$r(\phi^n(x), z) < \delta.$$

Since by (SS1) we have

$$[z, \phi^n(x)] = [[\phi^n(y), \phi^n(x)], \ \phi^n(x)] = [\phi^n(y), \phi^n(x)] = z,$$

it follows that $z \in V_\delta^u(\phi^n(x))$. Hence,

$$r(x, [y, x]) \leq \nu^n r(\phi^n(x), z) \leq \nu^n \delta.$$

The last inequality holds for any $n > 0$, this implies that

$$r(x, [y, x]) = 0, \text{ hence } x = [y, x].$$

By (1.45), for proper x, y we have

$$\phi([\phi^{-1}(x), \phi^{-1}(y)]) = [x, y],$$

this means that

$$\phi^{-1}[x, y] = [\phi^{-1}(x), \phi^{-1}(y)].$$

Thus, we can repeat the reasoning above (with $n < 0$) to show that $x = [x, y]$. Since our condition is symmetric, the same arguments show that

$$y = [x, y] = [y, x],$$

hence $x = y$.   □

Let X be a compact metrizable topological space and let ϕ be a homeomorphism of X. It is easy to see that if ϕ is a topologically Anosov homeomorphism with respect to some metric compatible with the topology of X, then it is topologically Anosov with respect to any such metric. The following result obtained independently by Ombach [Om1], Sakai [Sak4], and by Park, Lee, and Koo

[Par] shows that any topologically Anosov homeomorphism generates (in some sense) a Smale space.

Theorem 1.2.9. *Let ϕ be a homeomorphism of a compact metrizable space X. The following statements are equivalent.*

(1) ϕ is topologically Anosov (with respect to some metric);

(2) there exists a metric r such that (X, ϕ) is a Smale space with respect to r.

We refer the reader to [Om1, Sak4, Par] for the proof of this statement.

1.3 Shadowing for Mappings of Banach Spaces

In this section, we describe some "abstract" shadowing results for sequences of mappings of Banach spaces. In Subsect. 1.3.1, we prove a general "local" shadowing statement (Theorem 1.3.1) [Pi4].

The main difference between the situation described in Subsect. 1.3.1 and the one in the classical Shadowing Lemma is the possible nonuniqueness of a shadowing trajectory in the first case. Let us note that the nonuniqueness of a shadowing trajectory is not a "rare" property; for example, in the case of a structurally stable dynamical system a shadowing trajectory is, in general, not unique. On the other hand, it was mentioned in Sect. 1.1 (see the remark after Lemma 1.1.2) that "global shadowing with uniqueness" is not a common phenomenon.

Theorem 1.3.1 is one of the basic tools applied in this book to establish various shadowing properties ($\mathcal{L}_p$-shadowing, weighted shadowing, and shadowing for flows in Chap. 1, shadowing for structurally stable systems in Chap. 2, shadowing near the global attractor for a parabolic PDE in Chap. 3).

Of course, it is important to have a possibility to establish the uniqueness of a shadowing trajectory in the considered case of a sequence of mappings of Banach spaces. We give such a condition in Subsect. 1.3.2.

In Subsect. 1.3.3, we describe a general scheme of application of Theorem 1.3.1. Subsection 1.3.4 is devoted to two "abstract" theorems on shadowing proved by Chow, Lin, and Palmer [Cho1] and by Steinlein and Walther [Ste1]. We show that one can prove the first mentioned theorem applying results of Subsects. 1.3.1 and 1.3.2. It is also shown how to modify the method of proof of Theorem 1.3.1 to establish the LpSP near a "Steinlein-Walther hyperbolic set".

In [Pli2, PLi3], Pliss obtained necessary and sufficient conditions under which a family of linear systems of differential equations has uniformly bounded solutions. We "translate" these results into the shadowing language in Subsect. 1.3.5. It is shown (Theorem 1.3.6) that if a family $\{A_k\}$ of invertible linear mappings in $\mathbb{R}^n$ is piecewise hyperbolic, then the equations

$$u_{k+1} = A_k u_k + z_{k+1}$$

have a solution $\{u_k\}$ such that

$$||\{u_k\}|| \le M||\{z_k\}||,$$

where the constant M depends on the characteristics of the piecewise hyperbolicity of $\{A_k\}$. It follows from the results of [Pli3] that a converse (in some sense) statement is also true.

1.3.1 Shadowing for a Sequence of Mappings

Below we show that a lot of shadowing problems can be reduced to the following "abstract" shadowing problem.

Let H_k be a sequence of Banach spaces ($k \in \mathbb{Z}$ or $k \in \mathbb{Z}_+$), we denote by $|.|$ norms in H_k and by $||.||$ the corresponding operator norms for linear operators. Let us emphasize that the spaces H_k are not assumed to be isomorphic (and in Chap. 2 we apply the main result of this section in the case when H_k are Euclidean spaces of different dimensions).

Consider a sequence of mappings

$$\phi_k : H_k \to H_{k+1}$$

of the form

$$\phi_k(v) = A_k v + w_{k+1}(v), \tag{1.51}$$

where A_k are linear mappings.

It is assumed that the values $|\phi_k(0)|$ are uniformly small, say, $|\phi_k(0)| \le d$. We are looking for a sequence $v_k \in H_k$ such that $\phi_k(v_k) = v_{k+1}$ and the values $|v_k|$ are uniformly small, for example, the inequalities

$$\sup_k |v_k| \le Ld$$

hold with a constant L independent of d.

First we prove an auxiliary statement. Consider two Banach spaces of sequences $\bar{v} = \{v_k \in H_k\}$. Let $\mathcal{L}_\infty$ be the usual space with the norm

$$||\bar{v}||_\infty = \sup |v_k|$$

(we use the same notation in both cases $k \in \mathbb{Z}_+$ and $k \in \mathbb{Z}$), and let $\mathcal{B}$ be a Banach space such that its norm is *monotonous*, i.e., the inequalities $|v_k| \le |v'_k|$ imply $||\bar{v}||_\mathcal{B} \le ||\bar{v}'||_\mathcal{B}$ (it is also said in this case that the norm $||.||_\mathcal{B}$ is compatible with the lattice structure), and the inequalities

$$||\bar{v}||_\infty \le ||\bar{v}||_\mathcal{B} \tag{1.52}$$

hold.

Lemma 1.3.1. *Assume that for numbers $N_0, \kappa, \Delta > 0$*

(a) there exists a linear operator $\mathcal{G} : \mathcal{B} \to \mathcal{B}$ such that
(a1)
$$\|\mathcal{G}\| \leq N_0$$

(here $\|.\|$ is the operator norm generated by the norm $\|.\|_{\mathcal{B}}$ of $\mathcal{B}$);
    (a2) if $\bar{z} = \{z_k\} \in \mathcal{B}$, then the sequence $\bar{u} = \{u_k\}$ defined by $\bar{u} = \mathcal{G}\bar{z}$ satisfies the relations
$$u_{k+1} = A_k u_k + z_{k+1}; \tag{1.53}$$

(b) the inequalities
$$|w_{k+1}(v) - w_{k+1}(v')| \leq \kappa|v - v'| \ for \ |v|, |v'| \leq \Delta \tag{1.54}$$

and
$$\kappa N_0 < 1 \tag{1.55}$$

hold.
    Set
$$L = \frac{N_0}{1 - \kappa N_0}, \ d_0 = \frac{\Delta}{L}.$$

If
$$\|\{\phi_k(0)\}\|_{\mathcal{B}} \leq d \leq d_0, \tag{1.56}$$

then there exist $v_k \in H_k$ such that $\phi_k(v_k) = v_{k+1}$ and
$$\|\{v_k\}\|_{\mathcal{B}} \leq Ld. \tag{1.57}$$

Proof. We consider the case $k \in \mathbb{Z}_+$, the case $k \in \mathbb{Z}$ is considered similarly.
    Equalities
$$\phi_k(v_k) = v_{k+1}, \ k \geq 0, \tag{1.58}$$

are equivalent to
$$v_{k+1} = A_k v_k + w_{k+1}(v_k), \ k \geq 0. \tag{1.59}$$

For $\bar{v} \in \mathcal{B}$ set
$$\bar{w}(\bar{v}) = \{w_k(v_{k-1}) : k \geq 0\} \ with \ w_0 = 0$$

(in the case of $H_k, k \in \mathbb{Z}$, set $\bar{w}(\bar{v}) = \{w_k(v_{k-1})\}$).
    It follows from condition (a2) that if $\bar{v}$ is a solution of the equation
$$\bar{v} = \mathcal{G}\bar{w}(\bar{v}), \tag{1.60}$$

then relations (1.59) hold, hence $\bar{v}$ satisfies equalities (1.58). Thus, it remains to solve Eq. (1.60) and to estimate its solution.
    Assume that inequality (1.56) holds. Let D be the ball $\{\bar{v} \in \mathcal{B} : \|\bar{v}\|_{\mathcal{B}} \leq Ld\}$. Take $\bar{v}, \bar{v}' \in D$. Since $Ld \leq Ld_0 = \Delta$, we have $\|\bar{v}\|_{\mathcal{B}}, \|\bar{v}'\|_{\mathcal{B}} \leq \Delta$. Condition (1.52) implies that $\|\bar{v}\|_\infty, \|\bar{v}'\|_\infty \leq \Delta$. Since the norm of $\mathcal{B}$ is monotonous, we obtain from (1.54) the inequalities
$$\|\bar{w}(\bar{v}) - \bar{w}(\bar{v}')\|_{\mathcal{B}} \leq \kappa\|\bar{v} - \bar{v}'\|_{\mathcal{B}},$$

now it follows from condition (a1) that

$$\|\mathcal{G}\bar{w}(\bar{v}) - \mathcal{G}\bar{w}(\bar{v}')\|_{\mathcal{B}} \leq \kappa N_0 \|\bar{v} - \bar{v}'\|_{\mathcal{B}}. \tag{1.61}$$

By (1.56),

$$\|\{w_{k+1}(0)\}\|_{\mathcal{B}} = \|\{\phi_k(0)\}\|_{\mathcal{B}} \leq d,$$

hence for $\bar{v} \in D$ we have

$$\|\mathcal{G}\bar{w}(\bar{v})\|_{\mathcal{B}} \leq \|\mathcal{G}\bar{w}(0)\|_{\mathcal{B}} + \|\mathcal{G}\bar{w}(\bar{v}) - \mathcal{G}\bar{w}(0)\|_{\mathcal{B}} \leq N_0 d + N_0 \kappa L d = L d. \tag{1.62}$$

It follows from (1.55), (1.61), and (1.62) that $\mathcal{G}\bar{w}$ maps D into itself and contracts in D, hence there is a solution $\bar{v}$ of (1.60) that satisfies (1.57). This completes the proof. $\qquad\square$

Now we prove the main theorem of this section. This theorem is widely applied below to establish various shadowing properties.

Theorem 1.3.1. *Assume that*
(a) there exist numbers $\lambda \in (0,1), N \geq 1$, and projectors $P_k, Q_k : H_k \to H_k$ (we denote below $S_k = P_k H_k, U_k = Q_k H_k$) such that
(a1)

$$\|P_k\|, \|Q_k\| \leq N, \ P_k + Q_k = I;$$

(a2)

$$\|A_k\,|_{S_k}\,\| \leq \lambda, \ A_k S_k \subset S_{k+1}; \tag{1.63}$$

(b) if $U_{k+1} \neq \{0\}$, then there exist linear mappings $B_k : U_{k+1} \to H_k$ such that

$$B_k U_{k+1} \subset U_k, \ \|B_k\| \leq \lambda, \ A_k B_k\,|_{U_{k+1}} = I; \tag{1.64}$$

(c) there exist numbers $\kappa, \Delta > 0$ such that inequalities (1.54) and the inequality

$$\kappa N_1 < 1 \tag{1.65}$$

hold, where

$$N_1 = N \frac{1 + \lambda}{1 - \lambda}.$$

Set

$$L = \frac{N_1}{1 - \kappa N_1}, \ d_0 = \frac{\Delta}{L}.$$

If for a sequence of mappings (1.51) we have

$$|\phi_k(0)| \leq d \leq d_0 \ (k \geq 0 \text{ or } k \in \mathbb{Z}),$$

then there exist points $v_k \in H_k$ such that $\phi_k(v_k) = v_{k+1}$ and $|v_k| \leq L d$.

Proof. We consider the case $k \geq 0$. Take $\mathcal{B} = \mathcal{L}_\infty$, the norm of $\mathcal{B}$ is monotonous and condition (1.52) is trivially satisfied. We denote the norm of $\bar{v} \in \mathcal{B}$ simply by $\|\bar{v}\|$. Define a linear operator $\mathcal{G}$ on $\mathcal{B}$ by

$$\mathcal{G}\bar{z} = \bar{u}^1 + \bar{u}^2 + \bar{u}^3,$$

where

$$u_n^1 = P_n z_n, \ u_n^2 = \sum_{k=0}^{n-1} A_{n-1} \ldots A_k P_k z_k, \ u_n^3 = -\sum_{k=n}^{\infty} B_n \ldots B_k Q_{k+1} z_{k+1}.$$

Let us show that $\mathcal{G}$ maps $\mathcal{B}$ into $\mathcal{B}$ and estimate its norm. Take $\bar{z} \in \mathcal{B}$. Obviously,

$$||\bar{u}^1|| \leq N||\bar{z}||.$$

Since $P_k z_k \in S_k$, we obtain from (1.63) the estimate

$$|u_n^2| \leq N \sum_{k=0}^{n-1} \lambda^{n-k} |z_k| \leq N \frac{\lambda}{1-\lambda}||\bar{z}||.$$

Similarly, we apply (1.64) to show that the series defining u_n^3 converges, and the inequality

$$|u_n^3| \leq N \frac{\lambda}{1-\lambda}||\bar{z}||$$

holds. Finally, we arrive to the inequality

$$||\mathcal{G}|| \leq N + 2N \frac{\lambda}{1-\lambda} = N_1. \tag{1.66}$$

Let $\bar{u} = \mathcal{G}\bar{z}$. It follows from the equalities

$$A_n u_n^1 = A_n P_n z_n, \ A_n u_n^2 = \sum_{k=0}^{n-1} A_n \ldots A_k P_k z_k,$$

$$A_n u_n^3 = -(I - P_{n+1})z_{n+1} - \sum_{k=n+1}^{\infty} B_{n+1} \ldots B_k Q_{k+1} z_{k+1}$$

that

$$A_n u_n = u_{n+1} - z_{n+1},$$

hence $\bar{u}$ satisfies relations (1.53). Now our theorem follows from Lemma 1.3.1.
□

Remark. The case $k \in \mathbb{Z}$ is considered similarly, in the definition of $\mathcal{G}$ one has to take

$$u_n^2 = \sum_{k=-\infty}^{n-1} A_{n-1} \ldots A_k P_k z_k.$$

This does not change the estimates.

It follows from the remark after Theorem 1.2.3 that a neighborhood of a hyperbolic set has the following property. For any fixed $L' \geq L$ there exists $d_0(L')$ such that for any d-pseudotrajectory $\{x_k\}, k \in \mathbb{Z}$, with $d \leq d_0(L')$ there is a unique point x with the property

$$|\phi^k(x) - x_k| \le L'd.$$

Let us show that under the conditions of Theorem 1.3.1 the situation is qualitatively different, and the shadowing trajectory is not necessarily unique.

Example 1.12 Let $H_k = \mathbb{R}, k \in \mathbb{Z}$. Set $P_k = 0$ (so that $S_k = \{0\}, U_k = \mathbb{R}$) for $k < 0$, and $P_k = I$ (so that $S_k = \mathbb{R}, U_k = \{0\}$) for $k \ge 0$. Fix a number $\lambda \in (0,1)$ and set

$$A_k v = \lambda^{-1} v \text{ for } k < 0, \quad A_k v = \lambda v \text{ for } k \ge 0.$$

Obviously, conditions (1.63) and (1.64) are satisfied with $B_k v = \lambda v, k \le -2$. Take $d > 0$ and set $w_{k+1}(v) = d$ for all k, v. We can take $\kappa = 0$ in (1.54), so that condition (1.65) is automatically satisfied.

Fix arbitrary v_0. We deduce from the equalities $\phi_k(v_k) = v_{k+1}$ that

$$v_{k+1} = \lambda^{-1} v_k + d \text{ for } k < 0, \quad v_{k+1} = \lambda v_k + d \text{ for } k \ge 0.$$

It follows that

$$v_k = \lambda^{|k|} v_0 - d(\lambda + \ldots + \lambda^{|k|}) \text{ for } k < 0,$$
$$v_k = \lambda^k v_0 + d(1 + \lambda + \ldots + \lambda^{k-1}) \text{ for } k \ge 0.$$

Hence, for any v_0 with $|v_0| \le d$ we obtain a sequence $\bar{v} = \{v_k\}$ satisfying the equalities $\phi_k(v_k) = v_{k+1}$ and such that

$$\|\bar{v}\| \le \left(1 + \frac{1}{1-\lambda}\right) d.$$

Note that the described construction is a model of behavior typical for structurally stable systems. As the simplest example, take a diffeomorphism ϕ of the circle S^1 which has two hyperbolic fixed points, p with $D\phi(p) = \lambda < 1$, and q with $D\phi(q) = \lambda^{-1}$, and such that any trajectory of ϕ tends to $p \cup q$. Obviously, the behavior of pseudotrajectories of ϕ is similar to the one described in our example (we leave the details to the reader).

In addition, there exist interesting infinite-dimensional examples in which H_k are the same for all k, but S_k "grow" and U_k "decrease" at every step.

Example 1.13 Consider $H_k = l_\infty, k \in \mathbb{Z}$, where l_∞ is the usual Banach space of sequences $x = \{x_k : x_k \in \mathbb{R}, k \in \mathbb{Z}\}$, with

$$\|x\| = \sup_k |x_k|.$$

For $k \in \mathbb{Z}$ let $A_k : H_k \to H_{k+1}$ be defined as follows:

$$(A_k x)_m = \begin{cases} 1/2 x_m, & m \le k, \\ 2x_m, & m > k. \end{cases}$$

Obviously, conditions of Theorem 1.3.1 are satisfied in this example with $\lambda = 1/2, N = 1$,

$$(P_k x)_m = \begin{cases} x_m, & m \le k, \\ 0, & m > k. \end{cases}$$

1.3.2 Conditions of Uniqueness

Let us again consider a sequence of mappings

$$\phi_k : H_k \to H_{k+1}$$

which have form (1.51), where H_k are Banach spaces. We give sufficient conditions for the uniqueness of a sequence $\{v_k\}$ obtained in Theorem 1.3.1.

Theorem 1.3.2.   *Assume that condition (a) of Theorem 1.3.1 holds. Assume also that*
   (b')

$$A_k U_k \subset U_{k+1} \text{ and } \|A_k|_{U_k}\| \geq \frac{1}{\lambda}; \tag{1.67}$$

   (c') there exist numbers $\kappa_0, \Delta > 0$ such that

$$|w_{k+1}(v) - w_{k+1}(v')| \leq \kappa_0 |v - v'| \text{ for } |v|, |v'| \leq \Delta, \tag{1.68}$$

and for $\kappa = N\kappa_0$ the inequalities

$$\lambda + 2\kappa < 1, \quad \frac{1}{\lambda} - 2\kappa \geq \gamma > 1, \quad \frac{\lambda}{\gamma} + \frac{2\kappa}{\gamma} < 1 \tag{1.69}$$

are fulfilled.
   Then the relations

$$\phi_k(v_k) = v_{k+1}, \quad \phi_k(u_k) = u_{k+1}, \quad |v_k|, |u_k| \leq \Delta, \quad k \in \mathbb{Z}, \tag{1.70}$$

imply that $v_k = u_k, k \in \mathbb{Z}$.

Proof. Assume that there exist two sequences $\{v_k\}, \{u_k\}$ with properties (1.70). Set $z_k = v_k - u_k$. It follows from (1.68) that

$$z_{k+1} = A_k z_k + w'_{k+1}, \tag{1.71}$$

where

$$|w'_{k+1}| \leq \kappa_0 |z_k|. \tag{1.72}$$

   Represent

$$z_k = z_k^s + z_k^u,$$

where

$$z_k^s = P_k z_k \in S_k, \quad z_k^u = Q_k z_k \in U_k.$$

Since $|z_k| \leq 2\Delta$, we see that

$$|z_k^s|, |z_k^u| \leq 2N\Delta. \tag{1.73}$$

By (a),(b'), $A_k z_k^s \in S_{k+1}$, $A_k z_k^u \in U_{k+1}$, and it follows from the equalities

$$z_{k+1}^s = P_{k+1} A_k (z_k^s + z_k^u) + P_{k+1} w_{k+1}'$$

and from (1.72) that

$$|z_{k+1}^s| \leq |P_{k+1} A_k z_k^s| + |P_{k+1} w_{k+1}'| \leq \lambda |z_k^s| + \kappa |z_k|. \tag{1.74}$$

Similarly, we obtain the estimate

$$|z_{k+1}^u| \geq \frac{1}{\lambda} |z_k^u| - \kappa |z_k|. \tag{1.75}$$

Assume that there exists $l \in \mathbb{Z}$ such that $z_l \neq 0$. Consider two possible cases.

Case 1. There exists $m \leq l$ such that

$$|z_m^u| > |z_m^s|. \tag{1.76}$$

Then obviously $z_m^u \neq 0$.

Let us show that in this case

$$z_{m+1}^u \neq 0 \text{ and } |z_{m+1}^u| > |z_{m+1}^s|. \tag{1.77}$$

It follows from (1.76) that

$$|z_m| \leq |z_m^s| + |z_m^u| < 2|z_m^u|.$$

Now we deduce from (1.69) and (1.75) that

$$|z_{m+1}^u| \geq \frac{1}{\lambda} |z_m^u| - 2\kappa |z_m^u| \geq \gamma |z_m^u|,$$

this proves the first inequality in (1.77). It follows from the obtained inequality and from (1.74) that

$$\frac{|z_{m+1}^s|}{|z_{m+1}^u|} \leq \frac{\lambda |z_m^s| + 2\kappa |z_m^u|}{\gamma |z_m^u|} \leq \frac{\lambda}{\gamma} + \frac{2\kappa}{\gamma} < 1$$

(see (1.69)). This proves the second part of (1.77).

Repeating the same reasons, we see that the inequalities

$$|z_{m+k+1}^u| \geq \gamma |z_{m+k}^u|$$

hold for all $k \geq 0$. But these inequalities imply that

$$|z_{m+k}^u| \geq \gamma^k |z_m^u| \to \infty \text{ as } k \to \infty,$$

and this contradicts to (1.73). We proved that case 1 is impossible.

Case 2. For all $m \leq l$ we have

$$|z_m^u| \leq |z_m^s|. \tag{1.78}$$

Then
$$|z_m| \leq |z_m^s| + |z_m^u| \leq 2|z_m^s|, \ m \leq l.$$

It follows from (1.74) that

$$|z_{m+1}^s| \leq \lambda|z_m^s| + 2\kappa|z_m^s| \leq \lambda_1|z_m^s|, \ m \leq l, \ \text{where} \ \lambda_1 = \lambda + 2\kappa < 1.$$

This gives the inequalities

$$|z_l^s| \leq \lambda_1|z_{l-1}^s| \leq \ldots \leq \lambda_1^k|z_{l-k}^s| \leq \lambda_1^k N \Delta \tag{1.79}$$

for any $k \geq 0$ (we apply (1.73) here).

Since $k \geq 0$ is arbitrary in (1.79), we see that $z_l^s = 0$. But then (1.78) implies that also $z_l^u = 0$, hence $z_l = 0$. The obtained contradiction completes the proof. □

Remark. [1] An analogous statement can be proved in the same way for a sequence of mappings

$$\chi_k : H_k \to H_{k-1}$$

of the form

$$\chi_k(v) = B_k v + w_{k-1}(v)$$

under the conditions

$$B_k U_k \subset U_{k-1}, \ B_k S_k \subset S_{k-1}.$$

We leave the details to the reader.

Remark. [2] We also can obtain conditions of uniqueness for a "one-sided" sequence of mappings

$$\phi_k : H_k \to H_{k+1}, \ k \geq 0.$$

Assume that $H_k = U_k$ for $k \geq 0$, that condition (b') of Theorem 1.3.2 holds, that there exist numbers $\kappa_0, \Delta > 0$ such that (1.68) is fulfilled, and, finally, that

$$\frac{1}{\lambda} - \kappa_0 > 1.$$

Then the relations

$$\phi_k(v_k) = v_{k+1}, \ \phi_k(u_k) = u_{k+1}, \ |v_k|, |u_k| \leq \Delta, \ k \geq 0,$$

imply that $v_k = u_k, k \geq 0$.

We again take $z_k = u_k - v_k$. It follows that the inequalities

$$|z_{k+1}| \geq \frac{1}{\lambda}|z_k| - \kappa_0|z_k| \geq \gamma|z_k|, \ k \geq 0, \ \text{where} \ \gamma = \frac{1}{\lambda} - \kappa_0,$$

similar to (1.75), are true. If we assume that $z_l \neq 0$ for some $l \in \mathbb{Z}_+$, this will lead to

$$|z_k| \to \infty \ \text{as} \ k \to \infty,$$

and the obtained contradiction will prove our statement.

1.3.3 Application to the Classical Shadowing Lemma

We describe in this subsection a scheme of application of Theorem 1.3.1. The main application here is the Shadowing Lemma in the case $k \geq 0$ (Theorem 1.3.3 below). Similar arguments are applied in this book to study more complicated situations (such as shadowing for structurally stable systems in Chap. 2), so it is useful for the reader to see their application in the simplest case.

This scheme is applied literally in the same way to establish $\mathcal{L}_p$-shadowing and weighted shadowing (see Theorems 1.4.2 and 1.4.5 below). Let us describe here the common first step of the scheme (below we denote by SL, $\mathcal{L}_p$, and WS the cases of the Shadowing Lemma, of $\mathcal{L}_p$-shadowing, and of weighted shadowing, respectively).

The main idea of the proof is the following one. Let $\xi = \{x_k\}$ (where $k \in \mathbb{Z}$ or $k \in \mathbb{Z}_+$) be a sequence in a neighborhood W of an invariant set Λ for a diffeomorphism ϕ such that

$$|\phi(x_k) - x_{k+1}| \leq d.$$

We consider the sequence of Banach spaces $H_k = \mathbb{R}^n$ and the sequence of mappings $\phi_k(v) = \phi(x_k + v) - x_{k+1}$. The "extended" hyperbolic structure S, U given by Lemma 1.2.1 (for the SL and $\mathcal{L}_p$ cases) and a similar structure connected with the Sacker-Sell spectrum (for the WS case) allow us to show that we can represent ϕ_k in form (1.51) and then apply Theorem 1.3.1 (in the SL case) or its modifications (in other cases).

Thus, we consider a compact invariant set Λ for a diffeomorphism ϕ of $\mathbb{R}^n$.

If Λ is a hyperbolic set of ϕ, let C, λ_0 be the hyperbolicity constants of Λ. We take $\epsilon > 0, \lambda \in (\lambda_0, 1)$, and construct the corresponding Lyapunov norm and extended hyperbolic structure S', U' in a neighborghood W of Λ.

In the conditions of Theorem 1.4.5, let C, λ_0 be given by Theorem 1.4.4. We again take $\epsilon > 0, \lambda \in (\lambda_0, 1)$, and apply in this case Lemma 1.4.1 to construct a "Lyapunov" norm and an extended structure S', U' in a neighborghood W of Λ.

The same reasons as in the proof of Theorem 1.2.3 show that, without loss of generality, we may assume the Euclidean norm to be Lyapunov. We write below S, U instead of S', U'. Decreasing W, if necessary, we can find a constant $N > 0$ such that

$$||\Pi_x^s||, ||\Pi_x^u|| \leq N \tag{1.80}$$

for $x \in W$.

Obviously, we can find $\epsilon > 0$ such that for the number

$$\kappa_0 = \frac{1}{4N_1},$$

where $N_1 = N_1(N, \lambda)$ is defined in Theorem 1.3.1, the inequality

$$2N\epsilon < \kappa_0 \tag{1.81}$$

holds.

Now we take $d_1 > 0$ such that, for $x, y \in W$ with $|x - y| \leq d_1$, inequalities (1.10) and (1.11) of Lemma 1.2.1 (for the SL and $\mathcal{L}_p$ cases), and similar inequalities of Lemma 1.4.1 (for the WS case) are satisfied.

Take a sequence $\xi = \{x_k : k \geq 0\} \subset W$ with

$$|x_{k+1} - \phi(x_k)| < d \leq d_1, \; k \geq 0.$$

Set $H_k = \mathbb{R}^n, k \geq 0$. Define $\phi_k : H_k \to H_{k+1}$ by

$$\phi_k(v) = \phi(x_k + v) - x_{k+1}.$$

Since the neighborghood W is bounded, we can find $d_0 \in (0, d_1)$ such that, for the representation

$$\phi_k(v) = D\phi_k(0)v + \chi_{k+1}(v),$$

the inequalities

$$|\chi_{k+1}(v) - \chi_{k+1}(v')| \leq \kappa_0|v - v'| \text{ for } |v|, |v'| \leq d_0 \tag{1.82}$$

hold.

Denote $D_k = D\phi_k(0), S_k = S(x_k), P_k = \Pi_{x_k}^s, U_k = U(x_k), Q_k = \Pi_{x_k}^u$. Set, for $k \geq 0$,

$$A_k = A_k^s + A_k^u, \text{ where } A_k^s = P_{k+1}D_kP_k, \; A_k^u = Q_{k+1}D_kQ_k.$$

Let us estimate

$$\|A_k - D_k\| = \|A_k^s + A_k^u - D_k(P_k + Q_k)\| \leq$$

$$\leq \|(P_{k+1} - I)D_kP_k\| + \|(Q_{k+1} - I)D_kQ_k\|. \tag{1.83}$$

Take $v \in H_k$. Inequalities (1.80) imply that $|P_k v| \leq N|v|$. Since $P_k v \in S_k$, it follows from Lemma 1.2.1 (or from Lemma 1.4.1 for the WS case) that the estimate

$$|(P_{k+1} - I)D_kP_kv| = |Q_{k+1}D_kP_kv| \leq N\epsilon|v|$$

holds. Hence, the first term in (1.83) does not exceed $N\epsilon$. Similarly, the second term is also estimated by $N\epsilon$. By (1.81), this leads us to the inequality

$$\|A_k - D_k\| \leq 2N\epsilon < \kappa_0. \tag{1.84}$$

Now we represent

$$\phi_k(v) = A_kv + w_{k+1}(v),$$

where

$$w_{k+1}(v) = (D_k - A_k)v + \chi_{k+1}(v).$$

Note that

$$|w_{k+1}(0)| = |\phi_k(0)| < d.$$

We deduce from (1.82) and (1.83) that

$$|w_{k+1}(v) - w_{k+1}(v')| \le 2\kappa_0|v - v'| \text{ for } |v|, |v'| \le d_0. \tag{1.85}$$

This completes the description of the common part of the proofs for the SL, $\mathcal{L}_p$, and WS cases.

Now let us formulate and prove the analog of Theorem 1.2.3 for the LpSP$_+$. (Note that one can also apply Bowen's method described in Subsect. 1.2.4 to prove the statement below.)

Theorem 1.3.3. *If Λ is a hyperbolic set for a diffeomorphism ϕ, then there exists a neighborhood W of Λ such that ϕ has the LpSP$_+$ on W.*

Proof. It follows from Lemma 1.2.1 that

$$A_k S_k = P_{k+1} D_k S_k = S_{k+1}, \ A_k U_k = Q_{k+1} D_k U_k = U_{k+1},$$

and that for $A_k^s = A_k|_{S_k}$, $B_k = (A_k^u|_{U_k})^{-1}$ conditions (1.63),(1.64) are fulfilled.

Hence, we deduce from (1.85) and from the equality

$$2\kappa_0 N_1 = \frac{1}{2}$$

that Theorem 1.3.1 is applicable to $\xi = \{x_k : k \ge 0\}$ if

$$|\phi(x_k) - x_{k+1}| \le d \le d_0.$$

Thus, for the point $x = x_0 + v_0$ we have

$$|\phi^k(x) - x_k| \le Ld, \ k \ge 0.$$

This completes the proof. $\square$

1.3.4 Theorems of Chow-Lin-Palmer and Steinlein-Walther

Many shadowing results were established for mappings of Banach spaces. In this subsection, we describe in detail a variant of "abstract" shadowing lemma proved by Chow, Lin, and Palmer in [Cho1] and a theorem of Steinlein and Walther [Ste1]. A detailed treatment of exponential dichotomies and shadowing for mappings of Banach spaces is given by Henry [He3].

Chow, Lin, and Palmer studied in [Cho1] a sequence of C^1 mappings

$$\psi_k : H_k \to H_{k+1}, \ k \in \mathbb{Z},$$

of Banach spaces H_k under the following conditions:

(1) there exist subsets $T_k \subset H_k$ such that $\psi_k(T_k) \subset T_{k+1}$;

(2) for any $x \in T_k$ there is a continuous (in x) splitting

$$H_k = S_k(x) \oplus U_k(x) \tag{1.86}$$

such that

$$D\psi_k(x)S_k(x) \subset S_{k+1}(\psi_k(x)), \ D\psi_k(x)U_k(x) = U_{k+1}(\psi_k(x)),$$

and $D\psi_k(x) : U_k(x) \to U_{k+1}(\psi_k(x))$ is an isomorphism with a bounded inverse;

(3) there is a constant $N > 0$ such that if $P_k(x)$ $(Q_k(x))$ are the projectors in H_k onto $S_k(x)$ parallel to $U_k(x)$ (respectively, onto $U_k(x)$ parallel to $S_k(x)$), then

$$\|P_k(x)\|, \|Q_k(x)\| \leq N \text{ for all } x \in T_k, \ k \in \mathbb{Z};$$

(4) there exists $\lambda \in (0,1)$ such that, for any finite sequence $x_k, x_{k+1} = \psi_k(x_k), \ldots, x_{n+1} = \psi_n(x_n)$ with $x_k \in H_k$ and any integers $k \leq n$, the inequalities

$$\|D\psi_n(x_n) \ldots D\psi_k(x_k)P_k(x_k)\| \leq N\lambda^{n-k+1}, \tag{1.87}$$

$$\|D\psi_k(x_k)^{-1} \ldots D\psi_n(x_n)^{-1}Q_{n+1}(x_{n+1})\| \leq N\lambda^{n-k+1} \tag{1.88}$$

hold;

(5) there exists $\Delta > 0$ such that ψ_k and $D\psi_k$ are bounded and continuous in closed Δ-neighborghoods T'_k of T_k uniformly with respect to x in T'_k and $k \in \mathbb{Z}$.

The main result of [Cho1] states that there exists $\epsilon_0 > 0$ such that given $\epsilon \in (0, \epsilon_0)$ one can find $d > 0$ with the following property. If for a sequence $\{y_k \in T_k\}, k \in \mathbb{Z}$, the inequalities

$$|\psi_k(y_k) - y_{k+1}| \leq d, \ k \in \mathbb{Z}, \tag{1.89}$$

hold, then there exists a unique sequence $x_k \in H_k$, $k \in \mathbb{Z}$, such that

$$\psi_k(x_k) = x_{k+1} \text{ and } |x_k - y_k| \leq \epsilon, \ k \in \mathbb{Z}. \tag{1.90}$$

Let us show how to prove this statement applying Theorems 1.3.1 and 1.3.2. Fix a natural number ν such that

$$N\lambda^\nu \leq \lambda \tag{1.91}$$

and define, for $k \in \mathbb{Z}$,

$$\Psi_{k,\nu} = \psi_{k+\nu-1} \circ \ldots \circ \psi_k.$$

Set $G_m = H_{m\nu}, m \in \mathbb{Z}$. Take $p \in T_{m\nu}$ and denote

$$p_0 = p, \ p_1 = \psi_{m\nu}(p_0), \ \ldots, \ p_{m\nu+\nu-1} = \psi_{m\nu+\nu-2}(p_{m\nu+\nu-2}).$$

Then we can write, for $p \in T_{m\nu}$,

$$\Psi_{m\nu,\nu}(p+v) = \Psi_{m\nu,\nu}(p) + A_m v + w'_{m+1}(v),$$

where

$$A_m = D\psi_{m\nu+\nu-1}(p_{m\nu+\nu-1}) \ldots D\psi_{m\nu}(p_0).$$

It follows from the definition of A_m and from (1.87), (1.88), and (1.91) that for

$$W'_m = W_{m\nu}(p), \ W = S, U, P, Q; \ p \in T_{m\nu},$$

and for $A_m : G_m \to G_m$, analogs of conditions (a) and (b) in Theorem 1.3.1 are satisfied.

By condition (5) of [Cho1], given $\kappa > 0$ we can find $\Delta_0 \in (0, \Delta)$ (Δ is used in the definition of T'_k) such that for w'_{m+1} an analog of (1.54) holds (with Δ_0 instead of Δ). Take numbers d_0, L given by Theorem 1.3.1 for λ, N, Δ_0.

The same condition (5) implies the existence of a uniform Lipschitz constant $\mathcal{L}$ for ψ_k on the sets T'_k. Take $d_1 > 0$ such that

$$d_1 L_1 \leq d_0, \text{ where } L_1 = 1 + \mathcal{L} + \ldots + \mathcal{L}^{\nu-1},$$

and consider a sequence $y_k \in T_k$ such that (1.89) holds with $d \leq d_1$. Denote $\xi_m = y_{m\nu}$.

Set

$$\phi_m(v) = \Psi_{m\nu,\nu}(\xi_m + v) - \xi_{m+1}, \ m \in \mathbb{Z}.$$

By Lemma 1.1.3, the inequalities

$$|\phi_m(0)| = |\Psi_{m\nu,\nu}(\xi_m) - \xi_{m+1}| \leq dL_1 \leq d_0$$

hold.

We can represent

$$\phi_m(v) = A_m v + w_{m+1}(v),$$

where

$$w_{m+1}(v) = \Psi_{m\nu,\nu}(\xi_m) - \xi_{m+1} + w'_{m+1}(v).$$

Since

$$|w'_{m+1}(v) - w'_{m+1}(v')| = |w_{m+1}(v) - w_{m+1}(v')|,$$

inequalities (1.54) hold for $|v|, |v'| \leq \Delta_0$.

Applying Theorem 1.3.1, we see that there exists a sequence $v_m \in G_m$ with $|v_m| \leq LL_1 d$ such that for $z_m = \xi_m + v_m$ we have

$$\Psi_{m\nu,\nu}(z_m) = z_{m+1}, \ m \in \mathbb{Z}.$$

We obtain the estimates

$$|z_m - \xi_m| \leq LL_1 d, \ m \in \mathbb{Z}. \tag{1.92}$$

Represent arbitrary $k \in \mathbb{Z}$ in the form $k = k_0 \nu + k_1$, where $0 \leq k_1 \leq \nu - 1$, and set

$$x_k = \psi_{k-1} \circ \ldots \circ \psi_{k_0 \nu}(z_{k_0}).$$

It follows from Lemma 1.3.1 that $\psi_k(x_k) = x_{k+1}$ and

$$|x_k - y_k| \leq L_2 d,$$

where $L_2 = (\mathcal{L}^{\nu-1} L + 1) L_1$. Thus, if we take $d = \epsilon/L_2$, then the sequence $\{x_k\}$ satisfies the inequalities in (1.90).

To prove the uniqueness statement, take κ_0 such that for $\kappa = N\kappa_0$ inequalities (1.69) hold. Find $d' > 0$ such that for $|v|, |v'| \leq d'$ the inequality in (1.54) is fulfilled. It follows that if $d \leq d'$, then the sequence z_m that satisfies the inequalities

$$|z_m - \xi_m| \leq \epsilon_0, \ m \in \mathbb{Z},$$

with $\epsilon_0 = L_2 d$ is unique. Obviously, this implies the uniqueness of a sequence x_k for which (1.90) holds.

An "abstract" shadowing result based on another definition of a hyperbolic set was proved by Steinlein and Walther [Ste1]. This approach can be applied, for example, to study homoclinic trajectories for differential delay equations [Ste2]. We show that this result can be obtained as a corollary of a theorem similar to Theorem 1.3.1.

Let us begin with the definition of a hyperbolic set in the sense of Steinlein-Walther (our variant is a slight modification of the definition given in [Ste1]).

Let H be a Banach space and let $\phi : V \to H$ be a mapping of class C^1 on an open subset V of H. It is assumed that $D\phi$ is uniformly continuous on V and there exists a constant $K_0 \geq 1$ such that $\|D\phi(x)\| \leq K_0$ on V.

Definition 1.14 *We say that a set $T \subset V$ is a "Steinlein-Walther hyperbolic set" if*

(1) T is positively invariant for ϕ (i.e., $\phi(T) \subset T$) and there exists $r > 0$ such that $N_r(T) \subset V$;

(2) there exist complementary projectors $P(x), Q(x)$ on T (i.e., $P + Q = I$) and numbers $\lambda_0 \in (0,1)$, $K_1, C > 0$ such that P, Q are uniformly continuous on T, the inequalities $\|P(x)\|, \|Q(x)\| \leq K_1$ hold, and the spaces

$$S(x) = P(x)H, \ U(x) = Q(x)H$$

have the following properties:
(2.1)
$$D\phi(x)S(x) \subset S(\phi(x));$$

(2.2)
$$S(\phi(x)) + D\phi(x)U(x) = H;$$

(2.3)
$$|D\phi^k(x)v| \leq C\lambda_0^k|v| \ for \ v \in S(x), \ k \geq 0;$$

$$|Q(\phi^k(x))D\phi^k(x)v| \geq \frac{\lambda_0^{-k}}{C}|v| \ for \ v \in U(x), \ k \geq 0.$$

This definition of a hyperbolic set is more general than the usual ones (for example, than the one applied in [Cho1], see above), since it does not require the "unstable" spaces U to be $D\phi$-invariant. A detailed treatment of various definitions of hyperbolic sets for noninvertible mappings of Banach spaces can be found in [Lan1]. Below we prove the following statement.

Theorem 1.3.4. *If T is a Steinlein-Walther hyperbolic set for ϕ, then ϕ has the LpSP on T.*

Remark. The main result in [Ste1] states that under some additional smoothness assumption ϕ has the POTP and the SUP on T, here we prove only the existence statement.

Now we formulate and prove a result close to Theorem 1.3.1. Consider again a sequence of Banach spaces $H_k, k \in \mathbb{Z}$, and a sequence of mappings $\phi_k : H_k \to H_{k+1}$ of form (1.51), where $A_k = B_k + C_k$, and B_k, C_k are linear mappings.

Theorem 1.3.5. *Assume that*
(a) there exist numbers $\lambda \in (0,1), N \geq 1$, and projectors $P_k, Q_k : H_k \to H_k$ (we denote below $S_k = P_k H_k, U_k = Q_k H_k$) such that
(a1)
$$||C_k||, ||P_k||, ||Q_k|| \leq N, \ P_k + Q_k = I;$$

(a2)
$$B_k S_k \subset S_{k+1} \ and \ ||B_k|_{S_k}|| \leq \lambda; \tag{1.93}$$

(a3)
$$B_k U_k = U_{k+1} \ and \ ||(B_k)^{-1}|_{U_{k+1}}|| \leq \lambda; \tag{1.94}$$

(b)
$$C_k|_{S_k} = 0, \ C_k U_k \subset S_{k+1}; \tag{1.95}$$

(c) there exist numbers $\kappa, \Delta > 0$ such that inequality (1.54) and the inequality
$$\kappa N_2 < 1 \tag{1.96}$$
hold, where
$$N_2 = N \frac{1 + N^2 \lambda}{(1 - \lambda)^2}.$$

Then there exist constants $d_0, L > 0$ with the following property: if
$$|\phi_k(0)| \leq d \leq d_0 \ for \ k \in \mathbb{Z},$$
then there is a sequence $v_k \in H_k$ such that $\phi_k(v_k) = v_{k+1}$ and $|v_k| \leq Ld$.

Proof. The proof is similar to the one of Theorem 1.3.1. We take the space $\mathcal{B} = \mathcal{L}_\infty$, denote the norm of $\bar{v} \in \mathcal{B}$ by $||\bar{v}||$, and define a linear operator $\mathcal{G}$ on $\mathcal{B}$ as follows: we fix $\bar{z} \in \mathcal{B}$ and set
$$\mathcal{G}\bar{z} = \bar{u}^1 + \bar{u}^2 + \bar{y},$$
where we first define $\bar{y}$ by
$$y_n = -\sum_{k=n}^{\infty} B_n^{-1} \dots B_k^{-1} Q_{k+1} z_{k+1},$$

and then set

$$u_n^1 = P_n(z_n + C_{n-1}y_{n-1}), \ u_n^2 = \sum_{k=-\infty}^{n-1} B_{n-1}\ldots B_k(P_k z_k + C_{k-1}y_{k-1}).$$

Let us show that $\mathcal{G}$ maps $\mathcal{B}$ into $\mathcal{B}$ and estimate its norm. Take $\bar{z} \in \mathcal{B}$. Since $Q_{k+1}z_{k+1} \in U_{k+1}$, we have

$$|y_n| \le N \sum_{k=n}^{\infty} \lambda^{k-n+1} |z_{k+1}| \le N \frac{\lambda}{1-\lambda} ||\bar{z}||,$$

hence

$$||\bar{y}|| \le N \frac{\lambda}{1-\lambda} ||\bar{z}||.$$

This estimate and conditions (a1),(a2) imply the inequality

$$||\bar{u}^1|| \le N \left(1 + \frac{N^2 \lambda}{1-\lambda}\right) ||\bar{z}||.$$

By construction, $y_{k-1} \in U_{k-1}$. It follows from condition (b) that $C_{k-1}y_{k-1} \in S_k$. Since $P_k z_k \in S_k$, we see that

$$|u_n^2| \le N \sum_{k=-\infty}^{n-1} \lambda^{n-k} \left(1 + \frac{N\lambda}{1-\lambda}\right) ||\bar{z}|| \le \frac{N\lambda}{1-\lambda}\left(1 + \frac{N\lambda}{1-\lambda}\right)||\bar{z}||.$$

Finally, we arrive to the inequality

$$||\mathcal{G}|| \le \frac{N}{1-\lambda}\left(\lambda + 1 - \lambda + N^2\lambda + \lambda + \frac{N\lambda^2}{1-\lambda}\right) \le N_2$$

(here we take into account that $-\lambda^2 - N^2\lambda^2 + N\lambda^2 \le 0$).

Let $\bar{u} = \mathcal{G}\bar{z}$. It follows from the relations

$$B_n y_n = -Q_{n+1}z_{n+1} + y_{n+1},$$

$$B_n u_n^1 = B_n P_n z_n + B_n P_n C_{n-1} y_{n-1} = B_n P_n z_n + B_n C_{n-1} y_{n-1}, \ C_n u_n^1 = 0,$$

and

$$B_n u_n^2 = u_{n+1}^2 - B_n(P_n z_n + C_{n-1}y_{n-1}), \ C_n u_n^2 = 0,$$

that

$$(B_n + C_n)(y_n + u_n^1 + u_n^2) = -Q_{n+1}z_{n+1} + y_{n+1} + u_{n+1}^2 + C_n y_n =$$

$$= -z_{n+1} + P_{n+1}z_{n+1} + u_{n+1}^2 + y_{n+1} + P_{n+1}C_n y_n,$$

hence

$$u_{n+1} = (B_n + C_n)u_n + z_{n+1}, \ n \in \mathbb{Z}.$$

Now we set

$$L = \frac{N_2}{1 - \kappa N_2}, \quad d_0 = \frac{\Delta}{L}.$$

It follows from Lemma 1.3.1 that if $|\phi_k(0)| \le d \le d_0$, then there exists a sequence $v_k \in H_k$ such that $\phi_k(v_k) = v_{k+1}$ and $|v_k| \le Ld$.   □

Now let us reduce Theorem 1.3.4 to Theorem 1.3.5. First we note that condition (2.2) in Definition 1.14 implies the relation

$$Q(\phi(x))D\phi(x)U(x) = U(\phi(x)) \tag{1.97}$$

for $x \in T$.

Indeed, we have

$$U(\phi(x)) = Q(\phi(x))H = Q(\phi(x))(S(\phi(x)) + D\phi(x)U(x)) =$$

$$= Q(\phi(x))D\phi(x)U(x),$$

since $Q(\phi(x))S(\phi(x)) = \{0\}$.

To simplify our notation, we assume that $C = 1$ in condition (2.3) (this can be achieved passing from ϕ to ϕ^m with a proper natural m and applying Lemma 1.1.3). Set $K = \max(K_0, K_1)$.

Take a number $\lambda \in (\lambda_0, 1)$ and find $\epsilon > 0$ such that

$$\lambda_0 + K\epsilon < \lambda, \quad \frac{1}{\lambda_0} - K\epsilon > \frac{1}{\lambda}. \tag{1.98}$$

Set $N = K^3$, introduce $N_2 = N_2(N, \lambda)$ (see Theorem 1.3.5), and find $\kappa > 0$ such that inequality (1.96) holds. We assume that the inequality

$$2\epsilon K^2 < \kappa \tag{1.99}$$

is fulfilled.

Find $\Delta \in (0, r)$ (the number r is from condition (1) in Definition 1.14) such that the following statements hold:

(s1) if

$$\phi(x + v) = \phi(x) + D\phi(x)v + \chi(x, v),$$

then

$$|\chi(x, v) - \chi(x, v')| \le \frac{\kappa}{2} \text{ for } x \in T, \ |v|, |v'| \le \Delta;$$

(s2) for $x, y \in T$, the inequality $|x - y| \le \Delta$ implies that

(s2.1)

$$\|P(x) - P(y)\|, \|Q(x) - Q(y)\| < \epsilon,$$

and

(s2.2) if L is a linear subspace of H such that $Q(x)L = U(x)$, then $Q(y)L = U(y)$.

The existence of Δ follows from our assumptions on ϕ and T. Take a sequence $\xi = \{x_k : k \in \mathbb{Z}\} \subset T$ such that

$$|\phi(x_k) - x_{k+1}| \le d \le \Delta.$$

Denote

$$P_k = P(x_k), \ P'_k = P(\phi(x_k)), \ Q_k = Q(x_k), \ Q'_k = Q(\phi(x_k)),$$

and set $S_k = P_k H, U_k = Q_k H$.

Define linear operators

$$B_k = P_{k+1} D\phi(x_k) P_k + Q_{k+1} D\phi(x_k) Q_k, \ C_k = P_{k+1} D\phi(x_k) Q_k.$$

Take $v \in S_k$. Since $Q_k v = 0$ and $P_{k+1} H = S_{k+1}$, we deduce that $B_k v \in S_{k+1}$. It follows from (2.3), (s2.1), and (1.98) that

$$|B_k v| \le |P'_k D\phi(x_k) v| + |(P_{k+1} - P'_k) D\phi(x_k) v| \le$$

$$\le \lambda_0 |v| + \epsilon K |v| \le \lambda |v|.$$

This shows that B_k satisfies condition (a2) of Theorem 1.3.5.

Let $L = D\phi(x_k) U(x_k)$. Since $B_k U(x_k) = Q_{k+1} L$ and $Q'_k L = U(\phi(x_k))$ (see (1.97)), (s2.2) implies that $B_k U_k = U_{k+1}$.

Take $v \in U_k$. By (2.3), we have

$$|Q'_k D\phi(x_k) v| \ge \frac{1}{\lambda_0} |v|.$$

It follows from (1.98) and (s2.1) that

$$|B_k v| = |Q_{k+1} D\phi(x_k) v| \ge$$

$$\ge |Q'_k D\phi(x_k) v| - |(Q_{k+1} - Q'_k) D\phi(x_k) v| \ge \left(\frac{1}{\lambda_0} - K\epsilon \right) |v| \ge \frac{1}{\lambda} |v|.$$

Hence, B_k satisfies condition (a3) of Theorem 1.3.5.

Obviously, $\|C_k\| \le K^3 = N$. Since $K \ge 1$, the estimates $\|P_k\|, \|Q_k\| \le N$ are fulfilled.

Define, for $|v| \le \Delta$,

$$\phi_k(v) = [\phi(x_k) - x_{k+1}] + D\phi(x_k) v + \chi(x_k, v).$$

A sequence $\{v_k\}$ satisfies the relations

$$\phi(x_k + v_k) = x_{k+1} + v_{k+1}$$

if and only if $\phi_k(v_k) = v_{k+1}$. Represent $\phi_k(v)$ in form (1.51) with $A_k = B_k + C_k$ and

$$w_{k+1}(v) = [\phi(x_k) - x_{k+1}] + (D\phi(x_k) - A_k) v + \chi(x_k, v).$$

Let us estimate $\|(D\phi(x_k) - A_k)\|$. Note that since $Q'_k D\phi(x_k) P_k = \{0\}$ by condition (2.1) of Definition 1.14, we can write

$$C_k = P_{k+1}D\phi(x_k)Q_k + Q'_k D\phi(x_k)P_k.$$

It follows from $D\phi(x_k) = (P_{k+1} + Q_{k+1})D\phi(x_k)(P_k + Q_k)$ that

$$||(D\phi(x_k) - A_k)|| = ||(Q_{k+1} - Q'_k)D\phi(x_k)P_k|| \leq \epsilon K^2.$$

We see that if $d \leq \Delta$, then

$$|\phi_k(0)| \leq d,$$

and

$$|w_{k+1}(v) - w_{k+1}(v')| \leq \kappa \text{ for } |v|, |v'| \leq \Delta$$

(here we take into account (s1), the estimate above, and inequality (1.99)). Thus, it follows from Theorem 1.3.5 that if $d \leq d_0$, then for $x = x_0 + v_0$ we have $|\phi^k(x) - x_k| = |v_k|$. This completes the proof of Theorem 1.3.4.

1.3.5 Finite-Dimensional Case

In [Pli2, Pli3], Pliss studied conditions under which a family of linear systems of differential equations has uniformly bounded solutions.

We show in this subsection that these results "translated" into the shadowing language give necessary and sufficient conditions of Lipschitz shadowing for a sequence of mappings of $\mathbb{R}^n$.

Let

$$A_k : \mathbb{R}^n \to \mathbb{R}^n, \; k \in \mathbb{Z},$$

be a family of invertible linear mappings, consider the corresponding equations

$$u_{k+1} = A_k u_k + z_{k+1}, \; k \in \mathbb{Z}, \tag{1.100}$$

where the sequence $\{z_k\}$ is given and the sequence $\{u_k\}$ is unknown.

Let $\mathcal{B} = \mathcal{L}_\infty$ be the Banach space of sequences

$$\bar{v} = \{v_k \in \mathbb{R}^n : k \in \mathbb{Z}\}$$

with the usual norm

$$||\bar{v}|| = \sup_{k \in \mathbb{Z}} |v_k|.$$

We begin with some definitions. Let $< a, b >$ be a segment of $\mathbb{Z}$ defined as follows. If a, b are finite with $a < b$, then

$$< a, b >= \{k \in \mathbb{Z} : a \leq k \leq b\};$$

if a is finite, then

$$< a, \infty >= \{k \in \mathbb{Z} : a \leq k < \infty\}$$

and

$$< -\infty, a >= \{k \in \mathbb{Z} : -\infty < k \leq a\};$$

finally, $< -\infty, \infty >= \mathbb{Z}$.

Definition 1.15 *The family $\{A_k\}$ is "hyperbolic on a segment" $< a, b >$ with constants $N > 0, \lambda \in (0, 1)$ if for any $k \in < a, b >$ there exist complementary projectors $P(k)$ and $Q(k)$ of $\mathbb{R}^n$ such that*
    (1) $\|P(k)\|, \|Q(k)\| \leq N$;
    (2) the spaces $S(k) = P(k)\mathbb{R}^n$ and $U(k) = Q(k)\mathbb{R}^n$ have the properties

$$A_k T(k) = T(k + 1) \text{ for } T = S, U \text{ if } k, k + 1 \in < a, b >; \qquad (1.101)$$

(3) for $n > k$ such that $k, n \in < a, b >$, the following inequalities hold:

$$\|A_n \ldots A_k P(k)\| \leq N\lambda^{n-k+1} \qquad (1.102)$$

and

$$\|A_k^{-1} \ldots A_{n-1}^{-1} Q(n)\| \leq N\lambda^{n-k}. \qquad (1.103)$$

Consider two linear subspaces L_1 and L_2 of $\mathbb{R}^n$. As usual, we define the angle $\angle(L_1, L_2)$ as follows: let M_1 and M_2 be the orthogonal complements of $L_1 \cap L_2$ in L_1 and L_2, respectively, then

$$\angle(L_1, L_2) = \min_{v_i \in M_i} \angle(v_1, v_2).$$

Remark. It is assumed in Definition 1.15 that the norms of the projectors $P(k), Q(k)$ are uniformly bounded for $k \in < a, b >$. Let us show that it is possible to substitute this condition by the following one: the value $b - a$ is large enough and the norms of A_k are uniformly bounded.
    Indeed, fix $N > 0, \lambda \in (0, 1)$, and find a natural number T such that

$$\frac{\lambda^{-T}}{N} - N\lambda^T \geq 1.$$

Assume that $b - a > T$ and that $\|A_k\| \leq L$. Take $c \in < a, b >$ such that $c + T \leq b$. Fix vectors $v^s \in S(c), v^u \in U(c)$ with $|v^s| = |v^u| = 1$ and denote

$$\Delta(k) = A_{k-1} \ldots A_c(v^u - v^s);$$

according to this notation, $\Delta(c) = v^u - v^s$.
    Obviously, we have

$$|\Delta(k)| \leq L^{k-c}|\Delta(c)|. \qquad (1.104)$$

On the other hand, it follows from (1.102) and (1.103) that

$$|\Delta(k)| \geq |A_{k-1} \ldots A_c v^u| - |A_{k-1} \ldots A_c v^s| \geq \frac{\lambda^{c-k}}{N} - N\lambda^{k-c} \geq 1 \qquad (1.105)$$

if $k \in < a, b >, k = c + T$. Comparing (1.104) and (1.105), we see that if $c \in < a, b - T >$, then we have

$$|\Delta(c)| = |v^u - v^s| \geq L^{-T}$$

for any $v^s \in S(c), v^u \in U(c)$ with $|v^s| = |v^u| = 1$. It follows that there exists $\alpha_1 = \alpha_1(L, N, \lambda) > 0$ such that

$$\angle(S(c), U(c)) \geq \alpha_1 \text{ for } c \in < a, b - T >.$$

Since $\|A_k\| \leq L$ and relations (1.101) hold, the inequality above implies that we can find $\alpha_0 = \alpha_0(L, \alpha_1)$ such that

$$\angle(S(c), U(c)) \geq \alpha_0 \text{ for } c \in < b - T, b >.$$

Obviously, it follows from the last two inequalities that the norms of $P(c), Q(c)$ are bounded for $c \in < a, b >$ by a constant depending only on L, N, λ.

Definition 1.16 *The family $\{A_k\}$ is "piecewise hyperbolic" on $\mathbb{Z}$ with constants T, N, λ, α if there exist segments*

$$I_0 = < -\infty, t_1 >, \ I_1 = < t_1, t_2 >, \ \ldots, \ I_m = < t_m, \infty >$$

of $\mathbb{Z}$ (with finite $t_1, \ldots, t_m$) such that
(1) $t_{i+1} - t_i > T, i = 1, \ldots, m - 1;$
(2) the family $\{A_k\}$ is hyperbolic with constants N, λ on every segment I_j (we denote below by $P^j(k), Q^j(k), S^j(k), U^j(k)$ the corresponding projectors and spaces);
(3) $\dim U^j(t_{j+1}) > \dim U^{j+1}(t_{j+1}), j = 0, \ldots, m - 1;$
(4) the spaces $U^j(t_{j+1})$ and $S^{j+1}(t_{j+1})$ are transverse, and

$$\angle(U^j(t_{j+1}), S^j(t_{j+1})) \geq \alpha, \ j = 0, \ldots, m - 1.$$

In our notation, the main result of [Pli2] can be stated as follows.

Theorem 1.3.6. *There exist functions $T(N, \lambda, \alpha)$ and $M(N, \lambda, \alpha)$ such that if a family $\{A_k\}$ is piecewise hyperbolic on $\mathbb{Z}$ with constants $T(N, \lambda, \alpha), N, \lambda, \alpha$, then Eqs. (1.100) have a solution $\bar{u}$ such that*

$$\|\bar{u}\| \leq M(N, \lambda, \alpha)\|\bar{z}\|. \tag{1.106}$$

To prove this theorem, we first establish some auxiliary statements.

Lemma 1.3.2. *There exists a constant $L = L(N, \lambda)$ such that if a family $\{A_k\}$ is hyperbolic on a segment $< a, b >$ with constants N, λ, then one can find vectors $u_n, n \in < a, b >$, with the properties*

$$u_{n+1} = A_n u_n + z_{n+1}, \ n \in < a, b - 1 >, \tag{1.107}$$

and

$$|u_n| \leq L||\bar{z}||, \ n \in < a, b > . \tag{1.108}$$

Proof. Let $P(k), Q(k), k \in < a, b >$, be the projectors given by Definition 1.15. Set

$$u_n = u_n^1 + u_n^2 + u_n^3,$$

where

$$u_n^1 = P(n)z_n, \ u_n^2 = \sum_{k=a}^{n-1} A_{n-1} \ldots A_k P(k)z_k,$$

$$u_n^3 = - \sum_{k=n}^{b-1} A_n^{-1} \ldots A_k^{-1} Q(k+1)z_{k+1}.$$

The same reasons as in the proof of Theorem 1.3.1 show that relations (1.107) are satisfied, and inequalities (1.108) hold with

$$L = N\frac{1+\lambda}{1-\lambda}.$$

$\square$

The following statement is geometrically obvious.

Lemma 1.3.3. *There exists a function $\rho(\alpha) > 0$ defined for $\alpha \in (0, \pi/2)$ and having the following property. Assume that L_1 and L_2 are transverse linear subspaces of $\mathbb{R}^n$ such that*

$$\angle(L_1, L_2) \geq \frac{\alpha}{2}.$$

If $M_i = L_i + x_i$ and $|x_i| \leq c, i = 1, 2$, then there exists a vector $y \in M_1 \cap M_2$ such that $|y| \leq \rho(\alpha)c$.

Consider two linear subspaces S and U of $\mathbb{R}^n$ such that

$$\mathbb{R}^n = S \oplus U.$$

Fix nonnegative numbers R, l, g. We say that a set $D \subset \mathbb{R}^n$ is an (R, l, g)-*ball* with respect to (S, U) if

$$D = \{x + Fx : x \in U, |x| \leq R\} + p, \tag{1.109}$$

where F is a linear mapping such that $FU = S$, $||F|| \leq l$, and $p \in S, |p| \leq g$.

If $S_1 = S + x, Y_1 = Y + x$, where $x \in \mathbb{R}^n$, we say that a set $D_1 \subset \mathbb{R}^n$ is an (R, l, g)-ball with respect to (S_1, U_1) if $D = D_1 - x$ is an (R, l, g)-ball with respect to (S, U).

Now we consider linear subspaces S, U, S', U' of $\mathbb{R}^n$ such that

$$S \oplus U = S' \oplus U' = \mathbb{R}^n$$

and denote by P, Q, P', Q' the projectors such that

$$S = P\mathbb{R}^n, \ U = Q\mathbb{R}^n, \ S' = P'\mathbb{R}^n, \ U' = Q'\mathbb{R}^n.$$

Lemma 1.3.4. *Assume that for the spaces* S, U, S', U' *above*
(1) $\|P\|, \|Q\|, \|P'\|, \|Q'\| \leq N$*;*
(2) U *and* S' *are transverse, and*

$$\angle(U, S') \geq \alpha.$$

Then there exist positive constants R, γ, l_0, l_1, g *(depending on* N, α*) such that for any* $d > 0$ *the following holds: if* D *is a* $(\gamma Rd, l_0, d)$*-ball with respect to* (S, U)*, and* $S_1 = S' + x', U_1 = U' + x'$*, where* $|x'| \leq 2d$*, then* D *contains a subset* D_1 *being an* (Rd, l_1, gd)*-ball with respect to* (S_1, U_1)*.*

Remark. An analog of this lemma (for small disks in stable and unstable manifolds of trajectories in a small neighborhood of a hyperbolic set) was stated without proof in [Pi2] (see Lemma A.8).

Proof. Let L be a linear subspace of $\mathbb{R}^n$ of the form

$$L = \{x + Fx : x \in U\}, \tag{1.110}$$

where $FU = S$. Obviously, we can find a constant l_0 (depending only on α) such that if $\|F\| \leq l_0$, then

$$\angle(L, U) \leq \frac{\alpha}{2}. \tag{1.111}$$

Set

$$l_1 = \frac{N}{\sin(\alpha/2)}, \ g = 2(\rho(\alpha) + 1), \ \gamma = 2N(l_1 + 1), \ R = \frac{2\rho(\alpha)}{1 + l_1}.$$

Let D be a $(\gamma Rd, l_0, d)$-ball with respect to (S, U). Assume that D is given by a formula similar to (1.109). Consider the corresponding linear subspace L given by (1.110). Since $\dim L = \dim U$, it follows from the second condition of our lemma that there exists a linear subspace $L' \subset L$ such that

$$S' \oplus L' = \mathbb{R}^n.$$

Inequality (1.111) implies that

$$\angle(L', S') \geq \frac{\alpha}{2}. \tag{1.112}$$

Obviuosly, we can write

$$L' = \{x + F'x : x \in U'\},$$

where $F'U' = S'$.

Take a vector $v \in L'$, let $y = P'v \in S'$ and $x = Q'v \in U'$, then obviously $y = F'x$. It follows from (1.112) that

$$|x| \geq |v| \sin(\alpha/2),$$

hence

$$|y| \leq N|v| \leq l_1|x|.$$

We see that

$$||F'|| \leq l_1. \tag{1.113}$$

Set $L_1 = L' + p$ (p is the vector in the definition of D similar to (1.109)). Since $|p| \leq d$ and $S_1 = S' + x', |x'| \leq 2d$, it follows from (1.112) and from Lemma 1.3.3 that there is a vector $q \in L_1 \cap S_1$ such that

$$|q| \leq 2\rho(\alpha)d. \tag{1.114}$$

Now we have

$$|q - x'| \leq 2(\rho(\alpha) + 1)d = gd. \tag{1.115}$$

It follows from the inclusion $q \in S_1$ that $q - x' \in S'$, hence the set

$$D' = \{x + F'x : x \in U', |x| \leq Rd\} + q - x'$$

is an (Rd, l_1, gd)-ball with respect to (S', U') (here we take estimates (1.113) and (1.115) into account). This implies that the set

$$D_1 = D' + x'$$

is an (Rd, l_1, gd)-ball with respect to (S_1, U_1).

We claim that $D_1 \subset D$. Note that $L' + q = L' + p = L_1$. By construction,

$$D_1 \subset \{x + F'x : x \in U'\} + p - x' + x' = L' + q = L' + p \subset L + p. \tag{1.116}$$

It follows from (1.114) that

$$|Qq| \leq 2N\rho(\alpha)d$$

(recall that Q projects $\mathbb{R}^n$ to U). For any $v \in D_1$ we have

$$v - q = x + F'x,$$

where $x \in U', |x| \leq Rd$. Inequality (1.113) implies that

$$|Q(v - q)| \leq NR(1 + l_1)d.$$

Finally, it follows from the relation

$$2\rho(\alpha) = R(1 + l_1)$$

that

$$|Qv| \leq |Qq| + |Q(v - q)| \leq N[R(1 + l_1) + 2\rho(\alpha)]d =$$
$$= 2N(1 + l_1)Rd = \gamma Rd. \tag{1.117}$$

Now we deduce from (1.116) and (1.117) that $D_1 \subset D$. Our lemma is proved.

□

Now let us prove Theorem 1.3.6. Fix numbers N, λ, α and find the corresponding R, γ, l_0, l_1, g given by Lemma 1.3.4. Find a positive number T such that the following inequalities hold:

$$N\lambda^T g < 1, \tag{1.118}$$

$$N^2 \lambda^{2T} l_1 < l_0, \tag{1.119}$$

and

$$\lambda^{-T} > N\gamma. \tag{1.120}$$

Take a family $\{A_k\}$, let it be piecewise hyperbolic with constants T, N, λ, α on segments

$$I_0 = <-\infty, t_1>, \quad \ldots, \quad I_m = <t_m, \infty>.$$

Fix a sequence $\bar{z} \in \mathcal{L}_\infty$. By Lemma 1.3.2, there exist vectors $u_n^j, n \in I_j, j = 0, \ldots, m$, such that for u_n^j analogs of (1.107) and (1.108) hold on I_j. Set

$$d = L\|\bar{z}\|.$$

Let $S^j(k), U^j(k), k \in I_j$, be the subspaces given by Definition 1.16. Denote

$$\mathcal{S}^j(k) = S^j(k) + u_k^j, \quad \mathcal{U}^j(k) = U^j(k) + u_k^j.$$

To simplify notation, denote $t_1 = a, t_2 = b$.

Obviously, the set

$$D_0 = \{x \in U^0(a) : |x| \leq \gamma Rd\}$$

is a $(\gamma Rd, 0, 0)$-ball with respect to $(S^0(a), U^0(a))$, and hence the set

$$D_0^1 = D_0 + u_a^0$$

is a $(\gamma Rd, 0, 0)$-ball with respect to $(\mathcal{S}^0(a), \mathcal{U}^0(a))$.

Take a vector $u_a \in D_0^1$ and define $u_k, k \in \mathbb{Z}$, by (1.100). Set $v_k = u_k - u_k^0$ for $k \leq a$ (recall that $u_k^0, k \in I_0$, are given by Lemma 1.3.2). It follows from (1.100) and from an analog of (1.107) for u_k^0 that

$$v_{k-1} = A_{k-1}^{-1} v_k \text{ for } k \leq a,$$

hence

$$v_k = A_k^{-1} \ldots A_{a-1}^{-1} v_a \text{ for } k < a.$$

Since $v_a \in U^0(a)$, item (3) of Definition 1.15 implies that

$$|v_k| \leq N\lambda^{a-k}|v_a| \leq \gamma N Rd \text{ for } k < a.$$

It follows from the inequality $|u_k^0| \leq d$ that

$$|u_k| \leq (1 + \gamma N R)d \text{ for } k \leq a. \tag{1.121}$$

Introduce new coordinates in $\mathbb{R}^n$ moving the origin to u_a^0. Then we can write

$$\mathbb{R}^n = \mathcal{S}^0(a) \oplus \mathcal{U}^0(a).$$

Since $|u_a^0|, |u_a^1| \leq d$, items (2) and (4) of Definition 1.16 show that the conditions of Lemma 1.3.4 are satisfied for $(S, U) = (\mathcal{S}^0(a), \mathcal{U}^0(a))$ and $(S_1, U_1) = (\mathcal{S}^1(a), \mathcal{U}^1(a))$, hence there exists an (Rd, l_1, gd)-ball D^1 with respect to $(\mathcal{S}^1(a), \mathcal{U}^1(a))$ such that $D^1 \subset D_0^1$.

Denote $w_a^1 = \mathcal{S}^1(a) \cap D^1$ and define $w_k^1, k \in I_1$, by

$$w_{k+1}^1 = A_k w_k^1 + z_{k+1}.$$

Let $v_k^1 = w_k^1 - u_k^1$. Since D^1 is an (Rd, l_1, gd)-ball with respect to $(\mathcal{S}^1(a), \mathcal{U}^1(a))$, we have $|v_a^1| \leq gd$ and $v_a^1 \in S^1(a)$. Taking into account that $b - a > T$ and that

$$v_k^1 = A_{k-1} \dots A_a v_a^1 \text{ for } k \in I_1, \tag{1.122}$$

we deduce from (1.102) and (1.118) the estimate

$$|v_b^1| \leq N\lambda^{b-a}gd < d. \tag{1.123}$$

Now we take two vectors $w_a, w_a' \in D^1$ and define w_k, w_k', v_k, v_k' similarly to w_k^1, v_k^1. Represent $v_k = x_k + y_k, v_k' = x_k' + y_k'$, where $x_k, x_k' \in U^1(k)$, $y_k, y_k' \in S^1(k)$. Since the set $D^1 - u_a^1$ is an (Rd, l_1, gd)-ball with respect to $(S^1(a), U^1(a))$, we have $|y_a - v_a^1| \leq l_1|x_a|$, hence

$$|y_a| \leq gd + l_1 Rd.$$

Formulas similar to (1.122) show that

$$|y_k| \leq N\lambda^{k-a}|y_a| \leq N(l_1 R + g)d \text{ for } k \in I_1. \tag{1.124}$$

The same reasons show that

$$|y_k' - y_k| \leq N\lambda^{k-a}|y_a' - y_a|$$

and

$$|x_k' - x_k| \geq \frac{\lambda^{a-k}}{N}|x_a' - x_a| \tag{1.125}$$

for $k \in I_1$. Since

$$|y_a' - y_a| \leq l_1|x_a' - x_a|,$$

the relations above and (1.119) imply the inequality

$$|y_b' - y_b| \leq N^2\lambda^{2(b-a)}l_1|x_b' - x_b| \leq l_0|x_b' - x_b|. \tag{1.126}$$

If for v_a we have $|x_a| = Rd$, then it follows from (1.120) and (1.125) that

$$|x_b| \geq \frac{\lambda^{a-b}}{N}Rd \geq \gamma Rd. \tag{1.127}$$

Combining (1.123), (1.126), and (1.127), we deduce that the set

$$D' = A_{b-1} \dots A_a \left(D^1 - u_a^1 \right) + u_b^1$$

contains a subset D_0^2 being an $(\gamma R d, l_0, d)$-ball with respect to $(\mathcal{S}^1(b), \mathcal{U}^1(b))$.
Take $u_b \in D_0^2$, define u_k by (1.100), set $v_k = u_k - u_k^1$ for $k \in I_1$, and represent

$$v_k = \xi_k + \eta_k, \text{ where } \xi_k \in U^1(k), \ \eta_k \in S^1(k).$$

It follows from (1.101) that

$$\xi_k = A_k^{-1} \dots A_{b-1}^{-1} \xi_b \text{ for } k < b, \ k \in I_1,$$

hence we have

$$|\xi_k| \le N \lambda^{k-b} |\xi_b| \le \gamma N R d, \ k \in I_1. \tag{1.128}$$

The inclusion $D_0^2 \subset D'$ implies that

$$v_a = A_a^{-1} \dots A_{b-1}^{-1} v_b \in D^1 - u_a^1,$$

hence the values $|\eta_k|$ satisfy the same estimates (1.124) as $|y_k|$, so that

$$|\eta_k| \le N(l_1 R + g)d, \ k \in I_1.$$

Combine the last inequality with (1.128) to show that

$$|v_k| \le N(l_1 R + \gamma R + g)d, \ k \in I_1.$$

It follows that

$$|u_k| \le N(l_1 R + R + g + 1)d, \ k \in I_1. \tag{1.129}$$

Since $u_a \in D_1$, inequalities (1.121) imply that

$$|u_k| \le N'd \text{ for } k \le b,$$

where

$$N' = N(l_1 R + \gamma R + g + 1).$$

By Lemma 1.3.4, the set D_0^2 contains a subset D^2 being an (Rd, l_1, gd)-ball with
respect to $(\mathcal{S}^2(b), \mathcal{U}^2(b))$. Repeating the described procedure, we finally obtain
a set D^m being an (Rd, l_1, gd)-ball with respect to $(\mathcal{S}^m(t_m), \mathcal{U}^m(t_m))$.
    Take $u_{t_m} = \mathcal{S}^m(t_m) \cap D^m$, define $u_k, k \in \mathbb{Z}$, by (1.100), and set $v_k = u_k - u_k^m$.
Repeating the previous arguments, we show that

$$|v_k| \le N(l_1 R + g)d$$

and

$$|u_k| \le [N(l_1 R + g) + 1]d$$

for $k \ge t_m$, and that

$$|u_k| \le N'd \text{ for } k \le t_m.$$

Hence, the sequence $\{u_k\}$ has the desired property with $M = LN'$.   □

Let us prove an important corollary of Theorem 1.3.6 giving sufficient conditions of Lipschitz shadowing for a sequence of mappings of $\mathbb{R}^n$. Here we formulate only a "global" result assuming that the nonlinearities have small Lipschitz constants everywhere.

Theorem 1.3.7. *Assume that the family $\{A_k\}$ is piecewise hyperbolic on $\mathbb{Z}$ with constants T, N, λ, α, where $T = T(N, \lambda, \alpha)$ is given by Theorem 1.3.6. Fix $\kappa > 0$ such that $\kappa M < 1$, where M is given by Theorem 1.3.6, and assume that functions $w_k(v)$ satisfy the inequalities*

$$|w_k(v') - w_k(v)| \leq \kappa |v' - v|.$$

If

$$|w_k(0)| \leq d, \ k \in \mathbb{Z}, \tag{1.130}$$

with some $d > 0$, then there is a sequence $\bar{u}$ satisfying

$$u_{k+1} = A_k u_k + w_{k+1}(u_k) \tag{1.131}$$

and such that $\|\bar{u}\| \leq Ld$, where $L = M(1 - \kappa M)^{-1}$.

Proof. Fix $d > 0$ and assume that condition (1.130) is satisfied. Set $\epsilon = \kappa M$ and $\nu = Md$.

Construct a sequence $\bar{u}^j \in \mathcal{B}, j \geq 0$, as follows. Set $u_k^0 \equiv 0$. Let $\bar{u}^1$ be a solution of

$$u_{k+1}^1 = A_k u_k^1 + w_{k+1}(0)$$

such that

$$\|\bar{u}^1\| \leq M\|\bar{w}(0)\| \leq Md = \nu.$$

For $j \geq 1$ define inductively $\bar{u}^{j+1}$ as solutions of

$$u_{k+1}^{j+1} = A_k u_k^{j+1} + w_{k+1}\left(u_k^j\right)$$

with the following property: they satisfy the equations

$$u_{k+1}^{j+1} - u_{k+1}^j = A_k \left(u_k^{j+1} - u_k^j\right) + y_{k+1}^j,$$

where

$$y_{k+1}^j = w_{k+1}\left(u_k^j\right) - w_{k+1}\left(u_k^{j-1}\right),$$

and the estimates

$$\|\bar{u}^{j+1} - \bar{u}^j\| \leq M\|\bar{y}^j\|$$

hold (this is possible by Theorem 1.3.6).

Since

$$\|\bar{u}^1 - \bar{u}^0\| \leq Md = \nu \text{ and } \|\bar{u}^{j+1} - \bar{u}^j\| \leq \kappa M\|\bar{u}^j - \bar{u}^{j-1}\| = \epsilon\|\bar{u}^j - \bar{u}^{j-1}\|$$

for $j \geq 1$, we see that $||\bar{u}^j - \bar{u}^{j-1}|| \leq \epsilon^{j-1}\nu$, hence the sequence $\bar{u}^j$ converges to some $\bar{u}$. It easily follows that $\bar{u}$ satisfies (1.131). Since

$$||\bar{u}|| \leq ||\bar{u}^1 - \bar{u}^0|| + ||\bar{u}^2 - \bar{u}^1|| + \ldots \leq \nu + \epsilon\nu + \ldots = \frac{\nu}{1-\epsilon} = Ld,$$

our theorem is proved.   □

Consider the equations

$$u_{k+1} = (A_k + B_k)u_k + z_{k+1}, \tag{1.132}$$

where B_k are linear operators. Taking $w_{k+1}(u_k) = B_k u_k + z_{k+1}$, we represent the equations above in form (1.131). It follows from Theorem 1.3.7 that if the family $\{A_k\}$ satisfies the conditions of this theorem, then there exists $\epsilon = \epsilon(N, \lambda, \alpha)$ such that if $||B_k||, |z_k| < \epsilon$ for $k \in \mathbb{Z}$, then there is a sequence $\bar{u}$ satisfying (1.132) and such that $||\bar{u}|| < 1$.

Now we can formulate the necessity statement proved by Pliss in [Pli3].

Theorem 1.3.8. *There exist functions $N = N(C, \epsilon), \alpha = \alpha(C, \epsilon), \Theta = \Theta(C, \epsilon),$ and $\lambda = \lambda(C, \epsilon) \in (0, 1)$ defined for $C, \epsilon > 0$ and having the following property.*

Let $\{A_k : k \in \mathbb{Z}\}$ be a family of invertible linear mappings of $\mathbb{R}^n$ such that $||A_k||, ||A_k^{-1}|| \leq C$. If, for some $\epsilon > 0$ and any B_k, z_k with $||B_k||, |z_k| < \epsilon,$ Eqs. (1.132) have a solution $\bar{u}$ with $||\bar{u}|| < 1$, then the family $\{A_k\}$ is piecewise hyperbolic on $\mathbb{Z}$ with constants $\Theta, N, \lambda,$ and α.

The proof of Theorem 1.3.8 is rather complicated. We do not give it here and refer the reader to the original paper [Pli3].

1.4 Limit Shadowing

In the usual statement of the shadowing problem (see Sect. 1.1), the values $r(x_{k+1}, \phi(x_k))$ are assumed to be uniformly small. We can impose another condition on these values,

$$d_k = r(x_{k+1}, \phi(x_k)) \to 0 \text{ as } k \to \infty, \tag{1.133}$$

and look for a point x such that

$$h_k = r(\phi^k(x), x_k) \to 0 \text{ as } k \to \infty.$$

We study the introduced shadowing property (we call it *the limit shadowing property*) in Subsect. 1.4.1. It is shown that in a neighborhood of a hyperbolic set a diffeomorphism has this property (Theorem 1.4.1).

In Subsect. 1.4.2, we investigate the rate of convergence of the values h_k in terms of d_k. Theorem 1.4.2, the main result of this subsection, shows that if the sequence $\{d_k\}$ belongs to a Banach space $\mathcal{L}_p, p \geq 1$, then the sequence $\{h_k\}$ belongs to the same space.

Passing from the spaces $\mathcal{L}_p$ to their weighted analogs, the spaces $\mathcal{L}_{\bar{r},p}$, we obtain a possibility to establish the "$\mathcal{L}_{\bar{r},p}$"-shadowing in a neighborhood of an arbitrary compact invariant set (not necessarily hyperbolic) under the corresponding conditions on the weight sequence $\bar{r}$ (see Theorem 1.4.5 in Subsect. 1.4.3). These conditions are formulated in terms of the so-called Sacker-Sell spectrum [Sac].

Another possibility to establish the "$\mathcal{L}_{\bar{r},p}$"-shadowing is to assume that the weight sequence grows "fast enough" (see Theorems 1.4.6 and 1.4.8 in Subsect. 1.4.3).

Hirsch studied in [Hirs4] asymptotic pseudotrajectories, i.e., sequences $\{x_k\}$ such that

$$\overline{\lim}_{k \to \infty} d_k^{1/k} \leq \lambda, \text{ where } 0 < \lambda < 1.$$

He found conditions under which an asymptotic pseudotrajectory is asymptotically shadowed. The main shadowing result of [Hirs4] is described in Subsect. 1.4.4.

The main results of Subsects. 1.4.1-1.4.3 were obtained in [Ei2].

1.4.1 Limit Shadowing Property

Let ϕ be a dynamical system on a metric space (X, r).

Definition 1.17 *We say that ϕ has the "LmSP" (the "limit shadowing property") on $Y \subset X$ if for any sequence $\xi = \{x_k : k \in \mathbb{Z}\} \subset Y$ such that (1.133) holds there is a point x such that*

$$r(\phi^k(x), x_k) \to 0 \text{ as } k \to \infty.$$

If this property holds on $Y = X$, we say that ϕ has the LmSP.

From the numerical point of view, this property of a dynamical system ϕ means the following: if we apply a numerical method that approximates ϕ with "improving accuracy", so that one-step errors tend to zero as time goes to infinity, then the numerically obtained trajectories tend to real ones. Such situations arise, for example, when we are not so interested in the initial (transient) behavior of trajectories but want to get to areas where "interesting things" happen (e.g., neighborhoods of attractors), and then improve accuracy.

It is easy to see that there exist systems which do not have the LmSP.

Example 1.18 Consider the system ϕ on the circle S^1 with coordinate $x \in [0, 1)$ such that any point of S^1 is a fixed point of ϕ. It was shown in Sect. 1.1 that this system does not have the POTP. To show that it does not have the LmSP, consider the sequence $\{x_k : k \in \mathbb{Z}\}$ such that $x_0 = 0$ and

$$x_{k+1} = x_k + \frac{1}{k+1}(\text{mod } 1), \ k \geq 0.$$

Obviously, condition (1.133) is satisfied, but since the series

$$\sum_{k \geq 0} \frac{1}{k+1}$$

diverges, the sequence x_k does not tend to a fixed point of ϕ.

Now we give an example of a system that has the LmSP but does not have the POTP.

Example 1.19 Consider again the circle S^1 with coordinate $x \in [0,1)$. Let ϕ be a dynamical system on S^1 generated by a mapping $f : [0,1] \to [0,1]$ with the following properties:
- f is continuous and increasing;
- the set $\{f(x) = x\}$ coincides with $\{0, 1/3, 2/3, 1\}$;
- $f(x) > x$ for $x \in (0, 1/3) \cup (1/3, 2/3)$;
- $f(x) < x$ for $x \in (2/3, 1)$.

Thus, ϕ has three fixed points ($p = 0, q = 1/3, r = 2/3$) on S^1, the point r is asymptotically stable, the point p is completely unstable (i.e., it is asymptotically stable for ϕ^{-1}), and q is semi-stable.

Take a natural m and denote by V_s^m the $1/m$-neighborhoods of the points $s = p, q, r$. Set

$$W_1^m = (0, 1/3) \setminus (V_p^m \cup V_q^m), \ W_2^m = (1/3, 1) \setminus (V_p^m \cup V_q^m \cup V_r^m).$$

Note that there exist positive numbers a_m suchthat

$$f(x) \geq x + 2a_m \text{ for } x \in W_1^m, \ |f(x) - r| \leq |x - r| - 2a_m \text{ for } x \in W_2^m.$$

To show that ϕ has the LmSP on S^1, take a sequence $\xi = \{x_k : k \geq 0\} \subset S^1$ such that

$$d_k = |\phi^k(x) - x_{k+1}| \to 0 \text{ as } k \to \infty.$$

Obviously, there exist numbers m_0 and $b_1 > 0$ such that for $m \geq m_0$ the following holds. For any $x \in \phi(V_s^m)$ the inequality

$$\text{dist}(x, V_u^m) \geq b_1 \text{ for } s, u \in \{p, q, r\}, \ s \neq u,$$

is fulfilled. Find k_0 such that for $k \geq k_0$ we have $d_k < b_1$. Below we take $m \geq m_0$ and $k \geq k_0$. It follows from our choice that if $x_k \in V_s^m$, then x_{k+1} cannot "jump" into V_u^m with $u \neq s$.

Take an index m and find a number $b_2(m)$ which is less than any of the following three values:

$$f\left(\frac{1}{3} - \frac{1}{m}\right) - \left(\frac{1}{3} - \frac{1}{m}\right), f\left(\frac{2}{3} - \frac{1}{m}\right) - \left(\frac{2}{3} - \frac{1}{m}\right), \left(\frac{2}{3} + \frac{1}{m}\right) - f\left(\frac{2}{3} + \frac{1}{m}\right).$$

Now we claim that for any m there exists an index $s \in \{p, q, r\}$ and $k(m)$ such that

$$x_k \in V_s^m \text{ for } k \geq k(m). \tag{1.134}$$

It follows from our choice of m_0, k_0 that the fixed point s with the described property does not depend on m (since "jumps" from V_s^m into V_u^m with different s, u are impossible). Thus, if we prove this statement, this will establish the LmSP for ϕ.

The same reasons show that if $x_k \in V_p^m \cup V_q^m \cup V_r^m$ for $k \geq k_0$, then we have nothing to prove.

Thus, it remains to consider two possible cases. We fix m and take $l(m)$ such that, for $k \geq l(m)$, the inequalities $d_k < \min(a_m, b_2(m))$ are fulfilled.

Case 1. $x_{l(m)} \in W_2^m$. It follows that for $k \geq l(m)$ we have

$$|x_{k+1} - r| < |x_k - r| - a_m$$

while $x_k \in W_2^m$. By the choice of $b_2(m)$, there exists $k(m)$ such that (1.134) holds with $s = r$.

Case 2. $x_{l(m)} \in W_1^m$. It follows that for $k \geq l(m)$ we have

$$x_{k+1} > x_k + a_m$$

while $x_k \in W_1^m$. Hence, either there exists $k(m)$ such that (1.134) holds with $s = q$, or there exists k_1 such that $x_{k_1} \in V_2^m$, and case 2 is reduced to case 1 (note that we cannot "return" from V_q^m into W_1^m by the choice of $b_2(m)$).

It remains to show that ϕ does not have the POTP on S^1. Take arbitrary $d > 0$ and let

$$x_0 = \frac{1}{3} - \frac{d}{2}, \ x_1 = \frac{1}{3} + \frac{d}{2}, \ x_k = \phi^k(x_0), \ k < 0; \ x_k = \phi^{k-1}(x_1), \ k > 1.$$

Since $f(x_0) \in (x_0, 1/3)$, we see that $\xi = \{x_k\}$ is a d-pseudotrajectory for ϕ. Any trajectory of ϕ belongs to one of the sets $[0, 1/3], (1/3, 2/3], (2/3, 1)$, while ξ intersects small neighborhoods of the points $0, 1/3$, and $2/3$. This shows that ϕ does not have the POTP.

Let us formulate and prove the main statement of this subsection.

Theorem 1.4.1. *Let Λ be a hyperbolic set for a diffeomorphism ϕ of $\mathbb{R}^n$. There exists a neighbourhood W of Λ such that ϕ has the LmSP on W.*

Remark. For a fixed point Λ, this result was proved in [Ak] (Theorem 13 of Chap. 9). One can prove the stated result applying Proposition 11 of Chap. 10 in [Ak], but we prefer to give here a simple direct proof.

Proof. Take a neighborhood U of Λ and numbers ν, Δ given by Theorem 1.2.1. Let W_0 be a neighborhood of Λ on which ϕ has the LpSP with constants L, d_0'

(see Theorem 1.2.3). Set $W_1 = U \cap W_0$. Decreasing Δ, if necessary, we can find a neighborhood W of Λ such that the Δ-neighborhood of W belongs to W_1.

For a sequence $\xi = \{x_k, k \in \mathbb{Z}\} \subset W$, set

$$d_k = |x_{k+1} - \phi(x_k)| \tag{1.135}$$

and assume that

$$d_k \to 0 \text{ for } k \to \infty.$$

Take an integer $j_0 > 1$ such that $\Delta < L d_0 j_0$. For every $j \geq j_0$ find an index k_j such that

$$d_k < \frac{\Delta}{2Lj} < d_0' \text{ for } k > k_j.$$

Since ϕ has the LpSP on W_0, there exist points y_j such that

$$|\phi^k(y_j) - x_k| \leq \frac{\Delta}{2j} < \frac{\Delta}{2}, \; k > k_j, \; j \geq j_0. \tag{1.136}$$

It follows from the choice of j_0 that

$$\phi^k(y_j) \in W_1 \subset U \text{ for } k > k_j, \; j \geq j_0.$$

By (1.136),

$$|\phi^k(y_{j_0}) - \phi^k(y_j)| < \Delta \text{ for } k > k_j, \; j \geq j_0.$$

Theorem 1.2.1 implies that in this case

$$\phi^{k_j}(y_j) \in W_\Delta^s(\phi^{k_j}(y_{j_0})) \text{ for } k > k_j, \; j \geq j_0,$$

and hence

$$|\phi^k(y_{j_0}) - \phi^k(y_j)| \leq \nu^{k-k_j} \Delta \text{ for } k > k_j, \; j \geq j_0. \tag{1.137}$$

Take $x = y_{j_0}$. By (1.137), there exist numbers $l_j \geq k_j$ for $j \geq j_0$ such that

$$|\phi^k(x) - \phi^k(y_j)| < \frac{1}{j} \text{ for } k > l_j, \; j \geq j_0.$$

We deduce from this inequality and from (1.136) that given $\epsilon > 0$ we can find j such that

$$|x_k - \phi^k(x)| \leq |x_k - \phi^k(y_j)| + |\phi^k(x) - \phi^k(y_j)| \leq$$

$$\leq \frac{\Delta}{2j} + \frac{1}{j} < \epsilon \text{ for } k > l_j.$$

This means that

$$|x_k - \phi^k(x)| \to 0 \text{ as } k \to \infty.$$

$\square$

One can consider also the following "two-sided" variant of the LmSP on a set Y for a dynamical system ϕ on (X, r): given a sequence $\xi = \{x_k : k \in \mathbb{Z}\} \subset Y$ such that the analog of (1.133) holds for $|k| \to \infty$, to find a point x with the property

$$r(\phi^k(x), x_k) \to 0 \text{ as } |k| \to \infty.$$

We show in the next subsection that in a neighborghood of a hyperbolic set a diffeomorphism has this "two-sided" LmSP (see Theorem 1.4.3).

1.4.2 $\mathcal{L}_p$-Shadowing

It was shown in the previous subsection that in a neighborhood of its hyperbolic set a diffeomorphism ϕ has the following property. If for a pseudotrajectory $\xi = \{x_k\}$ of ϕ the "errors"

$$d_k = |x_{k+1} - \phi(x_k)|$$

tend to 0, then the pseudotrajectory tends to some trajectory $\{\phi^k(x)\}$ of ϕ. In this subsection, we study the rate of convergence for the values

$$|\phi^k(x) - x_k|.$$

For $1 \le p < \infty$, denote by $\mathcal{L}_p$ the Banach space of sequences $\bar{v} = \{v_k \in \mathbb{R}^n : k \in \mathbb{Z}\}$ with the norm

$$\|\bar{v}\|_p = \left(\sum_{k \in \mathbb{Z}} |v_k|^p\right)^{1/p}.$$

We consider also the Banach space $\mathcal{L}_{p,+}$ of sequences $\bar{v} = \{v_k \in \mathbb{R}^n : k \in \mathbb{Z}_+\}$ with the norm

$$\|\bar{v}\|_p = \left(\sum_{k \ge 0} |v_k|^p\right)^{1/p}.$$

Denote, as usual,

$$\|\bar{v}\|_\infty = \sup_{k \in \mathbb{Z}} |v_k|$$

and

$$\mathcal{L}_p(r) = \{\bar{v} \in \mathcal{L}_p : \|\bar{v}\| \le r\}.$$

Take a diffeomorphism ϕ of $\mathbb{R}^n$, a sequence $\xi = \{x_k \in \mathbb{R}^n\}$, and a point $x \in \mathbb{R}^n$, and denote

$$g_k(\xi) = x_{k+1} - \phi(x_k), \ g(\xi) = \{g_k(\xi)\},$$

$$h_k(x, \xi) = \phi^k(x) - x_k, \ h(x, \xi) = \{h_k(x, \xi)\}.$$

We use this notation in both cases $k \in \mathbb{Z}$ and $k \in \mathbb{Z}_+$.

The main result of this subsection is the following theorem (we consider the case $k \in \mathbb{Z}$).

Theorem 1.4.2. *Let Λ be a hyperbolic set for a diffeomorphism ϕ of $\mathbb{R}^n$. There exists a neighborhood W of Λ and numbers $L, d_0 > 0$ with the following property. If for $\xi = \{x_k\} \subset W$ and some $1 \le p < \infty$ the inequality*

$$\|g(\xi)\|_p \le d \le d_0$$

holds, then there is a unique point x such that

$$||h(x,\xi)||_p \leq Ld. \tag{1.138}$$

Remark. For the case $k \in \mathbb{Z}_+$, the statement of the theorem and its proof are similar, the only exception is that the shadowing point x is not necessarily unique.

Proof. We begin by repeating the scheme described in Subsect. 1.3.3. We take $\mathcal{B} = \mathcal{L}_p$, obviously, the norm $||.||_p$ is monotone and satisfies condition (1.52).

It follows from Lemma 1.2.1 that

$$A_k S_k = S_{k+1}, \ A_k U_k = U_{k+1},$$

and that for $A_k^s = A_k|_{S_k}$, $B_k = (A_k^u|_{U_k})^{-1}$ conditions (1.63) and (1.64) are fulfilled.

Now we fix $1 \leq p < \infty$ and define a linear operator $\mathcal{G}$ on $\mathcal{L}_p$ in the same way as in Theorem 1.3.1,

$$\mathcal{G}\bar{z} = \bar{u}^1 + \bar{u}^2 + \bar{u}^3,$$

where

$$u_n^1 = P_n z_n, \ u_n^2 = \sum_{k=-\infty}^{n-1} A_{n-1} \ldots A_k P_k z_k, \ u_n^3 = -\sum_{k=n}^{\infty} B_n \ldots B_k Q_{k+1} z_{k+1}.$$

Let us show that $\mathcal{G}$ maps $\mathcal{L}_p$ into $\mathcal{L}_p$ and estimate its norm. Take $\bar{z} \in \mathcal{L}_p$. By (1.80),

$$||\bar{u}^1||_p \leq N||\bar{z}||_p.$$

Since $P_k z_k \in S_k$, we deduce from (1.63) that

$$|u_n^2| \leq N \sum_{k=-\infty}^{n-1} \lambda^{n-k} |z_k|.$$

Let us estimate

$$||\bar{u}^2||_p = \left(\sum_{n \in \mathbb{Z}} |u_n^2|^p \right)^{1/p} \leq N \left(\sum_{n \in \mathbb{Z}} \left(\sum_{k=-\infty}^{n-1} \lambda^{n-k} |z_k| \right)^p \right)^{1/p}.$$

Now we apply the Minkowski inequality (see [Har], Theorem 165) in the following form: if $b_m, a_{m,n} \geq 0$ and $p \geq 1$, then

$$\left(\sum_n \left(\sum_m b_m a_{m,n} \right)^p \right)^{1/p} \leq \sum_m b_m \left(\sum_n a_{m,n}^p \right)^{1/p}.$$

We see that

$$N \left(\sum_{n \in \mathbb{Z}} \left(\sum_{k=-\infty}^{n-1} \lambda^{n-k} |z_k| \right)^p \right)^{1/p} = N \left(\sum_{n \in \mathbb{Z}} \left(\sum_{m=1}^{\infty} \lambda^m |z_{n-m}| \right)^p \right)^{1/p} \leq$$

$$\le N \sum_{m=1}^{\infty} \lambda^m \left(\sum_{n \in \mathbf{Z}} |z_{n-m}|^p \right)^{1/p} \le N \frac{\lambda}{1 - \lambda} \|\tilde{z}\|_p.$$

Similar estimates show that

$$\|\bar{u}^3\|_p \le N \frac{\lambda}{1 - \lambda} \|\tilde{z}\|_p.$$

It follows that $\|\mathcal{G}\|_p \le N_1$, where $N_1 = N_1(N, \lambda)$ is defined in Theorem 1.3.1. Now the existence statement of our theorem follows from Lemma 1.3.1 with

$$L = \frac{N_1}{1 - \kappa N_1}, \quad d_0 = \frac{d_1}{L}.$$

To prove the uniqueness of a point x for which (1.138) holds, note that conditions (a) of Theorem 1.3.1 and (b') of Theorem 1.3.2 are satisfied. Fix numbers κ_0, Δ given by Theorem 1.3.2. Take a point x that satisfies (1.138). Let $\bar{v}$ be the corresponding solution of Eq. (1.60). It follows from the equalities

$$\phi^k(x) - x_k = v_k$$

that $\|\bar{v}\|_p \le Ld$, hence $\|\bar{v}\|_\infty \le Ld$. Take d_0 such that the inequalities for w_{k+1} in (1.68) are satisfied for $|v|, |v'| \le Ld_0$. Now it follows from Theorem 1.3.2 that a sequence $\bar{v}$ solving (1.60) (and hence a point x with property (1.138)) is unique. Our theorem is proved.   □

Remark. In [Ea], Easton studied the following property of a dynamical system ϕ on a metric space (X, r): given $\epsilon > 0$ there exists $\delta > 0$ such that if a sequence of points $\{x_k : k \in \mathbf{Z}\}$ satisfies the inequality

$$\sum_{k \in \mathbf{Z}} r(x_{k+1}, \phi(x_k)) < \delta,$$

then there exists a point x such that

$$\sum_{k \in \mathbf{Z}} r(x_k, \phi^k(x)) < \epsilon.$$

This was called the *strong shadowing property*. In the case $p = 1$, our theorem shows that a diffeomorphism has the strong shadowing property in a neighborhood of a hyperbolic set.

In the case $p = \infty$, it is possible to consider two different shadowing properties. We obtain the first one taking the usual space $\mathcal{L}_\infty$ with the norm $\|\cdot\|_\infty$ instead of $\mathcal{L}_p$ in Theorem 1.4.2. It is easy to see that the corresponding statement coincides with the classical Shadowing Lemma.

The second one appears if we consider the space of bounded sequences with zero limits endowed with the standard supnorm topology.

Theorem 1.4.3. *There exists a neighbourhood W of Λ and numbers $L, d_0 > 0$ such that if for a sequence $\xi = \{x_k : k \in \mathbb{Z}\} \subset W$ we have $||g(\xi)||_\infty \leq d \leq d_0$ and $\lim_{|k|\to\infty} g_k(\xi) = 0$, then there exists a unique x such that*

$$||h(x, \xi)||_\infty \leq Ld \ and \ \lim_{|k|\to\infty} |h_k(x, \xi)| = 0.$$

The proof of this statement is similar to the proof of Theorem 1.4.2, and we omit it.

As an example of application of Theorems 1.4.1 and 1.4.2, we consider approximations of solutions for ordinary differential equations by one-step discretizations.

Example 1.20 Consider a system of differential equations

$$\dot{x} = F(t, x), \ x \in \mathbb{R}^n. \tag{1.139}$$

Assume that F and $\frac{\partial F}{\partial x}$ are continuous and bounded and that F is 1-periodic in t. Denote by $x(t, t_0, x_0)$ the solution of (1.139) with initial value $x(t_0) = x_0$, and let ϕ be the Poincaré mapping of (1.139), that is,

$$\phi(x_0) = x(1, 0, x_0).$$

Our assumptions on F imply that ϕ is a diffeomorphism of class C^1. Assume that ϕ has a hyperbolic invariant set Λ.

If F is smooth enough, any standard one-step discretization with time-step h (e.g., a Runge-Kutta method) gives a mapping

$$\Psi_h(t_0, x_0) = (t_0 + h, x(t_0 + h, t_0, x_0) + \Phi),$$

where $q \geq 1$ is the order of the method, and $|\Phi| \leq C_0 h^{q+1}$.

Take a point x_0 and an increasing sequence of natural numbers $\{n_k : k \geq 0\}$ such that $\lim_{k\to\infty} n_k = \infty$. Set $h_k = 1/n_k$ and define a sequence $\{x_k : k \geq 0\}$ by

$$\Psi_{h_k}^{n_k}(0, x_k) = (1, x_{k+1}).$$

Then $|\phi(x_k) - x_{k+1}| \leq C n_k^{-q}$. If we assume that $x_k \in W$ (W is given by Theorem 1.4.1), then it follows from Theorem 1.4.1 that there exists x such that $\lim_{k\to\infty} |\phi^k(x) - x_k| = 0$. To get the $\mathcal{L}_p$-shadowing described in Theorem 1.4.2, it suffices to have $n_k \geq n_0 + k^{\alpha/pq}$ with some $\alpha > 1$.

1.4.3 The Sacker-Sell Spectrum and Weighted Shadowing

We study in this subsection a shadowing property which is a "weighted" variant of the $\mathcal{L}_p$-shadowing described by Theorem 1.4.2. We fix a sequence $\bar{r} = \{r_k \geq 1 : k \geq 0\}$, a number $p \geq 1$, and consider instead of the space $\mathcal{L}_p$ (to be exact, instead of the space $\mathcal{L}_{p,+}$) the space $\mathcal{L}_{\bar{r},p}$ of sequences $\bar{v} = \{v_k \in \mathbb{R}^n : k \geq 0\}$ with the norm

$$||\bar{v}||_{\bar{r},p} = \left(\sum_{k\geq 0} r_k |v_k|^p\right)^{1/p}.$$

The shadowing problem is formulated similarly to the one in the previous subsection, i.e., given a dynamical system ϕ in $\mathbb{R}^n$ and a sequence $\xi = \{x_k \in \mathbb{R}^n : k \geq 0\}$ with small $||g(\xi)||_{\bar{r},p}$, to find a point x such that $||h(x,\xi)||_{\bar{r},p}$ is small (see the definitions of g, h before Theorem 1.4.2).

We will show that it is possible to establish this "$\mathcal{L}_{\bar{r},p}$"-shadowing in a neighborhood of an invariant set Λ of a diffeomorphism ϕ not assuming that Λ is hyperbolic.

To do this, we need the concept of Sacker-Sell spectrum for an invariant set [Sac]. Note that in [Sac] the spectrum is defined for a very general class of dynamical systems, while we describe it here only for diffeomorhisms of $\mathbb{R}^n$.

Let Λ be a compact invariant set for a diffeomorphism ϕ of $\mathbb{R}^n$. We identify the set of nonsingular $n \times n$ real matrices with the group $GL(n, \mathbb{R})$ of invertible linear transformations of $\mathbb{R}^n$.

Fix $\rho > 0$ and consider the mapping

$$\Phi_\rho : \Lambda \times \mathbb{Z} \to GL(n, \mathbb{R})$$

given by

$$\Phi_\rho(x, k) = \rho^k D\phi^k(x).$$

Definition 1.21 *We say that Φ_ρ has "exponential dichotomy" over a point $x \in \Lambda$ if there exists a projector $P = P(x)$ in $\mathbb{R}^n$ and numbers $K > 0, \alpha \in (0, 1)$ such that*

$$||\Phi_\rho(x, m)P\Phi_\rho^{-1}(x, l)|| \leq K\alpha^{m-l} \text{ for } l \leq m,$$

$$||\Phi_\rho(x, m)(I - P)\Phi_\rho^{-1}(x, l)|| \leq K\alpha^{l-m} \text{ for } m \leq l.$$

For a point $x \in \Lambda$, the resolvent $\mathcal{R}(x)$ is defined by

$$\mathcal{R}(x) = \{\rho : \Phi_\rho \text{ has exponential dichotomy over } x\}.$$

Now we define the spectrum $\Sigma(x)$ by

$$\Sigma(x) = (0, +\infty) \setminus \mathcal{R}(x),$$

the *Sacker-Sell spectrum* $\Sigma(\Lambda)$ by

$$\Sigma(\Lambda) = \cup_{x\in\Lambda}\Sigma(x),$$

and the resolvent $\mathcal{R}(\Lambda)$ by

$$\mathcal{R}(\Lambda) = (0, +\infty) \setminus \Sigma(\Lambda).$$

We need one more definition.

Definition 1.22 *We say that the set Λ is "invariantly connected" if it cannot be represented as the union of two disjoint nonempty compact invariant sets.*

The main result about the spectrum $\Sigma(\Lambda)$ of a compact invariant set we need is the following statement obtained by Sacker and Sell (it is a corollary of Lemmas 2, 4, 6, and Theorem 2 in [Sac]).

Theorem 1.4.4. *Assume that Λ is invariantly connected. Then*
(a) the spectrum $\Sigma(\Lambda)$ is the union of $k \leq n$ compact intervals,

$$\Sigma(\Lambda) = [a_1, b_1] \cup \ldots \cup [a_k, b_k];$$

(b) for any $\rho \in \mathcal{R}(\Lambda)$ there exist constants $C > 0, \lambda_0 \in (0,1)$, and a continuous family of linear subspaces $S(x), U(x) \subset \mathbb{R}^n$ for $x \in \Lambda$ such that
(b.1) $S(x) \oplus U(x) = \mathbb{R}^n$;
(b.2) $D\phi(x)S(x) = S(\phi(x)), \; D\phi(x)U(x) = U(\phi(x))$;
(b.3)

$$\rho^k |D\phi^k(x)v| \leq C\lambda_0^k |v| \text{ for } v \in S(x), \; k \geq 0, \tag{1.140}$$

$$\rho^{-k} |D\phi^{-k}(x)v| \leq C\lambda_0^k |v| \text{ for } v \in U(x), \; k \geq 0; \tag{1.141}$$

(c) if $\rho > b_k$, then $S(x) = \{0\}, U(x) = \mathbb{R}^n$ for $x \in \Lambda$.

Remark. In [Sac], the spectrum is defined in such a way that analogs of (1.140) and (1.141) hold for $\exp(-k\mu)|D\phi^k(x)v|$ (instead of $\rho^k|D\phi^k(x)v|$), hence $\Sigma(\Lambda)$ and the original Sacker–Sell spectrum are related by the transformation $\rho \mapsto \exp(-\mu)$. Of course, this does not change the geometry of the set $\Sigma(\Lambda)$.

The main result of this subsection is the following statement.

Theorem 1.4.5. *Let Λ be an invariantly connected compact invariant set for a diffeomorphism ϕ of $\mathbb{R}^n$. Assume that for some $r, p \geq 1$ we have*

$$\rho = r^{1/p} \in \mathcal{R}(\Lambda).$$

Then there exists a neighbourhood W of Λ and numbers $L, d_0 > 0$ such that if a sequence of points $\xi = \{x_k : k \geq 0\} \subset W$ satisfies the inequality

$$\|g(\xi)\|_{\bar{r},p} \leq d \leq d_0,$$

then there is a point x such that

$$\|h(x,\xi)\|_{\bar{r},p} \leq Ld, \tag{1.142}$$

where $\bar{r} = \{r^k, k \geq 0\}$.
   If $[\rho, \infty) \cap \Sigma(\Lambda) = \emptyset$, then we can find W and d_0 such that a point x with property (1.142) is unique.

In the proof, we use the lemma below.

Lemma 1.4.1.   *Assume that $\rho \in \mathcal{R}(\Lambda)$. Let $\lambda_0 \in (0,1)$ and C be given by Theorem 1.4.4. Then for any $\epsilon > 0, \lambda \in (\lambda_0, 1)$ there exists a neighborghood $W = W(\epsilon, \lambda)$ with the following properties. There exist positive constants N', δ, a C^∞ norm $|.|_x$ for $x \in W$, and continuous (not necessarily $D\phi$-invariant) extensions S', U' of S, U to W such that*

(1) $S'(x) \oplus U'(x) = \mathbb{R}^n, \ x \in W$;

(2) for $x, y \in W$ with $|y - \phi(x)| < \delta$, the mapping $\Pi_y^s D\phi(x) \ (\Pi_y^u D\phi(x))$ is an isomorphism between $S'(x)$ and $S'(y)$ (respectively, between $U'(x)$ and $U'(y)$), and the inequalities

$$\rho|\Pi_y^s D\phi(x)v|_y \le \lambda|v|_x, \ \rho|\Pi_y^u D\phi(x)v|_y \le \epsilon|v|_x, \ v \in S'(x); \tag{1.143}$$

$$\rho|\Pi_y^u D\phi(x)v|_y \ge \frac{1}{\lambda}|v|_x, \ \rho|\Pi_y^s D\phi(x)v|_y \le \epsilon|v|_x, \ v \in U'(x), \tag{1.144}$$

hold, where Π_x^s is the projection onto $S'(x)$ parallel to $U'(x)$, and $\Pi_x^u = I - \Pi_x^s$;

(3)

$$\frac{1}{N'}|v|_x \le |v| \le N'|v|_x \ for \ x \in W, \ v \in \mathbb{R}^n. \tag{1.145}$$

Proof. The proof is similar to the one of Lemma 1.2.1. For $\mu \in (\lambda_0, \lambda)$ we find a natural number ν such that

$$C \left(\frac{\lambda_0}{\mu} \right)^{\nu+1} < 1.$$

Take a point $p \in \Lambda$ and a vector $v \in \mathbb{R}^n$, represent $v = v^s + v^u \in S(p) \oplus U(p)$, and set

$$|v|_p^2 = (|v^s|_p^2 + |v^u|_p^2)^{1/2},$$

where

$$|v^s|_p = \sum_{j=0}^{\nu} \rho^j \mu^{-j} |D\phi^j(p)v^s|, \ |v^u|_p = \sum_{j=0}^{\nu} \rho^{-j} \mu^{-j} |D\phi^{-j}(p)v^u|.$$

For v^s we obtain the estimate

$$\rho|D\phi(p)v^s|_{\phi(p)} = \sum_{j=0}^{\nu} \rho^{j+1} \mu^{-j} |D\phi^j(\phi(p))D\phi(p)v^s| =$$

$$= \mu \left(\sum_{j=1}^{\nu} \rho^j \mu^{-j} |D\phi^j(p)v^s| + \rho^{\nu+1} \mu^{-\nu-1} |D\phi^{\nu+1}(p)v^s| \right) \le$$

$$\le \mu \left(\sum_{j=1}^{\nu} \mu^{-j} |D\phi^j(p)v^s| + \mu^{-\nu-1} C\lambda_0^{\nu+1} |v^s| \right) \le$$

(we applied (1.140) with $k = \nu + 1$ to estimate the second term)

$$\leq \mu |v^s|_x$$

(here we take into account the choice of ν). Similarly, for v^u we obtain the estimate

$$\rho |D\phi(p)v^u|_{\phi(p)} \geq \frac{1}{\mu}|v^u|_p.$$

The rest of the proof repeats the proof of Lemma 1.2.1.

Now we prove Theorem 1.4.5. We apply the scheme described in Subsect. 1.3.3 (for the WS case). Fix $r, p \geq 1$ and such that

$$\rho = r^{1/p} \in \mathcal{R}(\Lambda).$$

Let $\mathcal{B} = \mathcal{L}_{\bar{r},p}$, obviously the norm of $\mathcal{B}$ is monotonous and satisfies condition (1.52).

It follows from Lemma 1.4.1 that

$$A_k S_k = S_{k+1}, \ A_k U_k = U_{k+1}.$$

Now we define a linear operator $\mathcal{G}$ on $\mathcal{L}_{\bar{r},p}$ in the same way as in Theorem 1.3.1,

$$\mathcal{G}\bar{z} = \bar{u}^1 + \bar{u}^2 + \bar{u}^3,$$

where

$$u_n^1 = P_n z_n, \ u_n^2 = \sum_{k=-\infty}^{n-1} A_{n-1} \ldots A_k P_k z_k, \ u_n^3 = -\sum_{k=n}^{\infty} B_n \ldots B_k Q_{k+1} z_{k+1}.$$

Let us show that $\mathcal{G}$ maps $\mathcal{L}_{\bar{r},p}$ into $\mathcal{L}_{\bar{r},p}$ and estimate its norm. Take $\bar{z} \in \mathcal{L}_{\bar{r},p}$. By (1.80),

$$\|\bar{u}^1\|_{\bar{r},p} \leq N\|\bar{z}\|_{\bar{r},p}.$$

Let us write

$$r^n |u_n^2|^p = |\rho^n u_n^2|^p = \left| \sum_{k=0}^{n-1} (\rho A_{n-1}) \ldots (\rho A_k) P_k \rho^k z_k \right|^p.$$

Since $P_k z_k \in S_k$ and (1.143) holds, we obtain the inequality

$$r^n |u_n^2|^p \leq N^p \left(\sum_{k=0}^{n-1} \lambda^{n-k} \rho^k |z_k| \right)^p.$$

We apply the Minkowski inequality (see the proof of Theorem 1.4.1) and obtain the estimate

$$\|\bar{u}^2\|_{\bar{r},p} \leq N \left(\sum_{n=0}^{\infty} \left(\sum_{k=0}^{n-1} \lambda^{n-k} \rho^k |z_k| \right)^p \right)^{1/p} =$$

$$= N \left(\sum_{n=0}^{\infty} \left(\sum_{m=1}^{n} \lambda^m \rho^{n-m} |z_{n-m}| \right)^p \right)^{1/p} \leq N \sum_{m=1}^{\infty} \lambda^m \left(\sum_{k=m}^{\infty} \rho^{(k-m)p} |z_{k-m}|^p \right)^{1/p} =$$

$$= N \sum_{m=1}^{\infty} \lambda^m \left(\sum_{k=m}^{\infty} r^{k-m} |z_{k-m}|^p \right)^{1/p} \leq N \frac{\lambda}{1-\lambda} ||\bar{z}||_{\bar{r},p}^p.$$

Now let us write

$$r^n |u_n^3|^p = \left| \sum_{k=n}^{\infty} (\rho^{-1} A_n^{-1}) \dots (\rho^{-1} A_k^{-1}) Q_{k+1} \rho^{k+1} z_{k+1} \right|^p.$$

Since $Q_{k+1} z_{k+1} \in U_{k+1}$ and (1.144) holds, we obtain the inequality

$$r^n |u_n^3|^p \leq N^p \left(\sum_{k=n}^{\infty} \lambda^{k-n+1} \rho^{k+1} |z_{k+1}| \right)^p.$$

Applying again the Minkowski inequality, we see that

$$||\bar{u}^3||_{\bar{r},p} \leq N \frac{\lambda}{1-\lambda} ||\bar{z}||_{\bar{r},p}.$$

Finally we get the estimate

$$||\mathcal{G}\bar{z}||_{\bar{r},p} \leq N \left(1 + 2\frac{\lambda}{1-\lambda} \right) ||\bar{z}||_{\bar{r},p} \leq N_1 ||\bar{z}||_{\bar{r},p}, \tag{1.146}$$

where $N_1 = N_1(N, \lambda)$ is defined in Theorem 1.3.1.

Now the existence statement of our theorem follows from Lemma 1.3.1 with

$$L = \frac{N_1}{1 - \kappa N_1}, \quad d_0 = \frac{d_1}{L}.$$

Let us prove the uniqueness statement. Take a neighborhood W and the corresponding numbers L, d_0 given by the first part of the proof.

If $[\rho, \infty) \cap \Sigma(\Lambda) = \emptyset$, statement (c) of Theorem 1.4.4 and Lemma 1.4.1 imply that $U(x) = \mathbb{R}^n$ for $x \in W$. Denote

$$\psi(x, v) = \phi(x + v) - \phi(x) - D\phi(x)v$$

for $x \in W, v \in \mathbb{R}^n$.

It follows from Lemma 1.4.1 that in our case we can find numbers $d_0', \kappa > 0$ and a neighborhood W_0 of Λ such that the Ld_0'-neighborhood of W_0 is a subset of W, and the following inequalities hold:

$$\rho ||D\phi(x)|| \geq \frac{1}{\lambda} \text{ for } x \in W; \tag{1.147}$$

$$|\psi(x, v) - \psi(x, v')| \leq \frac{\kappa}{\rho} |v - v'| \text{ for } x \in W, |v|, |v'| \leq Ld_0'; \tag{1.148}$$

$$\gamma = \frac{1}{\lambda} - \kappa > 1.$$

We claim that, for a sequence $\xi = \{x_k : k \geq 0\}$, relation (1.142) cannot hold with $d \leq d_0'$ for two different points x, x'.

To obtain a contradiction, assume that (1.142) holds for different x and x'. Let

$$\phi^k(x) - x_k = v_k, \quad \phi^k(x') - x_k = w_k.$$

Since $||\bar{v}||_\infty \leq ||\bar{v}||_{\bar{r},p}$, we deduce from (1.142) that

$$|v_k|, |w_k| \leq Ld_0'. \tag{1.149}$$

It follows from our assumptions that $\{\phi^k(x)\}, \{\phi^k(x')\} \subset W$. Set $z_k = v_k - w_k$.

Comparing the relations

$$x_{k+1} + v_{k+1} = \phi(x_k + v_k) = \phi(x_k) + D\phi(x_k)v_k + \psi(x_k, v_k)$$

and

$$x_{k+1} + w_{k+1} = \phi(x_k + w_k) = \phi(x_k) + D\phi(x_k)w_k + \psi(x_k, w_k),$$

we see that

$$z_{k+1} = D\phi(x_k)z_k + \psi(x_k, v_k) - \psi(x_k, w_k).$$

Taking (1.147)-(1.149) into account, we obtain the inequality

$$\rho|z_{k+1}| \geq \frac{1}{\lambda}|z_k| - \kappa|v_k - w_k| \geq \gamma|z_k|.$$

Since $\rho|z_1| \geq \gamma|z_0|$, and

$$\rho^{k+1}|z_{k+1}| = \rho^k\rho|z_{k+1}| \geq \rho^k\gamma|z_k|,$$

we get by induction that

$$\rho^k|z_k| \geq \gamma^k|z_0| \text{ for } k \geq 0.$$

If $x \neq x'$, then $|z_0| \neq 0$, and it follows that for $\bar{z} = \{z_k\}$ we have

$$||\bar{z}||_{\bar{r},p} = \left(\sum_{k \geq 0} |\rho^k z_k|^p\right)^{1/p} \geq \left(\sum_{k \geq 0} |\gamma^k z_0|^p\right)^{1/p} = \infty.$$

The contradiction with the inequality

$$||\bar{z}||_{\bar{r},p} \leq ||\bar{v}||_{\bar{r},p} + ||\bar{w}||_{\bar{r},p} \leq 2Ld_0'$$

completes the proof.   □

Now let us show that if the weights "increase fast enough", then we can directly establish a "weighted" shadowing property with the uniqueness of the shadowing trajectory.

Theorem 1.4.6. *Let ϕ be a C^1-diffeomorphism of $\mathbb{R}^n$ and let U be a bounded set in $\mathbb{R}^n$. Assume that $p \geq 1$ and that the weight sequence $\bar{r} = \{r_k \geq 1 : k \geq 0\}$ satisfies the conditions*

$$r_{k+1} \geq \rho_0 r_k \ for \ k \geq k_0, \tag{1.150}$$

where

$$\rho = \rho_0^{1/p} > M = \max\left(1, \sup_{x \in U} \|D\phi(x)^{-1}\|\right). \tag{1.151}$$

Then there exist $L, d_0 > 0$ such that for a sequence $\xi = \{x_k : k \geq 0\} \subset U$ satisfying the condition $\|g(\xi)\|_{\bar{F},p} \leq d \leq d_0$ there exists a unique point x such that

$$\|h(x, \xi)\|_{\bar{F},p} \leq Ld. \tag{1.152}$$

Proof. First note that taking

$$c = \min_{0 \leq j \leq k_0} \rho_0^{-j} r_j,$$

we obtain the estimate

$$r_k \geq c\rho_0^k. \tag{1.153}$$

As in the proof of previous theorems, we are looking for a sequence $\bar{v} = \{v_k : k \geq 0\} \in \mathcal{L}_{\bar{F},p}$ that satisfies the equalities

$$\phi(x_k + v_k) = x_{k+1} + v_{k+1}, \ k \geq 0.$$

This means that we want to solve the equation

$$F(\bar{v}) = S\bar{v}, \tag{1.154}$$

where $(F\bar{v})_k = \phi(x_k + v_k) - x_{k+1}$, and S is the left shift,

$$S(v_0, v_1, \ldots) = (v_1, v_2, \ldots).$$

Let us show that the operator $S - DF(0)$ has a bounded right inverse in $\mathcal{L}_{\bar{F},p}$. First we establish the following elementary inequality: if $\{a_j : j \geq 0\}$ is a nonnegative sequence, and $p, \eta > 1$, then

$$\sum_{j \geq 0} a_j \leq C(\eta, p)\left(\sum_{j \geq 0} \eta^j a_j^p\right)^{1/p}, \tag{1.155}$$

where

$$C(\eta, p) = \left(\frac{\eta^{\frac{1}{p-1}}}{\eta^{\frac{1}{p-1}} - 1}\right)^{1 - \frac{1}{p}}.$$

Indeed, by Hölder's inequality we have

$$\sum_{j \geq 0} a_j = \sum_{j \geq 0} \eta^{j/p} a_j \eta^{-j/p} \leq \left(\sum_{j \geq 0} \eta^j a_j^p\right)^{1/p}\left(\sum_{j \geq 0}\left(\eta^{-j/p}\right)^{\frac{p}{p-1}}\right)^{\frac{p-1}{p}}.$$

This implies (1.155). For $p = 1$, inequality (1.155) holds trivially with $C(\eta, 1) = 1$.

Let us fix $\bar{w} \in \mathcal{L}_{\bar{r},p}$ and consider the equation $S\bar{v} - DF(0)\bar{v} = \bar{w}$ which is equivalent to

$$v_{k+1} = D\phi(x_k)v_k + w_k, \ k \geq 0. \tag{1.156}$$

To solve this last equation, set $D_{k,k} = I$ and

$$D_{j+1,k} = D\phi(x_j)D\phi(x_{j-1})\ldots D\phi(x_k).$$

We claim that the sequence $\bar{v} = \{v_k : k \geq 0\}$ defined by

$$v_k = -\sum_{j=k}^{\infty} D_{j+1,k}^{-1} w_j, \ k \geq 0, \tag{1.157}$$

satisfies (1.156). To show the convergence of (1.157), apply (1.151) to find

$$\tilde{\rho} \in (M^p, \rho_0)$$

and set

$$\eta = \frac{\tilde{\rho}}{M^p}.$$

It follows from (1.155) that

$$|v_k| \leq \sum_{j=k}^{\infty} |D_{j+1,k}^{-1} w_j| \leq \sum_{j=k}^{\infty} M^{j+1-k}|w_j| = M \sum_{j=0}^{\infty} M^j |w_{k+j}| \leq$$

$$\leq MC(\eta,p) \left(\sum_{j=0}^{\infty} \left(\frac{\tilde{\rho}}{M^p} \right)^j \left(M^j |w_{k+j}| \right)^p \right)^{1/p} =$$

$$= MC(\eta,p) \left(\sum_{j=k}^{\infty} \tilde{\rho}^{j-k} |w_j|^p \right)^{1/p}. \tag{1.158}$$

Since

$$\tilde{\rho}^{j-k} \leq \tilde{\rho}^j < \rho_0^j \leq \frac{r_j}{c},$$

we see that (1.158) does not exceed

$$Mc^{-1/p}C(\eta,p)\|\bar{w}\|_{\bar{r},p}.$$

Hence, the sums in (1.157) converge. Further, we have

$$v_{k+1} = -\sum_{j=k+1}^{\infty} D_{j+1,k+1}^{-1} w_j = -D_{k+1,k} \sum_{j=k+1}^{\infty} (D_{j+1,k+1}D_{k+1,k})^{-1} w_j =$$

$$= -D\phi(x_k) \sum_{j=k}^{\infty} D_{j+1,k}^{-1} w_j + w_k = D\phi(x_k)v_k + w_k,$$

i.e., $\bar{v}$ satisfies (1.156). We estimate the norm of $\bar{v}$ applying (1.158),

$$\sum_{k=0}^{\infty} r_k |v_k|^p \leq M^p C(\eta,p)^p \sum_{k=0}^{\infty} r_k \sum_{j=k}^{\infty} \tilde{\rho}^{j-k} |w_j|^p =$$

$$= M^p C(\eta,p)^p \sum_{j=0}^{\infty} \left(\sum_{k=0}^{j} \frac{r_k}{r_j} \tilde{\rho}^{j-k} \right) r_j |w_j|^p.$$

We can find $c_1 > 0$ such that $r_k \leq c_1 r_{k_0}$ for $0 \leq k \leq k_0 - 1$. Now we estimate

$$\sum_{k=0}^{j} \frac{r_k}{r_j} \tilde{\rho}^{j-k} \leq \sum_{k=0}^{k_0-1} \frac{r_{k_0}}{r_j} \tilde{\rho}^{j-k} + \sum_{k=k_0}^{j} \frac{r_k}{r_j} \tilde{\rho}^{j-k}.$$

The first sum does not exceed

$$\frac{c_1}{c_0} r_{k_0} \sum_{k=0}^{k_0-1} \frac{\tilde{\rho}^{j-k}}{\rho_0^j} \leq c_2 = \frac{c_1}{c_0} k_0 r_{k_0},$$

and the second one is estimated by

$$\sum_{k=k_0}^{j} \frac{\tilde{\rho}^{j-k}}{\rho_0^{j-k}} \leq c_3 = \frac{\rho_0}{\rho_0 - \tilde{\rho}}.$$

Hence, the equation $S\bar{v} - DF(0)\bar{v} = \bar{w}$ has a solution $\bar{v}$ depending linearly on $\bar{w}$ and such that

$$\|\bar{v}\|_{\bar{r},p} \leq MC(\eta,p)(c_2 + c_3)^{1/p} \|\bar{w}\|_{\bar{r},p},$$

i.e., the operator $S - DF(0)$ has a bounded right inverse. To prove that $S - DF(0)$ is invertible, it is enough to show that it is one-to-one. Assume that there exists $\bar{v}$ such that $S\bar{v} - DF(0)\bar{v} = 0$ and $v_m \neq 0$ for some m. Then for $k > m$ we have

$$v_k = D\phi(x_{k-1})v_{k-1}, \text{ hence } |v_k| \geq M^{-1}|v_{k-1}| \geq M^{m-k}|v_m|,$$

and

$$\sum_{k=m}^{\infty} r_k |v_k|^p \geq c|v_m|^p M^{mp} \sum_{k=m}^{\infty} \left(\frac{\rho_0}{M^p} \right)^k = \infty.$$

This shows that $\bar{v}$ does not belong to $\mathcal{L}_{\bar{r},p}$.

Hence, the operator $S - DF(0)$ has a bounded inverse in $\mathcal{L}_{\bar{r},p}$. We see that Eq. (1.154) is equivalent to

$$\bar{v} = (S - DF(0))^{-1}(F(\bar{v}) - DF(0)\bar{v}).$$

This last equation is solved exactly in the same way as in the proof of Theorem 1.2.3 (we take into account that $|v_k| \leq \|\bar{v}\|_{\bar{r},p}$). $\qquad \square$

To complete this subsection, we consider the following weighted analog of the space $\mathcal{L}_\infty$. Fix again a sequence $\bar{r} = \{r_k \geq 1 : k \geq 0\}$ and denote by $\mathcal{L}_{\bar{r},\infty}$ the Banach space of sequences $\bar{v} = \{v_k \in \mathbb{R}^n : k \geq 0\}$ with the norm

$$||\bar{v}||_{\bar{r},\infty} = \sup_{k\geq 0} r_k|v_k|.$$

The same reasons as in the proofs of Theorems 1.4.5 and 1.4.6 show that their analogs are true for the spaces $\mathcal{L}_{\bar{r},\infty}$. Let us explain only how to obtain the main estimates.

In the analog of Theorem 1.4.5, we take $\rho \in \mathcal{R}(\Lambda)$ and consider the space $\mathcal{L}_{\bar{r},\infty}$ with $r_k = \rho^k$. The basic step is to estimate $||\mathcal{G}||$.

For example, to estimate the second term in the expression defining the operator $\mathcal{G}$, let us write

$$\rho^n|v_n^2| = \left|\sum_{k=0}^{n-1}(\rho A_{n-1})\ldots(\rho A_k)P_k\rho^k w_k\right| \leq$$

$$\leq N\sum_{k=0}^{n-1}\lambda^{n-k}\sup_{k\geq 0}|\rho^k w_k| \leq N\frac{\lambda}{1-\lambda}||\bar{w}||_{\bar{r},\infty}.$$

Thus, we see that

$$||v^2||_{\bar{r},\infty} \leq N\frac{\lambda}{1-\lambda}||w||_{\bar{r},\infty}.$$

The remaining terms are estimated similarly.

In the analog of Theorem 1.4.6, we assume for simplicity that for the weight sequence $\bar{r}$ the inequalities

$$r_{k+1} \geq \rho r_k, \; k \geq 0,$$

are satisfied, and that condition (1.151) holds.

To estimate the norm of the sequence defined by (1.157), we write (taking into account that $r_k \leq \rho^{j-k}r_j$ for $j > k$ and putting $p = 1$ in the previous estimates)

$$r_k|v_k| \leq M\sum_{j=k}^{\infty}\frac{r_k}{r_j}\tilde{\rho}^{j-k}\sup_{j\geq 0}r_j|w_j| \leq$$

$$\leq M||\bar{w}||_{\bar{r},\infty}\sum_{j=k}^{\infty}\left(\frac{\tilde{\rho}}{\rho}\right)^{j-k} = M\frac{\rho}{\rho-\tilde{\rho}}||\bar{w}||_{\bar{r},\infty}.$$

We leave the remaining details to the reader.

Let us note that this last type of shadowing was first introduced by Fečkan in [Fe] (to be exact, two-sided sequences $\{v_k : k \in \mathbb{Z}\}$ and weights of the form $r_k = w^{-|k|}$ with $w > 1$ were considered; this approach does not seem to be of real interest for applications, since the "shadowing errors" $|\phi^k(x) - x_k|$ may grow to infinity).

Fečkan studied mappings $\phi : \mathbb{R}^n \to \mathbb{R}^n$ of special form,

$$\phi(x) = x + f(x), \tag{1.159}$$

under the condition

$$|f(x) - f(x')| \leq K|x - x'| \text{ with } K < 1. \tag{1.160}$$

The following statement is the main result of [Fe] (note that the Lipschitz dependence of the value

$$\sup_{k \in \mathbf{Z}} w^{-|k|} |\phi^k(x) - x_k|$$

on

$$\sup_{k \in \mathbf{Z}} w^{-|k|} |\phi(x_k) - x_{k+1}|$$

was not explicitly stated in [Fe]; in addition, we write condition (1.161) in the form of an inequality instead of the original equality. It is easily seen that the proof in [Fe] establishes the result formulated below).

Theorem 1.4.7. *Assume that*

$$w \geq \frac{1 + K}{1 - K}. \tag{1.161}$$

Then there is $L > 0$ with the following property: for any sequence $\{x_k \in \mathbb{R}^n : k \in \mathbf{Z}\}$ such that the value

$$\sup_{k \in \mathbf{Z}} w^{-|k|} |\phi(x_k) - x_{k+1}|$$

is finite, there exists a unique point x such that

$$\sup_{k \in \mathbf{Z}} w^{-|k|} |\phi^k(x) - x_k| \leq L \sup_{k \in \mathbf{Z}} w^{-|k|} |\phi(x_k) - x_{k+1}|.$$

One can apply reasons similar to the ones in the proof of Theorem 1.4.6 to obtain the formulated result.

Let us show that for the introduced space $\mathcal{L}_{\bar{r}, \infty}$ a similar result holds (under a condition weaker than inequality (1.161)). We give here a short direct proof close to the one in [Fe].

Theorem 1.4.8. *Assume that for a mapping ϕ given by (1.159) inequality (1.160) holds, and that*

$$\rho > \frac{1}{1 - K}. \tag{1.162}$$

Consider $\bar{r} = \{r_k \geq 1 : k \geq 0\}$ such that $r_{k+1} \geq \rho r_k, k \geq 0$. Then for any sequence $\xi = \{x_k \in \mathbb{R}^n : k \geq 0\}$ such that the value

$$\|g(\xi)\|_{\bar{r}, \infty}$$

is finite, there is a unique point x with the property

$$\|h(x, \xi)\|_{\bar{r}, \infty} \leq L \|g(\xi)\|_{\bar{r}, \infty}, \tag{1.163}$$

where

$$L = \frac{\rho}{\rho(1 - K) - 1}.$$

Proof. Let us write the main equation

$$\phi(x_k + v_k) = x_{k+1} + v_{k+1}$$

in the form

$$v_{k+1} - v_k = f(x_k + v_k) + x_k - \phi(x_k) + \phi(x_k) - x_{k+1},$$

or

$$\mathcal{A}\bar{v} = \mathcal{G}\bar{v} + \bar{w}, \tag{1.164}$$

where

$$(\mathcal{A}\bar{v})_k = v_{k+1} - v_k, \quad (\mathcal{G}\bar{v})_k = f(x_k + v_k) + x_k - \phi(x_k),$$

and

$$\bar{w}_k = \phi(x_k) - x_{k+1} = -g_k(\xi).$$

Set, for $\bar{v} \in \mathcal{L}_{\bar{r},\infty}$,

$$(\mathcal{B}\bar{v})_k = -\sum_{i=k}^{\infty} v_i.$$

To estimate $||\mathcal{B}||$, take $\bar{v} \in \mathcal{L}_{\bar{r},\infty}$ and set $\bar{v}' = \mathcal{B}\bar{v}$. Since $r_k \le \rho^{k-i} r_i$ for $i \ge k$, we obtain the inequalities

$$r_k|v_k'| \le \sum_{i=k}^{\infty} r_k|v_i| = \sum_{i=k}^{\infty} \left(\frac{r_k}{r_i}\right) r_i|v_i| \le$$

$$\le \sum_{i=k}^{\infty} \rho^{k-i} \sup_{i \ge 0} r_i|v_i| = \frac{\rho}{\rho-1}||\bar{v}||_{\bar{r},\infty}.$$

This shows that

$$||\mathcal{B}|| \le \frac{\rho}{\rho-1}. \tag{1.165}$$

It follows from the equality

$$(\mathcal{A}\mathcal{B}\bar{v})_k = -\sum_{i=k+1}^{\infty} v_i + \sum_{i=k}^{\infty} v_i = v_k$$

that $\mathcal{A}\mathcal{B} = I$. Similarly one shows that $\mathcal{B}\mathcal{A} = I$, hence $\mathcal{B} = (\mathcal{A})^{-1}$. Set

$$\mathcal{R}(\bar{v}) = \mathcal{B}(\mathcal{G}\bar{v} + \bar{w}).$$

Equation (1.164) is equivalent to

$$\bar{v} = \mathcal{R}(\bar{v}). \tag{1.166}$$

We easily deduce from (1.160) and (1.165) that for $\bar{v}, \bar{v}' \in \mathcal{L}_{\bar{r},\infty}$ we have

$$||\mathcal{R}(\bar{v}) - \mathcal{R}(\bar{v}')||_{\bar{r},\infty} \le K\frac{\rho}{\rho-1}||\bar{v} - \bar{v}'||_{\bar{r},\infty}.$$

By condition (1.162),

$$\rho - \rho K > 1, \text{ hence } K\frac{\rho}{\rho - 1} < 1.$$

This shows that $\mathcal{R}$ is a contraction on $\mathcal{L}_{\bar{r},\infty}$ with contraction constant

$$K\frac{\rho}{\rho - 1},$$

hence for any $\bar{w} \in \mathcal{L}_{\bar{r},\infty}$ there is a unique solution $\bar{v}$ of Eq. (1.166) (and of Eq. (1.164)). Since $\mathcal{G}(0) = 0$, we see that $||\mathcal{G}\bar{v}||_{\bar{r},\infty} \leq K||\bar{v}||_{\bar{r},\infty}$, and it follows from the inequalities

$$\frac{\rho}{\rho - 1}||\bar{w}||_{\bar{r},\infty} \geq ||\mathcal{B}\bar{w}||_{\bar{r},\infty} \geq ||\bar{v}||_{\bar{r},\infty} - ||\mathcal{B}\mathcal{G}\bar{v}||_{\bar{r},\infty} \geq$$

$$\geq \left(1 - K\frac{\rho}{\rho - 1}\right)||\bar{v}||_{\bar{r},\infty}$$

that our solution is estimated by

$$||\bar{v}||_{\bar{r},\infty} \leq \frac{\frac{\rho}{\rho - 1}}{1 - K\frac{\rho}{\rho - 1}}||\bar{w}||_{\bar{r},\infty} = \frac{\rho}{\rho(1 - K) - 1}||\bar{w}||_{\bar{r},\infty},$$

as was claimed.   □

Remark. If the function f in (1.159) is of class C^1, then condition (1.160) implies that ϕ is a diffeomorphism. Let us estimate $||(D\phi)^{-1}||$. Take $x, x' \in \mathbb{R}^n$, let $y = \phi(x), y' = \phi(x')$. It follows from the inequalities

$$|y' - y| \geq |x' - x| - |f(x') - f(x)| \geq (1 - K)|x' - x|$$

that

$$||(D\phi)^{-1}|| \leq \frac{1}{1 - K},$$

hence in the considered case condition (1.162) of Theorem 1.4.8 is similar to condition (1.151) of Theorem 1.4.6.

1.4.4 Asymptotic Pseudotrajectories

Hirsch studied asymptotic shadowing of asymptotic pseudotrajectories in [Hirs4] (see also [Ben]). Before stating his result, we give some definitions.

Consider a sequence $a = \{a_k \in \mathbb{R}^n : k \geq 0\}$. We denote

$$\mathcal{R}(a) = \overline{\lim}_{k\to\infty}|a_k|^{1/k},$$

where, as usual, $\overline{\lim}$ is the lim sup.

Definition 1.23 *A sequence $\xi = \{x_k : k \geq 0\}$ is called an "asymptotic pseudotrajectory" of exponent $\lambda < 1$ for a diffeomorphism ϕ of $\mathbb{R}^n$ if*

$$\mathcal{R}(g(\xi)) \leq \lambda$$

(the sequence $g(\xi)$ was defined in Subsect. 1.4.2).

Remark. In [Hirs4], a sequence with the property above is called a λ-pseudoorbit.

Let K be a compact invariant set for a diffeomorphism ϕ. Fix a point $x \in K$ and define the *expansion constant* of ϕ at x by the formula

$$EC(\phi, x) = \min_{|v|=1} |D\phi(x)v| = ||D\phi^{-1}(x)||^{-1}.$$

Now we define the number

$$EC(\phi, K) = \min_{x \in K} EC(\phi, x).$$

The main shadowing result in [Hirs4] can be stated as follows.

Theorem 1.4.9. *Assume that K is a compact invariant set for a diffeomorphism ϕ of $\mathbb{R}^n$. Let $\mu_0 = EC(\phi, K)$. If $\xi = \{x_k\} \subset K$ is an asymptotic pseudotrajectory of exponent λ for ϕ and*

$$0 < \lambda < \mu_0, \tag{1.167}$$

then there exists a unique point x such that

$$\mathcal{R}(h(x, \xi)) \leq \lambda$$

(the sequence $h(x, \xi)$ was defined in Subsect. 1.4.2).

Proof. Fix a number $\mu \in (\lambda, \mu_0)$. It is easy to see that there exists a number $\rho_0 > 0$ and a neighborhood U of the set K such that

$$N_{\mu\rho}(\phi(x)) \subset \phi(N_\rho(x)) \tag{1.168}$$

for $x \in K$ and $\rho \in (0, \rho_0)$, and

$$|\phi(x) - \phi(y)| \geq \mu|x - y|$$

for $x, y \in U$ with $|x - y| < \rho_0$.

We fix an asymptotic pseudotrajectory $\xi = \{x_k\} \subset K$ of exponent λ for ϕ such that (1.167) holds. For $d > 0$ we denote by $B_k(d)$ the closed ball of radius d^k centered at x_k.

Take a number d such that

$$\lambda < d < \min(1, \mu). \tag{1.169}$$

We claim that

$$B_{k+1}(d) \subset \phi(B_k(d)) \tag{1.170}$$

for large k. Fix a number ν such that

$$\lambda < \nu < d. \tag{1.171}$$

Obviously, we have

$$|g_k(\xi)| = |\phi(x_k) - x_{k+1}| \le \nu^k$$

for large k. Now it follows from (1.169) and (1.171) that

$$|\phi(x_k) - x_{k+1}| + d^{k+1} \le \nu^k + d^{k+1} = d^k \left[\left(\frac{\nu}{d} \right)^k + d \right] < \mu d^k$$

for large k.

The last inequality implies the inclusions

$$B_{k+1}(d) \subset N_{\mu d^k}(\phi(x_k)). \tag{1.172}$$

Since $x_k \in K$, we deduce from (1.168) that if $d^k < \rho_0$, then

$$N_{\mu d^k}(\phi(x_k)) \subset \phi(B_k(d)). \tag{1.173}$$

It follows from (1.172) and (1.173) that there exists m such that (1.170) holds for $k \ge m$.

Inclusions (1.170) imply that the set

$$Q_m = \bigcap_{k \ge 0} \phi^{-k}(B_{m+k}(d))$$

is not empty. Take a point $x \in \phi^{-m}(Q_m)$. Since $\phi^k(x) \in B_k(d)$ for $k \ge m$, we see that

$$|h_k(x, \xi)| = |\phi^k(x) - x_k| \le d^k, \ k \ge m, \tag{1.174}$$

hence

$$\mathcal{R}(h(x, \xi)) \le d.$$

Take two numbers d_1, d_2 satisfying inequalities analogous to (1.169). Let x_1, x_2 be the corresponding points for which analogs of (1.174) hold. Let us show that $x_1 = x_2$. Set $d = \max(d_1, d_2)$. Denote $z_k = |\phi^k(x_1) - \phi^k(x_2)|$, it follows that $z_k \le 2d^k$ for large k. There exists m such that

$$\phi^k(x_i) \in U, \ i = 1, 2, \ \text{and} \ z_k < \rho_0$$

for $k \ge m$. Since

$$z_{k+1} = |\phi^{k+1}(x_1) - \phi^{k+1}(x_2)| \ge \mu z_k$$

for $k \ge m$, we see that

$$2d^k \ge z_k \ge \mu^{k-m} z_m, \ k \ge m.$$

We obtain the inequalities

$$\left(\frac{d}{\mu}\right)^k \geq \frac{z_m}{2\mu^m}$$

which are contradictory for large k if $z_m \neq 0$. This shows that $x_1 = x_2$.

Now we fix a sequence of numbers $d_l > \lambda, l \geq 0$, such that $d_l \to \lambda$ as $l \to \infty$. It follows from the previous arguments that for large l there exists a point x independent of l and such that

$$\mathcal{R}(h(x,\xi)) \leq d_l.$$

Obviously, this point x has the desired property. $\qquad\square$

Remark. Let us show that the theorem above can be reduced to Theorem 1.4.6. Introduce the number

$$M_0 = \max_{x \in K} \|D\phi^{-1}(x)\|,$$

then obviously $\mu_0 M_0 = 1$. It follows from condition (1.167) that $\lambda M_0 < 1$, hence we can find a number $M > M_0$ such that $\lambda M < 1$ (below we assume that $M > 1$). Find a heighborhood U of K such that

$$\sup_{x \in U} \|D\phi^{-1}(x)\| \leq M.$$

Let $\xi = \{x_k : k \geq 0\}$ be an asymptotic pseudotrajectory of exponent λ for ϕ such that $\xi \subset U$.

Take sequences $\nu_l > \lambda$ and ρ_l such that $\nu_l \rho_l < 1$, $\nu_l \to \lambda$, and

$$\rho_l \to \frac{1}{\lambda} \qquad\qquad (1.175)$$

as $l \to \infty$. Since $\lambda M < 1$, we have

$$\rho_l > M \qquad\qquad (1.176)$$

for large l. Fix an index l such that the last inequality holds (for simplicity, below we write ν and ρ instead of ν_l and ρ_l). For large k we have $|g_k(\xi)| \leq \nu^k$, and since $\rho\nu < 1$, the series

$$\sum \rho^k |g_k(\xi)|$$

converges. It follows from (1.176) that we can apply Theorem 1.4.6 (with $p = 1$ and $r_k = \rho^k$). Find m such that

$$\sum_{k \geq m} \rho^k |g_k(\xi)| \leq d_0.$$

By Theorem 1.4.6, there exists a unique x such that

$$\sum_{k \geq m} \rho^k |h_k(x,\xi)| \leq L d_0$$

(and the same reasons as in Theorem 1.4.9 show that this x does not depend on l).

Since $\rho^k |h_k(x,\xi)| \leq 1$ for large k, we obtain the inequalities $|h_k(x,\xi)| \leq \rho^{-k}$. It follows that $\mathcal{R}(h(x,\xi)) \leq \rho_l^{-1}$. It remains to apply (1.175) to show that $\mathcal{R}(h(x,\xi)) \leq \lambda$.

1.5 Shadowing for Flows

In the case of dynamical systems with continuous time (flows), the shadowing problem becomes technically more complicated.

A lot of research was devoted to shadowing near a hyperbolic set of a flow. In this section, we consider a hyperbolic set Λ containing no rest points for an autonomous system of differential equations. Theorem 1.3.1 is applied to show that in a neighborhood of Λ pseudotrajectories are shadowed by real trajectories, and the shadowing is Lipschitz with respect to the "errors".

First shadowing results for flows were obtained by Bowen [Bo1]. He developed a method close to the one described in Subsect. 1.2.4 to approximate pseudotrajectories by real trajectories for the restriction of a flow to a basic set. After that many authors studied the shadowing problem for abstract flows and for differential equations (autonomous and nonautonomous). Let us mention Franke and Selgrade [Fr1], Bronshtein [Bro], Meyer and Sell [Me], Kato [Kato3], and Nadzieja [Na]. Note that in [Bro] and [Na] analogs of Theorem 1.2.5 for flows were proved.

The mentioned papers were mostly devoted to the case of a neighborhood of a hyperbolic set.

Komuro studied in [Ko2] the shadowing problem for a geometric model (introduced by Guckenheimer [Gu1]) of the famous Lorenz system [Lo],

$$\dot{x} = \sigma(y - x)$$
$$\dot{y} = rx - y - xz \tag{1.177}$$
$$\dot{z} = xy - \beta z.$$

This geometric model is characterized by its return map f. It is shown in [Ko2] that the geometric model does not have the property of weakly parametrized shadowing (see Definition 1.25 below) except a special case $f(0) = 0$ and $f(1) = 1$. These conditions correspond to the value $r = 24.06...$ in the Lorenz system under $(\sigma, b) = (10, 8/3)$.

In this section, we apply the techniques developed in Subsect. 1.3 to establish shadowing in a neighborhood of a hyperbolic set containing no rest points.

Consider an autonomous system of differential equations

$$\dot{x} = X(x), \; x \in \mathbb{R}^n. \tag{1.178}$$

We assume that $X \in C^1(\mathbb{R}^n)$. Let $\Xi(t, x)$ be the trajectory of system (1.178) such that $\Xi(0, x) = x$. Since we restrict ourselves to the case of a neighborhood of a compact invariant set for system (1.178), without loss of generality we may assume that any trajectory is defined for $t \in \mathbb{R}$. Thus, we study the flow $\Xi(t, x)$ of our system. We denote by $D\Xi(t, x)$ the corresponding variational flow, i.e.,

$$D\Xi(t, x) = \frac{\partial \Xi(t, x)}{\partial x}.$$

Let us define pseudotrajectories for flows.

Definition 1.24 *For $d, T > 0$ we say that a mapping*

$$\Psi : \mathbb{R} \to \mathbb{R}^n$$

is a "(d, T)-pseudotrajectory" of system (1.178) if, for any $\tau \in \mathbb{R}$,

$$|\Xi(t, \Psi(\tau)) - \Psi(t + \tau)| \le d, \ |t| \le T. \tag{1.179}$$

Remark. This definition is close to standard definitions of pseudotrajectories for flows (see [Kato3]). Note that we do not assume Ψ to be continuous.

Another possible definition of a pseudotrajectory for an autonomous system of differential equations is discussed in [Ano3]. We describe here a concept generalizing the one of [Ano3] (and close to it). Let us say that a mapping

$$\Psi^* : \mathbb{R} \to \mathbb{R}^n$$

is a (d)-*pseudosolution* of system (1.178) if there exists an increasing sequence $\{t_k \in \mathbb{R} : k \in \mathbb{Z}\}$ such that $\Psi^* \in C^1(t_k, t_{k+1})$,

$$\Psi^*(t_k) = \lim_{t \to t_k + 0} \Psi^*(t),$$

and the inequalities

$$\left| \frac{d}{dt} \Psi^*(t) - X(\Psi^*(t)) \right| \le d, \ t \in (t_k, t_{k+1}), \text{ and } |\Psi^*(t_k) - \Psi_-^*(t_k)| \le d \tag{1.180}$$

hold for any $k \in \mathbb{Z}$, where

$$\Psi_-^*(t_k) = \lim_{t \to t_k - 0} \Psi^*(t)$$

(of course, it is assumed that all the mentioned limits exist).

It is easy to see that if the right-hand side $X(x)$ of system (1.178) is Lipschitz (let $\mathcal{L}$ be its Lipschitz constant), and the values $t_{k+1} - t_k$ are separated from 0 (say, they satisfy the inequalities

$$0 < T_0 \le t_{k+1} - t_k$$

for $k \in \mathbb{Z}$), then the trajectory of a (d)-pseudosolution Ψ^* is an $(L_1 d, T)$-pseudotrajectory for the system, where

$$L_1 = (1 + T) \left(\sum_{i=1}^{m} \exp(i \mathcal{L} T) \right),$$

and m is the least natural number with the property

$$(m - 1) T_0 > T.$$

To prove this, we apply the following standard estimate (an easy consequence of Gronwall's lemma): if $\xi(t) \in C^1(a, b)$ satisfies the inequalites

$$|\dot{\xi}(t) - X(\xi(t))| \le d_1, \text{ and } |\xi(t') - x_0| \le d_2 \text{ for some } t' \in (a, b),$$

then

$$|\xi(t) - \Xi(t - t', x_0)| \le (d_2 + d_1|t - t'|) \exp(\mathcal{L}|t - t'|)$$

for $t \in (a, b)$.

Let us prove (1.179) for $0 \le t \le T$. Take $\tau \in [t_k, t_{k+1})$, it follows from the first inequality in (1.180) and from the estimate above that

$$|\Xi(t, \Psi^*(\tau)) - \Psi^*(t + \tau)| \le d|t| \exp(\mathcal{L}|t|) \le d(1 + T) \exp(\mathcal{L}T)$$

while $t + \tau \in [t_k, t_{k+1})$. Thus, if $T + \tau \in (t_k, t_{k+1})$, then our statement holds.

In the case $T + \tau \in [t_{k+1}, t_{k+2})$, the left-hand side of the previous estimate implies the inequality

$$|\Xi(t_{k+1} - \tau, \Psi^*(\tau)) - \Psi^*_-(t_{k+1})| \le dT \exp(\mathcal{L}T),$$

and hence the inequality

$$|\Xi(t, \Psi^*(\tau)) - \Psi^*(t + \tau)| \le$$

$$\le (d + dT \exp(\mathcal{L}T) + d(t + \tau - t_{k+1})) \exp(\mathcal{L}(t + \tau - t_{k+1})) \le$$

(we applied the second inequality in (1.180))

$$\le d(1 + T) \exp(\mathcal{L}T) + d(1 + T) \exp(2\mathcal{L}T)$$

while $t + \tau \in [t_{k+1}, t_{k+2})$. In this case, our statement is proved. Continuing this process, we obtain, for $t + \tau \in (t_{k+m-1}, t_{k+m})$, the estimate

$$|\Xi(t, \Psi^*(\tau)) - \Psi^*(t + \tau)| \le$$

$$\le d(1 + T)(\exp(\mathcal{L}T) + \ldots + \exp(m\mathcal{L}T)),$$

this gives the desired estimate, since by the choice of m we have

$$\tau + T < t_{k+1} + (m - 1)T_0 < t_{k+m}.$$

The case $t \in [-T, 0]$ is considered similarly.

Thus, we have proved that if the "jumps" t_k are not very frequent, then the trajectory of a (d)-pseudosolution is an $(\mathcal{L}_1, T)$-pseudotrajectory. Of course, since our Definition 1.24 requires nothing except (1.179), the inverse statement cannot be true.

Nevertheless, one can easily find, for a (d, T)-pseudotrajectory Φ, a (d)-pseudosolution Φ^* such that

$$|\Phi(t) - \Phi^*(t)| \le d \text{ for } t \in \mathbb{R}. \tag{1.181}$$

Indeed, take $t_k = kT, k \in \mathbb{Z}$, and set

$$\Phi^*(t) = \Xi(t - t_k, \Phi(t_k)) \text{ for } t \in [t_k, t_{k+1}).$$

Then obviously the first inequality in (1.180) holds with $d = 0$. Inequality (1.179) implies (1.181) and the second inequality in (1.180).

Thus, if a solution $\Xi(t, p)$ has the property

$$|\Xi(t, p) - \Phi^*(t)| \leq Ld,$$

then we have the estimate

$$|\Xi(t, p) - \Phi(t)| \leq (L + 1)d.$$

This means that in the problem of Lipschitz shadowing (and this is our main problem) the two definitions are equivalent.

Two types of shadowing for flows are usually considered. First let us define two sets of homeomorphisms (*reparametrizations*) $\alpha : \mathbb{R} \to \mathbb{R}$.

Define Rep as the set of increasing homeomorphisms α mapping $\mathbb{R}$ onto $\mathbb{R}$ and such that $\alpha(0) = 0$.

Fix $\epsilon > 0$ and define Rep(ϵ) as follows:

$$\mathrm{Rep}(\epsilon) = \left\{ \alpha \in \mathrm{Rep} : \left| \frac{\alpha(t) - \alpha(s)}{t - s} - 1 \right| \leq \epsilon \text{ for } t \neq s \right\}.$$

Definition 1.25 *System (1.178) has the "property of weakly parametrized shadowing" on a set $Y \subset \mathbb{R}^n$ if given $\epsilon > 0$ there exist $d, T > 0$ such that for any (d, T)-pseudotrajectory*

$$\Psi : \mathbb{R} \to Y$$

there is a point p and a reparametrization $\alpha \in$ Rep with the property

$$|\Psi(t) - \Xi(\alpha(t), p))| \leq \epsilon.$$

Definition 1.26 *System (1.178) has the "property of strongly parametrized shadowing" on a set $Y \subset \mathbb{R}^n$ if given $\epsilon > 0$ there exist $d, T > 0$ such that for any (d, T)-pseudotrajectory*

$$\Psi : \mathbb{R} \to Y$$

there is a point p and a reparametrization $\alpha \in$ Rep(ϵ) with the property

$$|\Psi(t) - \Xi(\alpha(t), p))| \leq \epsilon.$$

Relations between these (and some other) shadowing properties for flows were studied by Thomas [T1], Komuro [Ko1], and others.

Now let us define a hyperbolic set for system (1.178).

Definition 1.27 *A set Λ is called "hyperbolic" for system (1.178) if*
(h.1) Λ is compact and Ξ-invariant;

(h.2) there exist numbers $C > 0, \lambda_0 \in (0,1)$, and continuous families of linear subspaces $S(p), U(p)$ of $\mathbb{R}^n$, $p \in \Lambda$, such that
(h.2.1) the families S, U are $D\Xi$-invariant, i.e.,

$$D\Xi(t,p)T(p) = T(\Xi(t,p)) \text{ for } t \in \mathbb{R}, \ p \in \Lambda, \ T = S, U;$$

(h.2.2) for $p \in \Lambda$ we have

$$S(p) \oplus U(p) = \mathbb{R}^n \text{ if } X(p) = 0,$$

$$S(p) \oplus U(p) \oplus < X(p) >= \mathbb{R}^n \text{ if } X(p) \neq 0$$

(here $< X(p) >$ is the span of $X(p)$);
(h.2.3)

$$|D\Xi(t,p)v| \leq C\lambda_0^t|v| \text{ for } v \in S(p), \ t \geq 0;$$

$$|D\Xi(t,p)v| \leq C\lambda_0^{-t}|v| \text{ for } v \in U(p), \ t \leq 0.$$

We call C, λ_0 the *hyperbolicity constants* of Λ, and the families S, U are called the *hyperbolic structure* on Λ.

Now we state the main result of this subsection.

Theorem 1.5.1. *Assume that Λ is a hyperbolic set for system (1.178) and that $X(p) \neq 0, p \in \Lambda$. Then there exists a neighborhood W of Λ and numbers $d_0, L > 0$ such that for any $(d,1)$-pseudotrajectory $\Psi \subset W$ with $d \leq d_0$ there is a point p and a homeomorphism $\alpha \in \text{Rep}(Ld)$ such that*

$$|\Psi(t) - \Xi(\alpha(t),p)| \leq Ld, \ t \in \mathbb{R}. \tag{1.182}$$

Remark. The theorem states that the homeomorphism α is "close" to the identity in the sense of the inequality

$$\left|\frac{\alpha(t)}{t} - 1\right| \leq Ld.$$

Let us show that, in general, a similar statement with an estimate

$$|\alpha(t) - t| \leq D_1(d), \text{ where } \lim_{d \to 0} D_1(d) = 0, \tag{1.183}$$

is not true.

Assume that system (1.178) considered in $\mathbb{R}^2$ has a hyperbolic closed trajectory S corresponding to a 2π-periodic solution $\xi(t) = (\sin t, \cos t)$.

Consider for $d > 0$ a mapping Ψ given by

$$\Psi(t) = \xi(kd/2 + t), \ t \in [2\pi k, 2\pi(k+1)), \ k \in \mathbb{Z}.$$

Since $|\dot{\xi}(t)| = 1$, Ψ is a $(d,1)$-pseudotrajectory for system (1.178).

We claim that if d is small enough, then it is impossible to find a point p and an increasing homeomorphism $\alpha : \mathbb{R} \to \mathbb{R}$ such that estimate (1.183) holds together with

$$|\Xi(\alpha(t), p) - \Psi(t)| \leq D_2(d), \quad \text{where } \lim_{d \to 0} D_2(d) = 0. \qquad (1.184)$$

Note that the set $\{\Psi(2\pi k) : k \in \mathbb{Z}\}$ is $d/2$-dense in S (i.e., for any point $x \in S$ there is a point $x' = \Psi(2\pi k)$ such that $r(x, x') \leq d/2$, where $r(,)$ is the distance on S).

Since the trajectory S is hyperbolic, it is isolated (i.e., there is a neighborhood U of S such that S is the only complete trajectory of system (1.178) in U). It is easy to understand that if d is small enough, then a point p for which (1.184) is satisfied belongs to S.

Take $d > 0$ having the formulated property and such that

$$d < \pi, \quad D_1(d) < \frac{\pi}{4}, \quad D_2(d) < \frac{1}{2}.$$

Assume that for this d there exists a point $p \in S$ and a homeomorphism α for which the inequalities in (1.183) and (1.184) are fulfilled. Since $\Xi(2\pi k, p) = p$ for any $k \in \mathbb{Z}$, it follows from the inequality

$$r(\Xi(\alpha(2\pi k), p), \Xi(2\pi k, p)) \leq \frac{\pi}{4}$$

that

$$r(\Xi(\alpha(2\pi k), p), p) \leq \frac{\pi}{4}.$$

But it follows from (1.184) that

$$r(\Xi(\alpha(2\pi k), p), \Psi(2\pi k)) \leq \frac{\pi}{2} D_2(d) < \frac{\pi}{4},$$

hence for any point x of the set $\{\Psi(2\pi k)\}$ we have $r(x, p) \leq \pi/2$, and this set cannot be $\pi/2$-dense in S. The obtained contradiction proves our statement.

Now let us prove Theorem 1.5.1. Let Λ be a hyperbolic set of system (1.178) with hyperbolicity constants C, λ_0.

Find $T > 0$ such that

$$C\lambda_0^T \leq \lambda_0.$$

Denote $\phi(x) = \Xi(T, x)$.

We fix a bounded neighborhood W' of the set Λ and take a Lipschitz constant $\mathcal{L}$ for X in W'. We will take a neighborhood W in Theorem 1.5.1 such that

$$\Xi(t, x) \in W' \text{ for } x \in W, \ |t| \leq T + 1. \qquad (1.185)$$

The same reasons as in Lemma 1.1.3 show that there is a constant $K = K(T, \mathcal{L})$ such that if a $(d', 1)$-pseudotrajectory Ψ of system (1.178) belongs to W, then Ψ is a (Kd', T)-pseudotrajectory. We will show that there exist constants d_0, L

such that for a (d, T)-pseudotrajectory $\Psi \subset W$ with $d \leq d_0$ we can find a point p and a homeomorphism α with the properties described in our theorem. Obviously, this proves the theorem for $(d, 1)$-pseudotrajectories with $d_0' = d_0/K$ and $L' = KL$.

The first step of the proof is to extend the hyperbolic structure to a neighborhood of Λ.

Lemma 1.5.1. *Given $\epsilon > 0, \lambda \in (\lambda_0, 1)$ there exists a neighborghood $W = W(\epsilon, \lambda)$, a positive constant δ, and continuous (not necessarily $D\phi$-invariant) extensions S', U' of S, U to W such that*

(1)

$$S'(x) \oplus U'(x) \oplus < X(x) >= \mathbb{R}^n, x \in W$$

(we denote below by Π_x^s, Π_x^u, and Π_x^0 the complementary projectors onto $S'(x)$, $U'(x)$, and $< X(x) >$ generated by this representation of $\mathbb{R}^n$);

(2) for $x, y \in W$ with $|y - \phi(x)| < \delta$, the mapping $\Pi_y^s D\phi(x)$ $(\Pi_y^u D\phi(x))$ is an isomorphism between $S'(x)$ and $S'(y)$ (respectively, between $U'(x)$ and $U'(y)$), and the inequalities

$$|\Pi_y^s D\phi(x)v| \leq \lambda|v|, \ |\Pi_y^u D\phi(x)v| \leq \epsilon|v|, \ v \in S'(x); \tag{1.186}$$

$$|\Pi_y^u D\phi(x)v| \geq \frac{1}{\lambda}|v|, \ |\Pi_y^s D\phi(x)v| \leq \epsilon|v|, \ v \in U'(x), \tag{1.187}$$

hold;

(3) if $\xi(x) = \Xi(T', x)$ with $|T'-T| \leq 1$, then, for $x, y \in W$ with $|y-\xi(x)| < \delta$ and for $v \in S'(x) \oplus U'(x)$, the inequality

$$|\Pi_y^0 D\xi(x)v| \leq \epsilon|v| \tag{1.188}$$

holds.

Proof. We first extend S, U to continuous (not necessarily $D\phi$-invariant) families S', U' on a closed neighborghood W_0 of Λ so that statement (1) holds.

It follows immediately from Definition 1.27 that for points $x \in \Lambda, y = \phi(x)$ the mapping $\Pi_y^s D\phi(x)$ $(\Pi_y^u D\phi(x))$ is an isomorphism between $S(x)$ and $S(y)$ (respectively, between $U(x)$ and $U(y)$), and the relations

$$||\Pi_y^s D\phi(x)|_{S(x)}|| \leq \lambda_0, \ \Pi_y^u D\phi(x)|_{S(x)} = 0,$$

$$||\Pi_y^u D\phi(x)|_{U(x)}|| \geq 1/\lambda_0, \ \Pi_y^s D\phi(x)|_{U(x)} = 0$$

hold. For points $x \in \Lambda, y = \xi(x)$, we have

$$\Pi_y^0 D\xi(x)|_{S(x)\oplus U(x)} = 0.$$

Since $\Pi_x^s, \Pi_x^u, \Pi_x^0, \phi(x)$, and $D\phi(x)$ are uniformly continuous on Λ, and ξ is a shift along trajectories of system (1.178) for a bounded time, given arbitrary $\epsilon > 0$ we obviously can find a neighborhood $W(\epsilon, \lambda) \subset W_0$ and a number δ with the desired properties.   □

Below we write S, U instead of S', U'. Take $\lambda \in (\lambda_0, 1)$ and find the corresponding neighborhood $W = W(1, \lambda)$. We will decrease W in the proof. Since $W \subset W_0$ and W_0 is compact, there exists $N > 0$ such that

$$|X(x)|, \|\Pi_x^s\|, \|\Pi_x^u\| \leq N \text{ for } x \in W, \tag{1.189}$$

and

$$\|D\Xi(t, x) - D\Xi(t', x)\| \leq N|t - t'| \text{ for } x \in W, |t|, |t'| \leq T + 1, \tag{1.190}$$

For $p \in W$ set
$$Z(p) = S(p) + U(p).$$

We will denote in this proof by $\Sigma(p)$ the $(n-1)$-dimensional hyperplane through the point p parallel to $Z(p)$. Note that this hyperplane is transverse to the vector $X(p)$. Let us formulate two auxiliary statements (Lemmas 1.5.2 and 1.5.3). In these statements, $\text{dist}(x, \Sigma(z))$ is the distance between a point x and a hyperplane $\Sigma(z)$, and $\Sigma(b, x)$ is the ball in $\Sigma(x)$ of radius $b > 0$ centered at x.

Lemma 1.5.2.   *There exist constants* $d_1 > 0, \epsilon_0 < 1, K_1 \geq 1$ *such that if* $z, q \in W$ *and* $|z - q| \leq d_1$, *then there is a unique scalar function* $f(x)$ *in the* d_1-*neighborhood of* q *such that* $|f(x)| \leq \epsilon_0$ *and*

$$\Xi(f(x), x) \in \Sigma(z).$$

This function f *is of class* C^1 *and has the properties*

$$|f(x)| \leq K_1 \text{dist}(x, \Sigma(z)) \text{ and } |z - \Xi(f(x), x)| \leq K_1|x - z|.$$

Remark.   The geometric sense of the function f is obvious. For a point x in a small neighborhood of z, $f(x)$ is the time at which the trajectory through x reaches its nearest point of intersection with $\Sigma(z)$.

Proof.   If we assume, without loss of generality, that z is the origin, and that $\Sigma(z)$ is given by the relation

$$y = Ls, \ s \in \mathbb{R}^{n-1},$$

then we can write the equation for f in the form

$$F(x, f, s) = \Xi(f, x) - Ls = 0. \tag{1.191}$$

Since $F(0, 0, 0) = 0$, and the matrix

$$A = \frac{\partial F}{\partial(f, s)}(0, 0, 0) = [X(0), -L]$$

is nonsingular, it follows from the implicit function theorem that Eq. (1.191) has a unique small solution $f(x), s(x)$ for small x. Since our neighborhood W is a subset of a compact set W_0 on which statement (1) of Lemma 1.5.1 holds, for any $z \in W$ the angle between $\Sigma(z)$ and $X(z)$ and the norm $|X(z)|$ are separated from 0 by a constant independent of z. It follows from standard estimates in the implicit function theorem that f has the properties described in our lemma with constants d_1, ϵ_0, K_1 independent of z.

Lemma 1.5.3. *There exist positive constants $d_2 \leq d_1$ and b with the following property.*

If $p, z \in W$ and $|\phi(p) - z| \leq d_2$, then there is a unique function $\tau(x)$ of class C^1 on $\Sigma(b, p)$ such that

$$\psi(x) = \Xi(\tau(x), x) \in \Sigma(z) \text{ and } |\tau(x) - T| \leq \epsilon_0$$

(d_1 and ϵ_0 are from Lemma 1.5.2).

Remark. We do not prove this lemma here, its proof is similar to the proof of the preceding lemma. We will study in detail properties of the functions f, τ in Subsect. 2.2.1, Chap. 2, investigating equations analogous to Eq. (1.191).

Take $N_1 = N_1(N, \lambda)$ given by Theorem 1.3.1, fix the corresponding κ so that inequality (1.65) holds, and find $\epsilon > 0$ such that

$$\epsilon < \frac{\kappa}{12K_1 N}. \tag{1.192}$$

Below we work with a neighborhood $W = W(\epsilon, \lambda)$ (increasing it, if necessary, to satisfy (1.185)).

The main technical part in the proof of Theorem 1.5.1 is the lemma below. In this lemma, we consider points $p, z \in W$ with $|z - \phi(p)| \leq d_2$, so that $\tau(x), \psi(x)$ are defined on $\Sigma(b, p)$.

Lemma 1.5.4. *There exists $d_3 \leq d_2$ such that for any $p, z \in W$ with $|z - \phi(p)| < d \leq d_3$ and for any $v \in Z(p)$ with $|v| = 1$ we have*

$$|(D\psi(p) - D\phi(p))v| < \frac{\kappa}{4}.$$

Proof. Take $v \in Z(p)$ with $|v| = 1$. By Lemma 1.5.3, we can define $\tau(x)$ on $\Sigma(b, p)$. Denote $\tau_0 = \tau(p)$ and define $\xi(x) = \Xi(\tau_0, x)$. Denote $q = \phi(p), w = \xi(p)$, and let

$$v_1 = D\phi(p)v, \quad v_2 = D\xi(p)v, \quad v_3 = D\psi(p)v.$$

Take

$$d_3' = \min\left(d_2, \frac{\epsilon_0}{K_1}, \frac{\kappa}{8K_1 N}\right),$$

where ϵ_0 and K_1 are given by Lemma 1.5.2, and d_2 is given by Lemma 1.5.3. If $|z - q| < d \leq d'_3$, then obviously $\text{dist}(q, \Sigma(z)) < d$, hence $|f(q)| \leq K_1 d$ by Lemma 1.5.2, and $|f(q)| \leq \epsilon_0$ by the choice of d'_3. The point w belongs to $\Sigma(z)$. It follows from the equalities

$$w = \Xi(\tau_0, p) = \Xi(\tau_0 - T, \Xi(T, p)) = \Xi(\tau_0 - T, q),$$

from the estimate $|\tau_0 - T| \leq \epsilon_0$, and from the uniqueness of f that $\tau_0 - T = f(q)$. Since $\epsilon_0 < 1$, we deduce from estimate (1.190) that

$$\|D\phi(p) - D\xi(p)\| = \|D\Xi(T, p) - D\Xi(\tau_0, p)\| \leq N|T - \tau_0|.$$

This leads to the inequality

$$|v_1 - v_2| \leq N|f(q)| \leq NK_1 d < \frac{\kappa}{8}. \tag{1.193}$$

We want to estimate $|v_2 - v_3|$. Let $\gamma(s) = p + vs, s \geq 0$, be a ray in $Z(p)$. Define

$$\gamma_1(s) = \xi(\gamma(s)) = \Xi(\tau_0, \gamma(s)),$$
$$\gamma_2(s) = \psi(\gamma(s)) = \Xi(\tau(\gamma(s)), \gamma(s))$$

for $s \geq 0$ with $\gamma(s) \in \Sigma(b, p)$.

Now we take

$$d''_3 = \min\left(d_2, \frac{d_1}{2}, \frac{\delta}{2}\right),$$

where δ is given by Lemma 1.5.1. If $|z - q| < d \leq d''_3$, then obiously there exists $s_0 > 0$ such that, for $s \in [0, s_0)$, the function $\tau^*(s) = f(\gamma_1(s))$ is defined. It follows from the equalities

$$\Xi(f(\gamma_1(s)), \gamma_1(s)) = \Xi(f(\gamma_1(s)) + \tau_0, \gamma(s)) = \Xi(\tau(\gamma(s)), \gamma(s))$$

that $\tau(\gamma(s)) = \tau_0 + \tau^*(s)$, hence

$$v_2 = \frac{d\gamma_1}{ds}(0) = \frac{d}{ds}\Xi(\tau_0, \gamma(s))\,|_{s=0}, \tag{1.194}$$

$$v_3 = \frac{d\gamma_2}{ds}(0) = \frac{d}{ds}\Xi(\tau_0 + \tau^*(s), \gamma(s))\,|_{s=0} = X(w)\frac{d\tau^*}{ds}(0) + v_2. \tag{1.195}$$

We can write, for small $s > 0$,

$$\gamma_1(s) = w + sv_2 + o(s),$$

where, as usual,

$$\frac{o(s)}{s} \to 0 \text{ as } s \to 0.$$

By statement (3) of Lemma 1.5.1,

$$|\Pi_z^0 v_2| = |\Pi_z^0 D\xi(p)v| \leq \epsilon,$$

hence
$$\operatorname{dist}(\gamma_1(s), \Sigma(z)) \leq \epsilon s + o(s),$$

and consequently (see Lemma 1.5.2),
$$|\tau^*(s)| = |f(\gamma_1(s))| < \epsilon K_1 s + o(s).$$

This leads to the estimate
$$\left| \frac{d}{ds} \tau^*(0) \right| \leq \epsilon K_1.$$

It follows from (1.189) that
$$\left| X(w) \frac{d}{ds} \tau^*(0) \right| \leq \epsilon N K_1 < \frac{\kappa}{8}.$$

Comparing (1.194) and (1.195), we see that
$$|v_2 - v_3| < \frac{\kappa}{8}.$$

This inequality and inequality (1.193) prove that for $|z - \phi(p)| < d \leq d_3$, where $d_3 = \min(d_3', d_3'')$, we have
$$|v_1 - v_3| < \frac{\kappa}{4}.$$

This completes the proof.   □

Now we take a (d, T)-pseudotrajectory Ψ for system (1.178) such that $\Psi(t) \in W$ for $t \in \mathbb{R}$ and $d \leq d_3$. Denote $x_k = \Psi(kT)$ for $k \in \mathbb{Z}$. We choose coordinates in $\Sigma(x_k)$ so that x_k is the origin of $\Sigma(x_k)$, this identifies $Z(x_k)$ with $\Sigma(x_k)$. Let $H_k = Z(x_k)$. Denote $P_k = \Pi^s_{x_k}|_{Z(x_k)}, Q_k = \Pi^u_{x_k}|_{Z(x_k)}$, and $S_k = P_k H_k, U_k = Q_k H_k$. Obviously,
$$S_k = S(x_k), \quad U_k = U(x_k).$$

Fix a pair x_k, x_{k+1} and denote $p = x_k, z = x_{k+1}$. Let $\phi_k(x) = \psi(x - p)$ for $x \in \Sigma(b, p)$ (the mapping ψ and the number b are given by Lemma 1.5.3).

We can write
$$\phi_k(v) = D_k v + \chi_{k+1}(v),$$

where $D_k = D\phi_k(0), \chi_{k+1}(0) = \phi_k(0), D\chi_{k+1}(0) = 0$. Since the mapping ϕ_k is a shift along trajectories of system (1.178) for time not exceeding $T + 1$, there exists $d_4 < d_3$ such that
$$|\chi_{k+1}(v) - \chi_{k+1}(v')| \leq \frac{\kappa}{2}|v - v'| \text{ for } v, v' \in Z(p), |v|, |v'| \leq d_4. \tag{1.196}$$

Now we set
$$A_k = A_k^s + A_k^u, \text{ where } A_k^s = \Pi_z^s D\phi(p)\Pi_p^s, \ A_k^u = \Pi_z^u D\phi(p)\Pi_p^u.$$

Represent $\phi_k(v)$ in form (1.51), where

$$w_{k+1}(v) = (D_k - A_k)v + \chi_{k+1}(v).$$

It follows from Lemma 1.5.1 that A_k^s maps $S(p) = S_k$ to $S(z) = S_{k+1}$, and that

$$||A_k|_{S_k}|| = ||A_k^s|_{S_k}|| \le \lambda. \tag{1.197}$$

Similar reasons show that $A_k^u|_{U_k} = A_k|_{U_k}$ maps isomorphically U_k to U_{k+1}, and for $B_k = (A_k^u|_{U_k})^{-1}$ we have

$$||B_k|| \le \lambda. \tag{1.198}$$

Now let us estimate $|(D_k - A_k)v|$ for $v \in H_k, |v| = 1$. Since

$$D_k = D\phi_k(0) = D\psi(p)$$

and $v = (\Pi_p^s + \Pi_p^u)v$, we can write

$$(D_k - A_k)v = (D\psi(p) - D\phi(p))v+$$

$$+(\Pi_z^s + \Pi_z^u + \Pi_z^0)D\phi(p)(\Pi_p^s + \Pi_p^u)v - \Pi_z^s D\phi(p)\Pi_p^u v - \Pi_z^u D\phi(p)\Pi_p^s v.$$

It follows that

$$|(D_k - A_k)v| \le |(D\psi(p) - D\phi(p))v|+$$

$$+|\Pi_z^0 D\phi(p)v| + |\Pi_z^s D\phi(p)\Pi_p^u v| + |\Pi_z^u D\phi(p)\Pi_p^s v|. \tag{1.199}$$

By Lemma 1.5.4, the first term on the right in (1.199) does not exceed $\kappa/4$. Since $v \in S_k + U_k$, (1.188) implies that the second term does not exceed $\epsilon \le \kappa/12$. Since $\Pi_p^u v \in U_k, \Pi_p^s v \in S_k$ and $|\Pi_p^u v|, |\Pi_p^s v| \le N$, it follows from the second inequalities in (1.186) and (1.187) that the third and the fourth terms in (1.199) are each estimated by $N\epsilon \le \kappa/12$. Hence, if $d \le d_3$, then

$$|(D_k - A_k)v| \le \frac{\kappa}{2}.$$

Now it follows from this inequality and from (1.196) that

$$|\chi_{k+1}(v) - \chi_{k+1}(v')| \le \kappa|v - v'| \text{ for } v, v' \in H_k, |v|, |v'| \le d_4. \tag{1.200}$$

Finally, we deduce from (1.197)-(1.200) and from Theorem 1.3.1 that there exist constants $d', L' > 0$ such that if

$$|\phi_k(0)| \le d'' \le d',$$

then one can find points $v_k \in H_k$ with the properties

$$\phi_k(v_k) = v_{k+1} \text{ and } |v_k| \le L'd''. \tag{1.201}$$

Since any ϕ_k is a shift along a trajectory of system (1.178), we see that all the points $p_k = x_k + v_k$ belong to the same trajectory $\Xi(t, p_0)$. It follows from Lemma 1.5.2 and from the definition of ϕ_k that if Ψ is a (d, T)-pseudotrajectory with $d \le d_2$, then

$$|\phi_k(0)| = |z - \Xi(f(\phi(p)), \phi(p))| \leq K_1|\phi(p) - z| \leq K_1 d,$$

hence we deduce from (1.201) that if

$$d \leq d_0 = \min\left(d_4, \frac{d'}{K_1}, \frac{d_1}{1 + L_0 \exp(\mathcal{L}T)}\right),$$

then

$$|p_k - x_k| \leq L_0 d, \qquad (1.202)$$

where $L_0 = L'K_1$ (recall that $\mathcal{L}$ is a Lipschitz constant for X in the neighborhood W').

Define numbers t_k by $p_{k+1} = \Xi(t_k, p_k)$. Let us estimate

$$|\Xi(T, p_k) - z| \leq |\Xi(T, p) - z| + |\Xi(T, p_k) - \Xi(T, p)|$$

(recall that $p = x_k, z = x_{k+1}$). The first summand does not exceed d, the second one does not exceed $L_0 d \exp(\mathcal{L}T)$.

Since by the choice of d_0 we have the inequality

$$d(1 + L_0 \exp(\mathcal{L}T)) \leq d_1,$$

it follows from Lemma 1.5.2 and from the equality $p_{k+1} = \Xi(t_k, p_k)$ that $|t_k - T| \leq L_1 d$, where $L_1 = K_1(1 + L_0 \exp(\mathcal{L}T))$.

Set $\tau_0 = 0, \tau_k = t_1 + \ldots + t_k$ for $k > 0$, and $\tau_k = t_k + \ldots + t_{-1}$ for $k < 0$. Now we define $\alpha : \mathbb{R} \to \mathbb{R}$. For $k \in \mathbb{Z}$ we set

$$\alpha(Tk) = \tau_k,$$

for $t \in [Tk, T(k+1)]$ we set

$$\alpha(t) = \tau_k + (t - Tk)\frac{t_{k+1}}{T}.$$

Obviously, α is an increasing homeomorphism mapping $\mathbb{R}$ onto $\mathbb{R}$.

Take $t \in [Tk, T(k+1)]$ for $k > 0$, set $t' = t - Tk$. Since $|t'| \leq T$, we obtain the estimate

$$|\alpha(t) - \tau_k - t'| = \left|t'\left(\frac{t_{k+1}}{T} - 1\right)\right| \leq L_1 d. \qquad (1.203)$$

Now let us show that $\alpha \in \text{Rep}(L_2 d)$, where $L_2 = L_1/T$. Take $t, s \in \mathbb{R}$ such that $t \neq s$. Assume that $t < s$ and that

$$t \in [Tk, T(k+1)), \quad s \in [Tm, T(m+1)).$$

Since

$$\frac{t_{k+1}}{T} = 1 + \frac{t_{k+1} - T}{T} \leq 1 + \frac{L_1 d}{T} = 1 + L_2 d,$$

we obtain the inequalities

$$\alpha(T(k+1)) - \alpha(t) \leq (1 + L_2 d)(T(k+1) - t),$$

$$\alpha(T(k+2)) - \alpha(T(k+1)) \le (1 + L_2 d)T,$$

$$\cdots$$

$$\alpha(s) - \alpha(Tm) \le (1 + L_2 d)(s - Tm).$$

Adding these inequalities and dividing the sum by $s - t$, we prove the estimate

$$\frac{\alpha(s) - \alpha(t)}{s - t} \le 1 + L_2 d.$$

Similar reasons establish an analogous estimate from below. This proves that $\alpha \in \text{Rep}(L_2 d)$.

Set $p = p_0$. Obviously, $\Xi(\alpha(Tk), p) = \Xi(\tau_k, p) = p_k$. For $t \in [Tk, T(k+1)]$ we have

$$|\Xi(\alpha(t), p) - \Psi(t)| \le |\Xi(\alpha(t) - \tau_k, p_k) - \Xi(t', p_k)| +$$

$$+ |\Xi(t', p_k) - \Xi(t', x_k)| + |\Psi(Tk + t') - \Xi(t', \Psi(Tk))|$$

(note that $\Xi(\alpha(t), p) = \Xi(\alpha(t) - \tau_k, p_k)$, $x_k = \Psi(Tk)$, $\Psi(t) = \Psi(Tk + t')$).

The first term on the right does not exceed $N L_1 d$ (we apply (1.203) and the inequality $|X| \le N$). The second term does not exceed

$$|p_k - x_k| \exp(\mathcal{L}T) \le L_0 \exp(\mathcal{L}T)d,$$

and the third term is not more than d (since Ψ is a (d, T)-pseudotrajectory). This leads to the inequality

$$r(\Xi(\alpha(t), p), \Psi(t)| \le L_3 d,$$

where $L_3 = N L_1 + L_0 \exp(\mathcal{L}T) + 1$.

Take $L = \max(L_2, L_3)$ to complete the proof of Theorem 1.5.1.   □

Remark. Analyzing the proof of Theorem 1.5.1, it is easy to understand that a point p and a homeomorphism α such that (1.183) holds are not unique. This nonuniqueness is due to the absence of hyperbolicity "along the flow".

Nevertheless, since the "linear parts" of the mappings ϕ_k introduced in the proof obviously satisfy the conditions of Theorem 1.3.2, the following "uniqueness" statement for the considered problem holds.

We can take d_0 so small that, for $d \le d_0$, a sequence $v_k \in Z(x_k)$ that satisfies (1.201) is unique.

By Lemma 1.5.2, this means that there exist numbers d^* and L^* such that for $d \le d^*$ a point $p = p_0 \in \Sigma(x_0)$ with the properties

(1) $p_k \in \Sigma(x_k), |p_k - x_k| \le L^* d$;
(2) $p_{k+1} = \Xi(\tau_k, p_k)$;
(3) $|\tau_k - T| \le L^* d$
is unique.

Another way to formulate the uniqueness of a shadowing trajectory in a neighborhood of a hyperbolic set for a flow is described in [Coo3].

2. Topologically Stable, Structurally Stable, and Generic Systems

Everywhere in this chapter, M is a C^∞ smooth n-dimensional closed (i. e., compact and boundaryless) manifold, and r is a Riemannian metric on M.

2.1 Shadowing and Topological Stability

This section is devoted to relations between shadowing and topological stability. Following Walters [Wa2], we show that a topologically stable homeomorphism has the POTP. It is also shown that if an expansive homeomorphism has the POTP, then it is topologically stable.

Various authors used different definitions of topological stability. In this book, we work with the following one.

We consider the space $Z(M)$ of discrete dynamical systems on M with the metric ρ_0 defined by the formula

$$\rho_0(\phi, \psi) = \max_{x \in M}(r(\phi(x), \psi(x)), r(\phi^{-1}(x), \psi^{-1}(x))).$$

Definition 2.1 *We say that a dynamical system $\phi \in Z(M)$ is "topologically stable" if given $\epsilon > 0$ there exists a neighborhood W of ϕ in $Z(M)$ such that for any system $\psi \in W$ there is a continuous mapping h of M onto M having the following properties:*
  (a) $r(x, h(x)) < \epsilon$ for $x \in M$;
  (b) $\phi \circ h = h \circ \psi$.

This definition of topological stability is close to the corresponding ones in [Wa1, Y1]. In [Ni2], the defined property is called the C^0 lower semi-stability in the strong sense.

Walters [Wa2] and Morimoto [Morim1] showed that topological stability implies the the POTP. Walters proved Theorem 2.1.1 for $n=\dim M \geq 2$, the proof given by Morimoto is valid for $n \geq 1$.

Theorem 2.1.1. *If a system $\phi \in Z(M)$ is topologically stable, then ϕ has the POTP.*

Walters also proved the following statement.

Theorem 2.1.2. *Let $\phi \in Z(M)$ be expansive with expansivity constant α. If ϕ has the POTP, then given $\epsilon > 0$ with $3\epsilon < \alpha$ there exists $d > 0$ with the following property. If $\psi \in Z(M)$ and $\rho_0(\phi, \psi) < d$, then there is a unique continuous mapping $h : M \to M$ such that $h \circ \psi = \phi \circ h$ and $r(x, h(x)) < \epsilon$ for all $x \in M$. If ϵ is small enough, then h maps M onto M.*

Remark. This last result shows that a topologically Anosov homeomorphism is topologically stable (and even has a stronger property, the corresponding mapping h close to identity is unique).

We follow [Wa2] in the proofs of Theorems 2.1.1 and 2.1.2. The proof of Theorem 2.1.1 is given here only in the case $n = \dim M \geq 2$ (the remaining case $\dim M = 1$ is simple but requires special treatment, the reader can find the corresponding proof in Chap. 2 of [Pi2]).

Let us establish some preliminary results. First we prove a "C^0-closing" statement obtained by Nitecki and Shub in [Ni3].

Lemma 2.1.1. *Assume that $\dim M \geq 2$. Consider a finite collection*

$$\{(p_i, q_i) \in M \times M : i = 1, \ldots, k\}$$

such that
(a) $p_i \neq p_j, q_i \neq q_j$ for $1 \leq i < j \leq k$;
(b) $r(p_i, q_i) < d$ for $i = 1, \ldots, k$, with small positive d.
Then there exists a diffeomorphism f of M with the following properties:
(a) $\rho_0(f, \mathrm{id}) < 2d$ (here id is the identity mapping of M);
(b) $f(p_i) = q_i$ for $i = 1, \ldots, k$.

Proof. Take the circle S^1 with coordinate $\theta \in [0, 1)$. Consider the following system of differential equations on $M \times S^1$ with coordinates (p, θ):

$$\dot{p} = 0, \quad \dot{\theta} = 1.$$

Its vector field is $X = (0, 1)$, and its flow is given by

$$\Xi(t, p, \theta) = (p, \theta + t(\mathrm{mod}\ 1)).$$

Obviously, $\Xi(1, p, \theta)$ takes $M \times 0$ to itself and induces the identity mapping there.

Given the points $p_i, q_i \in M$, we consider the points $(p_i, \frac{1}{4}), (q_i, \frac{3}{4})$ in $M \times S^1$. We take for each i a curve $\gamma_i(t)$ in $M, 0 \leq t \leq \frac{1}{2}$, of constant speed, joining p_i to q_i and having length less than d. We can change parameter t on $\gamma_i(t)$ so that

$$\dot{\gamma}_i(0) = \dot{\gamma}_i(\frac{1}{2}) = 0, \quad |\dot{\gamma}_i(t)| < 2d. \tag{2.1}$$

Consider the curves g_i given by

$$g_i(t) = (\gamma_i(t), \frac{1}{4} + t), \ 0 \leq t \leq \frac{1}{2}.$$

The curves g_i are one-dimensional, and $\dim(M \times S^1) \geq 3$. It follows from condition (a) and from the transversality theorem (see [Hirs1]) that we can slightly perturb the curves $\gamma_i(t)$ so that (2.1) holds, and

$$g_i(t) \neq g_j(t) \text{ for } 0 \leq t \leq \frac{1}{2}, \ 1 \leq i < j \leq k.$$

Hence, we can find tubular neighborhoods N_i of g_i in $M \times S^1$ such that

$$\overline{N}_i \cap \overline{N}_j = \emptyset, \ 1 \leq i < j \leq k.$$

Take the tubular neighborhood N_i of g_i, extend g_i beyond its endpoints, and then extend N_i to be a tubular neighborhood of the extended curve (let this extended neighborhood be N_i^*). This can be done so that

$$\overline{N}_i^* \cap \overline{N}_j^* = \emptyset, \ 1 \leq i < j \leq k. \tag{2.2}$$

Let Y_i be the vector field $Y_i = (\dot{\gamma}_i(t), 1)$ on g_i (we can take $Y_i = (0, 1)$ on the extended part of g_i). Extend Y_i to N_i^* making it constant along fibers. Now take a "bump" function Θ which is 1 on $g_i(t), 0 \leq t \leq \frac{1}{2}$, and 0 off N_i^*. We obtain the vector field

$$Z_i = X + \Theta(Y_i - X)$$

such that $Z_i = (\dot{\gamma}_i(t), 1) = \dot{g}_i(t)$ on $g_i(t), 0 \leq t \leq \frac{1}{4}$, hence $g_i(t)$ is a trajectory of the system generated by Z_i, and

$$|Z_i - X| < 2d. \tag{2.3}$$

It follows from (2.2) that we can obtain a vector field Z which coincides with X on $M \setminus (N_1^* \cup \ldots \cup N_k^*)$, and with Z_i on $N_i^*, i = 1, \ldots, k$. Let $\Psi(t, p, \theta)$ be the flow of the corresponding system of differential equations. Since $X = (0, 1)$, and inequalities (2.3) hold for $i = 1, \ldots, k$, we deduce from standard estimates for differential equations that for any $(p, \theta) \in M \times S^1$ we have

$$r(\Psi(t, p, \theta), \Xi(t, p, \theta)) < 2d$$

for $|t| \leq 1$. Let f be the time-one mapping of the flow Ψ. It follows from our construction that $f(p_i) = q_i, i = 1, \ldots, k$, and $\rho_0(f, \text{id}) < 2d$. $\qquad\square$

The following statement enables us to "improve" a finite segment of a d-pseudotrajectory.

Lemma 2.1.2. *Let $\xi = \{x_k\}$ be a d-pseudotrajectory for a dynamical system ϕ. Let $m \geq 0$ be an integer, and let $\eta > 0$ be given. Then there exists a set of points $\{y_0, \ldots, y_m\}$ such that*

(a) $r(x_k, y_k) < \eta$, $0 \le k \le m$;
(b) $r(\phi(y_k), y_{k+1}) < 2d$, $0 \le k \le m - 1$;
(c) $y_i \ne y_j$ for $0 \le i < j \le m$.

Proof. We use induction on m. For $m = 0$ our lemma is obviously true. We assume that the lemma is true for $m - 1$, and prove it for m. Let $\eta > 0$ be given, we consider $\eta < d$. Choose $\lambda \in (0, \eta)$ such that the inequality $r(x, y) < \lambda$ implies $r(\phi(x), \phi(y)) < d$. By our assumption, we can choose $\{y_0, \ldots, y_{m-1}\}$ so that

$$r(y_k, x_k) < \lambda, \; 0 \le k \le m - 1; \; r(\phi(y_k), y_{k+1}) < 2d, \; 0 \le k \le m - 2,$$

and

$$y_i \ne y_j \text{ for } 0 \le i < j \le m - 1.$$

Since

$$r(\phi(y_{m-1}), x_m) \le r(\phi(y_{m-1}), \phi(x_{m-1})) +$$

$$+ r(\phi(x_{m-1}), x_m) < 2d,$$

we can find y_m such that $y_m \ne y_i$ for $0 \le i \le m - 1$, $r(y_m, x_m) < \eta$, and $r(\phi(y_{m-1}), y_m) < 2d$. $\qquad \square$

Now we show that it is possible to approximate a finite segment of a d-pseudotrajectory for ϕ by a trajectory of a close dynamical system.

Lemma 2.1.3. *Let* $\dim M \ge 2$ *and let* ϕ *be a dynamical system. Given a natural* m *and a positive* Δ, *there exists* $d > 0$ *with the following property. If* $\xi = \{x_k\}$ *is a* d-*pseudotrajectory for* ϕ, *then there exists a system* $\psi \in Z(M)$ *and a point* $y \in M$ *such that*

$$\rho_0(\phi, \psi) < \Delta$$

and

$$r(\psi^k(y), x_k) < \Delta \text{ for } k = 0, \ldots, m. \tag{2.4}$$

Proof. Fix a system ϕ, a natural number m, and $\Delta > 0$. Find $d_1 \in (0, \Delta)$ such that the inequality $r(x, y) < d_1$ implies $r(\phi^{-1}(x), \phi^{-1}(y)) < \Delta$. Take $d = d_1/4$ and consider a d-pseudotrajectory ξ for ϕ. By Lemma 2.1.2, there exists a set $\{y_0, \ldots, y_m\}$ such that

$$r(\phi(y_k), y_{k+1}) < d_1/2, \; k = 0, \ldots, m - 1;$$

$$d(y_k, x_k) < \Delta, \; k = 0, \ldots, m; \tag{2.5}$$

and

$$y_i \ne y_j, \; 0 \le i < j \le m$$

(hence, $\phi(y_i) \ne \phi(y_j)$, $0 \le i < j \le m$).

It follows from Lemma 2.1.1 that there exists a diffeomorphism f of M having the following properties:

$$\rho_0(f, \text{ id}) < d_1$$

and

$$f(\phi(y_k)) = y_{k+1} \text{ for } k = 0, \ldots, m - 1. \tag{2.6}$$

Consider $\psi = f \circ \phi$ and take $y = y_0$. Obviously, (2.5) and (2.6) imply (2.4).

For any $x \in M$ we have

$$r(\psi(x), \phi(x)) = r(f(\phi(x)), \phi(x)) < d_1 < \Delta$$

and

$$r(\psi^{-1}(x), \phi^{-1}(x)) = r(\phi^{-1}(f^{-1}(x)), \phi^{-1}(x)) < \Delta$$

(we take into account that

$$r(f^{-1}(x), x) \leq \rho_0(f, \text{ id}) < d_1).$$

This completes the proof.   □

Now we can prove Theorem 2.1.1 (let us remind that we consider the case $\dim M \geq 2$). Let ϕ be topologically stable.

Fix arbitrary $\epsilon > 0$. Find $\Delta \in (0, \epsilon)$ such that for any system $\psi \in Z(M)$ with $\rho_0(\phi, \psi) < \Delta$ there exists a mapping $h : M \to M$ having the properties described in Definition 2.1. Find $d > 0$ such that the statement of Lemma 2.1.3 is true.

Let ξ be a d-pseudotrajectory for ϕ. Fix a natural number m and apply Lemma 2.1.3 to find a dynamical system ψ and a point y such that $\rho_0(\phi, \psi) < \Delta$ and

$$r(\psi^k(y), x_k) < \Delta, \ k = 0, \ldots, m.$$

Take $x = h(y)$, then $\phi^k(x) = h(\psi^k(y)), k \in \mathbb{Z}$. Since

$$r(\phi^k(x), \psi^k(y)) < \epsilon \text{ for } k \in \mathbb{Z},$$

we see that

$$r(\phi^k(x), x_k) \leq \epsilon \text{ for } k = 0, \ldots, m.$$

The number $\epsilon > 0$ is arbitrary. It follows from part (a) of Lemma 1.1.1 that ϕ has the POTP. This completes the proof of Theorem 2.1.1.

Remark. Let us emphasize that topological stability is a really stronger property than the POTP. A generic continuous dynamical system has the POTP (see Sect. 2.4), while topological stability is not a generic property [Hu]. Yano constructed a simple example of a dynamical system on the circle S^1 which has the POTP but fails to be topologically stable [Y1]. An explanation of the example of Yano is given in Sect. 3.1, after the proof of Theorem 3.1.1.

Now we prove Theorem 2.1.2.

Lemma 2.1.4. *Assume that a dynamical system ϕ is expansive with expansivity constant α. Given $\lambda > 0$ there exists $N \geq 1$ such that if $r(\phi^k(x), \phi^k(y)) \leq \alpha$ for all k with $|k| < N$, then $r(x, y) < \lambda$.*

Proof. Let $\lambda > 0$ be given. If no N can be chosen with the property stated, then for each $N \geq 1$ there exist points x_N, y_N such that $r(\phi^k(x_N), \phi^k(y_N)) \leq \alpha$ for all k with $|k| < N$, and $d(x_N, y_N) \geq \lambda$. Choose subsequences N_i and points x, y with $x_{N_i} \to x, y_{N_i} \to y$ as $i \to \infty$. Then $r(x, y) \geq \lambda$, and $r(\phi^k(x), \phi^k(y)) \leq \alpha$ for all k. This contradicts the expansivity of ϕ.   □

To prove Theorem 2.1.2, we fix $\epsilon > 0$ such that $3\epsilon < \alpha$ and choose $d > 0$ corresponding to ϵ by Definition 1.3. Take $\psi \in Z(M)$ such that $\rho_0(\phi, \psi) < d$. Then any trajectory of ψ is a d-pseudotrajectory of ϕ. By part (b) of Lemma 1.1.2, ϕ has the SUP with constant $\alpha/2$.

Hence, for any $x \in M$ there is a unique point y such that

$$r(\phi^k(y), \psi^k(x)) < \epsilon \text{ for } k \in \mathbb{Z}.$$

This defines a mapping $h : M \to M$, $h(x) = y$. Take $k = 0$ in the inequality above to show that $r(x, h(x)) < \epsilon$. Since

$$r(\phi^k(h(\psi(x))), \psi^{k+1}(x)) < \epsilon, \ k \in \mathbb{Z},$$

and

$$r(\phi^k(\phi(h(x))), \psi^{k+1}(x)) = r(\phi^{k+1}(h(x)), \psi^{k+1}(x)) < \epsilon, \ k \in \mathbb{Z},$$

we see that the d-pseudotrajectory $\{x_k = \psi^{k+1}(x) : k \in \mathbb{Z}\}$ is (ϵ, ϕ)-shadowed by the points $h(\psi(x))$ and $\phi(h(x))$. Since x is arbitrary, the SUP implies that $h \circ \psi = \phi \circ h$.

Now let us show that h is continuous. Let $\lambda > 0$ be given. Apply Lemma 2.1.4 to find N such that if $r(\phi^k(u), \phi^k(v)) < \alpha$ for $|k| \leq N$, then $r(u, v) < \lambda$. Choose $\eta > 0$ such that $r(x, y) < \eta$ implies $r(\psi^k(x), \psi^k(y)) < \alpha/3$ for $|k| \leq N$. Then for x, y with $r(x, y) < \eta$ we have

$$r(\phi^k(h(x)), \phi^k(h(y))) = r(h(\psi^k(x)), h(\psi^k(y))) \leq$$

$$\leq r(h(\psi^k(x)), \psi^k(x)) + r(\psi^k(x), \psi^k(y)) +$$

$$+ r(\psi^k(y), h(\psi^k(y))) < \epsilon + \frac{\alpha}{3} + \epsilon < \alpha$$

for $|k| \leq N$. Therefore, the inequality $r(x, y) < \eta$ implies $r(h(x), h(y)) < \lambda$, and the continuity of h is proved.

The mapping h is the only one with $h \circ \psi = \phi \circ h$ and $r(x, h(x)) < \epsilon$ since if l is another one, then

$$r(\phi^k(l(x)), \phi^k(h(x))) = r(l(\psi^k(x)), h(\psi^k(x))) \leq$$

$$\leq r(l(\psi^k(x)), \psi^k(x)) + r(\psi^k(x), h(\psi^k(x))) < 2\epsilon < \alpha$$

for any $k \in \mathbb{Z}$, and it follows that $l(x) = h(x)$.

It is well known that if M is a compact manifold and $r(x, h(x)) < \epsilon$ for any $x \in M$ with ϵ small enough, then h maps M onto M. This completes the proof. $\square$

We consider here only the case of discrete dynamical systems. The same problem for flows was studied by Thomas [T2].

2.2 Shadowing in Structurally Stable Systems

Structurally stable systems (both flows and diffeomorphisms) were the main objects of interest in the global qualitative theory of dynamical systems in the last 30 years. Now we know that structural stability is equivalent to Axiom A (for diffeomorphisms) or Axiom A' (for flows) combined with the strong transversality condition, see [Robb, Robi2, Ma] in the case of diffeomorphisms, and [Robi1, Hay, We] in the case of flows.

Various approaches were applied to show that a structurally stable diffeomorphism has the POTP [Robi3, Mori2, Saw].

In Subsect. 2.2.1, we prove that a structurally stable flow has a Lipschitz shadowing property [Pi4]. We begin to work with a flow since this case is technically more difficult than the case of a diffeomorphism (mostly due to the possible coexistence of rest points and nonwandering trajectories that are not rest points). The main statement (Theorem 2.2.3) is reduced to shadowing results for sequences of mappings of Banach spaces with noninvertible "linear parts" (see Sect. 1.3). It was an intention of the author to make the presentation of Theorem 2.2.3 maximally "self-contained". Due to this reason, we give a detailed proof of the existence of Robinson's "compatible extensions of stable and unstable bundles" [Robi1] (see Lemma 2.2.9).

Shadowing for structurally stable diffeomorphisms is studied in Subsect. 2.2.2. It is shown that a structurally stable diffeomorphism has the LpSP. This result was first published in [Pi2] with a proof based on another approach. Here we show that a method similar to the one applied in Subsect. 2.2.1 gives a shorter and a more clear proof.

Sakai noted that the POTP is "uniform" in a C^1-neighborhood of a structurally stable diffeomorphism [Sak1] and that the C^1 interior of the set of diffeomorphisms with the POTP consists of structurally stable diffeomorphisms [Sak2]. We prove (Theorem 2.2.8) that the LpSP is also uniform [Be] and that a diffeomorphism in the C^1 interior of diffeomorphisms with uniform LpSP is structurally stable.

2.2.1 The Case of a Flow

In this subsection, we consider autonomous systems of differential equations generated by vector fields of class C^1 on M. Below we identify a vector field X

on M and the system

$$\dot{x} = X(x) \tag{2.7}$$

generated by X.

To define structural stability for the flows of considered systems, we introduce the C^1-topology on the set of vector fields on M as follows.

By the Whitney theorem [Hirs1], for any closed smooth n-dimensional manifold M there exists an embedding

$$f : M \to \mathbb{R}^{2n+1}.$$

Let us fix such an embedding f. Thus, we can consider M as a smooth submanifold of $\mathbb{R}^{2n+1}$. For a point $p \in M$ we identify the tangent space $T_p M$ with a linear subspace of $\mathbb{R}^{2n+1}$ and denote by $|.|$ the norm in $T_p M$ generated by the Euclidean norm of $\mathbb{R}^{2n+1}$.

Now we can consider a vector field X on M as a mapping that takes a point $p \in M$ to $X(p) \in T_p M \subset \mathbb{R}^{2n+1}$.

For two vector fields X and Y on M, we define

$$\rho_0(X, Y) = \max_{p \in M} |X(p) - Y(p)|.$$

In this setting, $DX(p)$ is considered as the derivative of the mapping above, and for $p \in M$ we can define

$$\|DX(p) - DY(p)\| = \max_{v \in T_p M, |v|=1} |DX(p)v - DY(p)v|.$$

Now we define the C^1-distance ρ_1 between two vector fields X and Y on M by the formula

$$\rho_1(X, Y) = \rho_0(X, Y) + \max_{p \in M} \|DX(p) - DY(p)\|.$$

We denote by $\mathcal{X}^1(M)$ the space of vector fields of class C^1 of M with the topology induced by ρ_1.

Remark. Of course, one can consider metrics generated by coverings of M with local charts (see Chap. 0 in [Pi1]).

Let $\Xi(t, x)$ be the flow generated by system (2.7), i.e., $\Xi(t, x)$ is the trajectory of (2.7) such that $\Xi(0, x) = x$. We denote by $D\Xi(t, x)$ the corresponding variational flow,

$$D\Xi(t, x) = \frac{\partial \Xi(t, x)}{\partial x}.$$

Let $H(t, x)$ be the flow of a system

$$\dot{x} = Y(x). \tag{2.8}$$

Definition 2.2 *A homeomorphism h of M is called a "topological equivalence" of the flows Ξ and H if h maps oriented trajectories of (2.7) onto oriented trajectories of (2.8).*

Now we define structural stability for flows.

Definition 2.3 *We say that the flow Ξ of system (2.7) is "structurally stable" if given $\epsilon > 0$ there is $\delta > 0$ such that for any $Y \in \mathcal{X}^1(M)$ with $\rho_1(X, Y) < \delta$ there is a topological equivalence h of the flows Ξ and H with the property*

$$r(x, h(x)) < \epsilon \text{ for } x \in M.$$

Remark. Another possible definition of the structural stability for the flow Ξ requires the existence of a number $\delta > 0$ such that for any $Y \in \mathcal{X}^1(M)$ with $\rho_1(X, Y) < \delta$ there is a topological equivalence h of the flows Ξ and H (and h is not required to be close to the identity).

It follows from [Hay, We] that these two definitions are equivalent.

Before we state necessary and sufficient conditions for structural stability, we give some related definitions and discuss the appearing structures.

We will apply the usual notation, for a set $V \subset M$ we denote

$$\Xi(t, V) = \bigcup_{x \in V} \Xi(t, x).$$

Definition 2.4 *A point $p \in M$ is called "nonwandering" for the flow Ξ if for any neighborhood V of p and for any T there exists $t > T$ such that*

$$\Xi(t, V) \cap V \neq \emptyset.$$

It is easy to show that the set of nonwandering points of Ξ (we denote this set $\Omega(\Xi)$) is compact and Ξ-invariant.

Smale introduced in [Sm2] the following property of Ξ.

Axiom A'. *The nonwandering set $\Omega(\Xi)$ of the flow Ξ is hyperbolic and can be represented as the union of two disjoint compact sets, Ω' and Ω^0, such that*

(1) Ω^0 consists of a finite number of rest points of Ξ;
(2) $X(x) \neq 0$ for $x \in \Omega'$;
(3) closed trajectories of Ξ are dense in Ω'.

Remark. In the case of the flow Ξ, we apply Definition 1.27 of a hyperbolic set with the following obvious modification. Everywhere in this definition, we take T_pM instead of $\mathbb{R}^n$, and, for $v \in T_pM$, $|v|$ is the norm generated by the

Riemannian metric r. For a point $p \in \Omega(\Xi)$ we denote by $S^0(p)$ and $U^0(p)$ the corresponding subspaces of the hyperbolic structure on $\Omega(\Xi)$.

Take a point $p \in M$ and define its *stable and unstable manifolds* as follows:

$$W^s(p) = \{x \in M : r(\Xi(t, x), \Xi(t, p)) \to 0 \text{ as } t \to +\infty\},$$

$$W^u(p) = \{x \in M : r(\Xi(t, x), \Xi(t, p)) \to 0 \text{ as } t \to -\infty\}.$$

For a trajectory γ of Ξ we define the *stable manifold*

$$W^s(\gamma) = \{x \in M : \text{dist}(\Xi(t, x), \gamma) \to 0 \text{ as } t \to +\infty\}$$

and the *unstable manifold*

$$W^u(\gamma) = \{x \in M : \text{dist}(\Xi(t, x), \gamma) \to 0 \text{ as } t \to -\infty\}.$$

Let us also introduce the following notation. For a set $V \subset M$ we denote

$$O^+(V) = \bigcup_{t \geq 0} \Xi(t, V), \quad O^-(V) = \bigcup_{t \leq 0} \Xi(t, V),$$

$$O(V) = O^+(V) \cup O^-(V).$$

According to this notation, $O(x)$ is the trajectory of a point $x \in M$.

The following statement is a corollary of the stable manifold theorem for flows [Sm1] for the hyperbolic set $\Omega(\Xi)$.

Theorem 2.2.1.
(1a) For a point $p \in \Omega(\Xi)$, the stable manifold $W^s(p)$ and the unstable manifold $W^u(p)$ are the images of Euclidean spaces $\mathbb{R}^k$ under immersions $\mathbb{R}^k \to M$ of class C^1. The equalities

$$T_p W^s(p) = S^0(p), \quad T_p W^u(p) = U^0(p)$$

hold;
(1b) for any point $x \in W^\sigma(p)$, the tangent space $T_x W^\sigma(p)$ has the property

$$D\Xi(t, x) T_x W^\sigma(p) = T_y W^\sigma(r), \quad t \in \mathbb{R}, \quad \sigma = s, u,$$

where $y = \Xi(t, x), r = \Xi(t, p)$ (below we use the notation

$$S^0(x) = T_x W^s(p), \quad U^0(x) = T_x W^u(p));$$

(1c) the stable manifolds $W^s(p)$ have the following property: if $p_m, p \in \Lambda$, $p_m \to p$ as $m \to \infty$, and for $x_m \in W^s(p_m)$ we have $x_m \to x \in W^s(p)$, then $S^0(x_m) \to S^0(x)$ (and the unstable manifolds have the same property);
(2) for a trajectory $\gamma \subset \Omega(\Xi)$, the sets $W^s(\gamma), W^u(\gamma)$ are Ξ-invariant images under immersions $N \to M$ of class C^1, where N is $\mathbb{R}^k$ if the trajectory γ

is not closed, and N is a fiber bundle over the circle S^1 with fiber $\mathbb{R}^k$ if γ is a closed trajectory. If $p \in \gamma$, then

$$\mathcal{S}(p) := T_p W^s(\gamma) = S^0(p) + <X(p)>, \qquad (2.9)$$

$$\mathcal{U}(p) := T_p W^u(\gamma) = U^0(p) + <X(p)>; \qquad (2.10)$$

(3) for $p \in \Omega(\Xi)$ we have

$$W^\sigma(O(p)) = \bigcup_{r \in O(p)} W^\sigma(r), \quad \sigma = s, u.$$

To prove the main result about shadowing in structurally stable flows, we construct a special geometric structure connected with a structurally stable flow ("weakly upper semicontinuous" subspaces $S(p), U(p)$ for $p \in M$, see Lemma 2.2.10). To do this, we need a group of auxiliary results. A part of these results (Lemmas 2.2.1 and 2.2.2) are well-known, their original proofs are detailed or reproduced in books (see, for example, [Pi1] for proofs of Lemmas 2.2.1 and 2.2.2). In contrast, the proof of Lemma 2.2.9 on the existence of "compatible extensions of stable and unstable bundles" (very important for us) is only sketched in the fundamental paper [Robi1], thus we give here a detailed proof of this statement.

Lemma 2.2.1 [Sm2]. *There exists a unique decomposition*

$$\Omega(\Xi) = \Omega_1 \cup \ldots \cup \Omega_m,$$

where Ω_i are compact disjoint invariant sets, and each Ω_i contains a dense trajectory.

The sets Ω_i are called *basic sets*. For a basic set Ω_i we define the stable and unstable sets

$$W^s(\Omega_i) = \{x \in M : \text{dist}(\Xi(t, x), \Omega_i) \to 0 \text{ as } t \to +\infty\}$$

and

$$W^u(\Omega_i) = \{x \in M : \text{dist}(\Xi(t, x), \Omega_i) \to 0 \text{ as } t \to -\infty\}.$$

The following statement easily follows from Definition 2.4 and Lemma 2.2.1.

Lemma 2.2.2 [Sm2].

$$M = \bigcup_{1 \le i \le m} W^s(\Omega_i) = \bigcup_{1 \le i \le m} W^u(\Omega_i).$$

Below we formulate a very important property of basic sets.

Lemma 2.2.3 [Hirs1]. *For a basic set Ω_i we have*

$$W^s(\Omega_i) = \bigcup_{\gamma \subset \Omega_i} W^s(\gamma), \quad W^u(\Omega_i) = \bigcup_{\gamma \subset \Omega_i} W^u(\gamma).$$

Remark. In [Hirs1], semi-invariant disk families are applied to prove Lemma 2.2.3. Let us show that one can prove this lemma using an approach based on shadowing. We give here only a sketch of the proof, leaving the details to the reader.

By Lemma 2.2.1, we can find a neighborghood U of Ω_i such that

$$U \cap (\Omega(\Xi) \setminus \Omega_i) = \emptyset.$$

Fix a point $x \in W^s(\Omega_i)$. We can take U so small that an analog of Theorem 1.4.3 for Ξ holds in U (one can prove it using the same ideas as in the proof of Theorem 1.5.1). Take a small $d > 0$ and find t_0 such that

$$r(x_1, \Omega_i) < d \text{ and } \Xi(t, x_1) \in U \text{ for } t \geq 0,$$

where $x_1 = \Xi(t_0, x)$. Find a point $y \in \Omega_i$ such that $r(x_1, y) < d$, and consider the mapping $\Psi : \mathbb{R} \to M$ defined by

$$\Psi(t) = \begin{cases} \Xi(t, x_1), & t \geq 0, \\ \Xi(t, y), & t < 0. \end{cases}$$

Define a sequence $g = \{g_k : k \in \mathbb{Z}\}$ by

$$g_k = r(\Xi(1, \Psi(k-1)), \Psi(k)).$$

Obviously, we have

$$\|g\|_\infty < d \text{ and } |g_k| \to 0 \text{ as } |k| \to \infty.$$

An analog of Theorem 1.4.3 for flows implies that there exists a point p close to x_1 (so that we may assume p to be in U) and a reparametrization α such that

$$r(\Xi(\alpha(t), p), \Psi(t)) \to 0 \text{ as } |t| \to \infty.$$

Obviously, it follows that $x_1 \in W^s(O(p))$, and hence $x \in W^s(O(p))$. Thus, it remains to show that $p \in \Omega_i$. Of course, if Ω_i is a rest point, Lemma 2.2.3 is trivial, so we consider a basic set Ω_i such that closed trajectories are dense in it. The trajectory $O(p)$ belongs to a small neighborghood of a hyperbolic set Ω_i, hence it has properties similar to the ones described in Theorem 1.2.1. Take an arbitrary neighborhood V of p. The same reasons as in the proof of Lemma 2.2.1 (see also the proof of Lemma B.6 in [Pi2]) show that there exist closed trajectories $p_1, p_2 \subset \Omega_i$ with the following properties:

 - $W^u(p_1)$ and $W^s(p_2)$ have a point q_1 of transverse intersection;
 - $W^s(p_1)$ and $W^u(p_2)$ have a point q_2 of transverse intersection such that $q_2 \in V$.

It follows that the set $O(q_1) \cup O(q_2)$ is a transverse homoclinic contour [Pi1], hence the point q_2 is nonwandering [Pi1]. Since V is arbitrary and the set $\Omega(\Xi)$ is closed, we have $p \in \Omega(\Xi)$. By the choice of U, this proves that $p \in \Omega_i$.

By Lemma 2.2.2, any trajectory of the flow Ξ tends to a basic set as $|t| \to \infty$. By Lemma 2.2.3, for any trajectory γ of Ξ there exist nonwandering trajectories γ_1 and γ_2 such that γ belongs to the intersection of $W^s(\gamma_1)$ an $W^u(\gamma_2)$.

It follows now from our definition of the stable and unstable manifolds for a trajectory of Ξ that

$$W^s(\gamma) = W^s(\gamma_1) \text{ and } W^u(\gamma) = W^u(\gamma_2).$$

Since the stable and unstable manifolds for nonwandering trajectories are the images under immersions of class C^1 of smooth manifolds, for any point $x \in M$ we can define the following subspaces of $T_x M$:

$$S(x) = T_x W^s(O(x))$$

and

$$\mathcal{U}(x) = T_x W^s(O(x)).$$

Now we formulate the second main definition in the theory of structural stability.

Definition 2.5 We say that the flow Ξ satisfies the "geometric STC" (the "geometric strong transversality condition") if, for any trajectories $\gamma_1, \gamma_2 \subset \Omega(\Xi)$, the stable and unstable manifolds $W^s(\gamma_1)$ and $W^u(\gamma_2)$ are transverse.

Theorem 2.2.2 [Robi1, Hay, We]. *The flow Ξ of system (2.7) is structurally stable if and only if Ξ satisfies Axiom A' and the geometric STC.*

Remark. Robinson proved in [Robi] that the conditions of Theorem 2.2.2 are sufficient for the structural stability of Ξ, Hayashi and Wen established their necessity.

We give a definition of a (d, T)-pseudotrajectory for Ξ similar to Definition 1.24. A mapping

$$\Psi : \mathbb{R} \to M$$

is called a (d, T)-pseudotrajectory for Ξ if

$$r(\Xi(t, \Psi(\tau)), \Psi(t + \tau)) \leq d, \ |t| \leq T, \tag{2.11}$$

for any $\tau \in \mathbb{R}$.

Let us state the main result of this subsection.

Theorem 2.2.3. *Assume that the flow Ξ of system (2.7) is structurally stable. Then there exist numbers $d_0, L > 0$ such that for any $(d, 1)$-pseudotrajectory Ψ*

*of Ξ with $d \leq d_0$ there is a point $p \in M$ and a homeomorphism $\alpha \in \mathrm{Rep}(Ld)$
such that*

$$r(\Xi(\alpha(t), p), \Psi(t)) \leq Ld \ for \ t \in \mathbb{R} \tag{2.12}$$

(the set $\mathrm{Rep}(\epsilon)$ is defined in Sect. 1.5).

Remark. If $\Omega' = \emptyset$ (i.e., the nonwandering set consists only of rest points), one can take $\alpha(t) = t$.

Now we describe some additional properties of structurally stable flows. Below the flow Ξ is assumed to be structurally stable.

It is shown in [Sm2] that we can choose the indices of the basic sets so that

$$W^s(\Omega_i) \cap W^u(\Omega_j) \neq \emptyset \text{ with } i \neq j$$

implies the inequality $i > j$.

Lemma 2.2.4 [Pu]. *There exists a smooth function $G : M \to [0, m] \subset \mathbb{R}$ (a Lyapunov function) such that*
   (1) $\Omega_i \subset G^{-1}(i), 1 \leq i \leq m$;
   (2) if $x \notin \Omega(\Xi)$, then the function $g(t) = G(\Xi(t, x))$ is increasing.

The following statement is a consequence of Theorem 1.1 in [Hirs1]. We prefer to give here a simple direct proof based on the existence of a Lyapunov function described in Lemma 2.2.4.

Lemma 2.2.5. *Fix $0 < d < 1$ and define the set*

$$\mathcal{D} = G^{-1}(i - d) \cap W^s(\Omega_i).$$

For any neighborhood Q of $\mathcal{D}$ in $H := G^{-1}(i-d)$, the set $O^+(Q) \cup W^u(\Omega_i)$ contains a neighborhood of Ω_i.

Proof. Fix a neighborhood Q of the set $\mathcal{D}$ in H and assume that the statement of our lemma does not hold. Then there exists a sequence of points $\{x_m\}$ such that

$$x_m \to \Omega_i \text{ as } m \to \infty, \tag{2.13}$$

$$x_m \notin W^u(\Omega_i), \tag{2.14}$$

and

$$x_m \notin O^+(Q). \tag{2.15}$$

It follows from (2.14) that, for any m,

$$\Xi(t, x_m) \to \Omega_j \text{ as } t \to -\infty$$

with $j \neq i$. By (2.13), for large m we have the inequality

$$G(x_m) < i + 1,$$

hence it follows from Lemma 2.2.4 that $j < i$.

On the other hand, we have

$$i - d < G(x_m)$$

for large m, hence for any such m there exists a number $t_m > 0$ such that

$$y_m = \Xi(-t_m, x_m) \in H.$$

The set H is compact, let y be a limit point of the set $\{y_m\}$. If

$$y \in \mathcal{W}^s(\Omega_i), \tag{2.16}$$

then $y_m \in Q$ for large m, hence $x_m \in O^+(Q)$, and this contradicts to (2.15).

Thus, to prove our lemma it remains to establish inclusion (2.16). If it does not hold, then we deduce from the inequality $0 < d < 1$ and from Lemma 2.2.4 that $y \in \mathcal{W}^s(\Omega_k)$ with $k > i$. Then there exists $T > 0$ such that

$$G(\Xi(T, y)) > i + \frac{1}{2}.$$

Since G is continuous and $t_m \to \infty$ as $m \to \infty$ (this follows immediately from (2.13) and from the compactness of Ω_i), for large m we have

$$G(\Xi(T, y_m)) = G(\Xi(T - t_m, x_m)) > i + \frac{1}{2}.$$

But since $T - t_m < 0$ for large m, the last inequality contradicts to the relations

$$G(\Xi(T - t_m, x_m)) < G(x_m) \to i \text{ as } m \to \infty.$$

This contradiction completes the proof.   □

We will need also the following simple statement.

Lemma 2.2.6. *For any neighborghood V of Ω_i there exists a number $0 < d < 1$ such that, for any $d' \in (0, d)$, the set*

$$G^{-1}(i - d') \cap \mathcal{W}^s(\Omega_i)$$

is a subset of V.

Proof. Consider the set

$$H_1 = G^{-1}(i - \frac{1}{2}) \cap \mathcal{W}^s(\Omega_i).$$

First we show that the set H_1 is compact. Indeed, consider a sequence $\{x_m\} \subset H_1$, and let x be a limit point of this sequence. Since G is continuous, we have

$$G(x) = i - \frac{1}{2}.$$

Since G increases along trajectories, the relation $x \notin \mathcal{W}^s(\Omega_i)$ implies that there is an index $l > i$ such that $x \in \mathcal{W}^s(\Omega_l)$. Then we can find $T > 0$ such that

$$G(\Xi(T, x)) > i + \frac{1}{2}.$$

Since G and Ξ are continuous, it follows that for large m we have

$$G(\Xi(T, x_m)) > i + \frac{1}{2},$$

and this contradicts to the inclusions $x_m \in \mathcal{W}^s(\Omega_i)$. Hence, H_1 is compact.

Since any trajectory through H_1 tends to Ω_i, standard reasons show that there is a number $T > 0$ such that $\Xi(t, H_1) \subset V$ for $t \geq T$. Obviously, we can find a number $d > 0$ with the property

$$G(\Xi(t, x)) \leq i - d \text{ for } x \in H_1, \ t \in [0, T].$$

Since any trajectory on $\mathcal{W}^s(\Omega_i) \setminus \Omega_i$ intersects H_1 at exactly one point, we see that this number d has the desired property. □

Consider a family $E = \{E(x) : x \in M\}$, where any $E(x)$ is a linear subspace of $T_x M$.

Definition 2.6 *A family E as above is called "lower semicontinuous" at $p \in M$ if for any linear subspace $L \subset E(p)$ there exists a neighborghood U of p and a continuous subbundle F of $TM|_U$ such that $F(p) = L$, and $F(x) \subset E(x)$ for $x \in U$.*

A family E is called "lower semicontinuous" if it is lower semicontinuous at every $p \in M$.

Remark. Let E be a lower semicontinuous family as above and let L be a continuous subbundle of TM on an open set $U \subset M$. It is easy to see that if

$$E(x) + L(x) = T_x M$$

for some $x \in U$, then

$$E(y) + L(y) = T_y M$$

for all y close to x.

It is known [Nil] that on M there exists a C^∞ Riemannian metric r' such that for the hyperbolic set Ω we can take $C = 1$ in item (h.2.3) of Definition 1.27 (for norms generated by r'). Of course, this metric is an analog of the metric generated by a Lyapunov norm (see Lemma 1.2.1), and we call it a Lyapunov metric. The same reasons as in the proof of Theorem 1.2.3 show that if we prove

Theorem 2.2.3 for a Lyapunov metric r', then it is true for any metric r (with another constants d_0, L). Thus, below we assume that r is a Lyapunov metric.

Let Ω_i be a basic set. Fix a number $d \in (0,1)$, consider the set

$$\mathcal{D} = G^{-1}(i - d) \cap W^s(\Omega_i),$$

and let Q be a neighborhood of $\mathcal{D}$ such that $Q \cap W^u(\Omega_i) = \emptyset$. Assume that $\mathcal{V}$ is a continuous subbundle of $TM|_Q$ such that

$$\mathcal{S}(x) \oplus \mathcal{V}(x) = T_x M \text{ for } x \in \mathcal{D} \tag{2.17}$$

(the spaces $\mathcal{S}(x)$ are defined in item (2) of Theorem 2.2.1 and before Definition 2.5). Let us extend the subbundle $\mathcal{V}$ to the set $O(Q) \cup W^u(\Omega_i)$ as follows:
- for $y = \Xi(t, x)$, where $t \in \mathbb{R}, x \in Q$, we set

$$\mathcal{V}(y) = D\Xi(t, x)\mathcal{V}(x);$$

- for $y \in W^u(\Omega_i)$, we set

$$\mathcal{V}(y) = U^0(y)$$

(the spaces $U^0(x)$ are elements of the hyperbolic structure for $x \in \Omega(\Xi)$, for $x \notin \Omega(\Xi)$ they are defined in item (1b) of Theorem 2.2.1).

Lemma 2.2.7 [Robb]. *There exists a neighborhood Q of $\mathcal{D}$ such that the subbundle $\mathcal{V}$ constructed as above is continuous on the set $O(Q) \cup W^u(\Omega_i)$.*

Remark. Statements analogous to Lemmas 2.2.7 and 2.2.8 were established in [Robb] for a diffeomorphism, our proofs for a flow are based on similar arguments.

In the proofs of Lemmas 2.2.7-2.2.9, for a basic set Ω_i we fix a small neighborhood V_i and introduce "coordinates" in $TM|_{V_i}$ in the following way. We take a neighborhood V of Ω_i and fix continuous subbundles U' and S' of $TM|_V$ such that

$$U'(x) = U^0(x) \text{ for } x \in W^u(\Omega_i) \cap V$$

and

$$S'(x) = S(x) \text{ for } x \in W^s(\Omega_i) \cap V$$

(the continuity of U^0 on $W^u(\Omega_i)$ follows from item (1c) of Theorem 2.2.1, the continuity of S on $W^s(\Omega_i)$ follows from the same statement of Theorem 2.2.1 and from formulas (2.9) and (2.10)). Note that, for $x \notin W^s(\Omega_i)$, $S'(x)$ does not necessarily include $< X(x) >$. In addition, if the basic set Ω_i is a rest point, then $S'(x) = S^0(x)$ for $x \in W^s(\Omega_i)$.

Since for $x \in \Omega_i$ we have

$$U^0(x) \oplus S(x) = T_x M,$$

we can take the neighborghood V so small that

$$U'(x) \oplus S'(x) = T_x M \text{ for } x \in V. \tag{2.18}$$

We denote a neighborhood of Ω_i with this property by V_i. By (2.18), U' and S' introduce "coordinates" in $T_x M, x \in V_i$.

Now we prove Lemma 2.2.7.

Proof. Since $\mathcal{V}$ is continuous on Q, it follows from the construction of $\mathcal{V}$ on $O(Q)$ that it is continuous on $O(Q)$. It remains to show that $\mathcal{V}$ is continuous on $W^u(\Omega_i)$. For this purpose, we apply a variant of argument used in [Pali1] to prove the so-called λ-Lemma (see also [Pi1]).

First let us assume that our basic set Ω_i is not a rest point. Since the set Ω_i is compact, there exists a positive constant μ_0 such that

$$|X(x)| \leq \mu_0 |X(y)| \text{ for } x, y \in \Omega_i. \tag{2.19}$$

Now we fix a constant $N \geq 1$ such that

$$||\Pi_x^s||, ||\Pi_x^u||, ||\Pi_x^0|| \leq N \text{ for } x \in \Omega_i, \tag{2.20}$$

where $\Pi_x^s, \Pi_x^u, \Pi_x^0$ are the complementary projectors onto the spaces $S^0(x), U^0(x)$, $< X(x) >$ of the hyperbolic structure on Ω_i. Set

$$\mu = N(1 + \mu_0),$$

obviously, $\mu > 1$.

Find positive numbers T and Δ such that

$$\rho = \frac{2\mu}{R - \Delta} < 1, \text{ where } R = \frac{1}{2}\lambda_0^{-T}. \tag{2.21}$$

Take a vector $v \in T_x M, x \in V_i$, and represent it in the form

$$v = v^s + v^u, \text{ where } v^s \in S'(x), v^u \in U'(x)$$

(here we refer to the "coordinates" corresponding to (2.18)). If $v^u \neq 0$, we can define the inclination α of v by the formula

$$\alpha = \frac{|v^s|}{|v^u|}$$

(in the original paper [Pali1], inclinations were denoted by λ, and due to this reason the result concerning them was called the λ-Lemma; in this book, we denote by λ characteristics of hyperbolicity, and we do not want to give other meanings to this symbol).

Denote $x_1 = \Xi(T, x)$ and consider the operator

$$F(x) : T_x M \to T_{x_1} M, \; F(x)v = D\Xi(T, x)v.$$

If $x_1 \in V_i$, then we can represent $F(x)$ in the form

$$\begin{bmatrix} F_{ss} & F_{us} \\ F_{su} & F_{uu} \end{bmatrix}$$

according to (2.18) (i.e., $F_{ss}v^s \in \mathcal{S}'(x_1)$ etc).

First let us consider a point $x \in \Omega_i$. Take $v = v^s + v^u \in T_x M$ and represent $v^s = v^{ss} + v^0$, where $v^{ss} \in S^0(x)$ and $v^0 \in < X(x) >$. It follows from the definition of hyperbolic structure that

$$F(x)v^{ss} \in S^0(x_1) \text{ and } |F(x)v^{ss}| \leq \lambda_0^T |v^{ss}| \leq |v^{ss}|,$$

$$F(x)v^u \in U^0(x_1) \text{ and } |F(x)v^u| \geq \lambda_0^{-T} |v^u|.$$

On the other hand, it is well known that if $v^0 = cX(x)$, then $F(x)v^0 = cX(x_1)$ (see [Pi1], Chap. 5), hence $|F(x)v^0| \leq \mu_0 |v^0|$. Since $|v^{ss}|, |v^0| \leq N|v^s|$, we see that

$$\|F_{ss}\| \leq \mu, \ F_{us} = 0, \ F_{su} = 0, \ |F_{uu}v^u| \geq 2R|v^u|.$$

Take a positive δ such that

$$\delta < \frac{\Delta R(1-\rho)}{4}, \ \rho\Delta + \frac{\delta}{R-\delta} < \Delta.$$

It follows from the estimates above that we can take the neighborghood V_i such that, for $x \in V_i$, the inequalities

$$\|F_{ss}\| \leq 2\mu, \ \|F_{us}\|, \|F_{su}\| < \delta, \ |F_{uu}v^u| \geq R|v^u| \qquad (2.22)$$

hold.

Denote $\mathcal{D}_1 = \Xi([0,T], \mathcal{D})$. It follows from condition (2.17) that there exists a constant K such that, for $v \in \mathcal{V}(x), x \in \mathcal{D}_1$, the inclination α of v does not exceed K.

Take a point $x_0 \in \mathcal{D}_1$. Denote $x_{k+1} = \Xi(T, x_k)$ for $k \geq 0$. Since $W^s(\Omega_i)$ is Ξ-invariant and $G(x_{k+1}) < G(x_k)$, the choice of d implies that $x_k \in W^s(\Omega_i) \cap V_i$. Let α_k be the inclination of $v_k = F^k(x)v_0$, represent $v_k = v_k^s + v_s^u$ according to (2.18). Let us estimate

$$|v_1^s| \leq |F_{ss}v_0^s| + |F_{us}v_0^u| \leq 2\mu|v_0^s| + \delta|v_0^u|. \qquad (2.23)$$

For $x \in W^s(\Omega_i)$ we have $\mathcal{S}'(x) = \mathcal{S}(x)$. Since the family $\mathcal{S}$ is $D\Xi$-invariant, for $v = v^s \in \mathcal{S}(x)$ its image $F(x)v \in \mathcal{S}(x_1)$, hence $F_{su} = 0$. This proves the equality $v_1^u = F_{uu}v_0^u$ and the inequality

$$|v_1^u| \geq R|v_0^u|.$$

This inequality, inequality (2.21), and estimate (2.23) imply that

$$\alpha_1 \leq \frac{2\mu|v_0^s| + \delta|v_0^u|}{R|v_0^u|} \leq \rho\alpha_0 + \frac{\delta}{R}.$$

Continuing this process, we show that for any m we have

$$\alpha_m \leq \rho^m \alpha_0 + \frac{\delta}{R}(1 + \rho + \ldots + \rho^{m-1}) \leq \rho^m \alpha_0 + \frac{\delta}{R(1-\rho)}.$$

It follows from the choice of δ that we can find m_0 such that, for any

$$x \in \mathcal{D}' = \Xi(m_0 T, \mathcal{D}_1)$$

and any $v \in \mathcal{V}$, the inclination α of v does not exceed $\Delta/2$. Since the family $\mathcal{V}$ is continuous on $O(Q)$, there exists a neighborghood Q' of $\mathcal{D}'$ such that the inclination

$$\alpha \leq \Delta \text{ for any } v \in \mathcal{V}(x), \ x \in Q'. \tag{2.24}$$

Take a point $x = x_0 \in Q'$, a vector $v = v_0 \in \mathcal{V}(x_0)$, and consider the point $x_1 = \Xi(T, x_0)$ and the vector $v_1 = F(x)v_0$. Let us estimate the inclination α_1 of the vector v_1. The value $|v_1^s|$ is estimated by (2.23). Since now we cannot refer to the equality $F_{su} = 0$, we give a new estimate for $|v_1^u|$. We obtain the inequality

$$|v_1^u| \geq |F_{uu}v_0^u| - |F_{su}v_0^s| \geq R|v_0^u| - \delta|v_0^s|.$$

Combined with (2.23), this inequality gives

$$\alpha_1 \leq \frac{2\mu|v_0^s| + \delta|v_0^u|}{R|v_0^u| - \delta|v_0^s|} \leq \rho\alpha_0 + \frac{\delta}{R - \delta} < \Delta$$

(we take into account the inequality $\alpha_0 \leq \Delta$ and the second condition on δ).

Now we decrease the neighborghood Q chosen to define $\mathcal{V}$ so that $\Xi(m_0 T, Q) \subset Q'$. In this case, for any $x \in \Xi(t, Q)$ with $t \geq m_0 T$ and any $v \in \mathcal{V}(x)$, the inclination α of v does not exceed Δ. We also assume that, for $x \in Q$, the inequalities

$$G(x) < i - \frac{d}{2} \tag{2.25}$$

hold.

Apply an analog of Lemma 2.2.6 for $\mathcal{W}^u(\Omega_i)$ to find a number $d_1 > 0$ such that

$$G^{-1}(i + d') \cap \mathcal{W}^u(\Omega_i) \subset V_i^0$$

for $d' \in (0, d_1]$. Set $H_2 = G^{-1}(i + d_1) \cap \mathcal{W}^u(\Omega_i)$.

To show that $\mathcal{V}$ is continuous on $\mathcal{W}^u(\Omega_i)$, it is enough to check the continuity of $\mathcal{V}$ on H_2 and on Ω_i (this second case is treated similarly, and we leave it to the reader).

Take a point $y \in H_2$ and consider a sequence of points $y_p \in O(Q)$ such that $y_p \to y$ as $p \to \infty$. Since

$$i < G(\Xi(t, y)) \leq i + d_1 \text{ for } t \in (-\infty, 0],$$

and similar inequalities are true for $G(\Xi(t, y_p))$ on time segments with lengths tending to infinity, it follows from (2.25) that we can write $y_p = \Xi(m(p)T, x_p)$, where $x_p \in Q$, and $m(p) \to \infty$ as $p \to \infty$.

For $x \in W^u(\Omega_i) \cap V_i$ we have $U'(x) = U^0(x)$, the same reasons as above show that in this case $F_{us} = 0$. Fix arbitrary $\epsilon > 0$ and find a neighborghood $V(\epsilon)$ of the set $W^u(\Omega_i) \cap V_i$ such that for $x \in V(\epsilon)$ we have

$$||F_{us}|| < \epsilon_1 = \frac{\epsilon(R - \delta)(1 - \rho)}{2}.$$

Since $O^-(y) \subset V(\epsilon)$, there exist numbers $m_1(p)$ such that $\Xi(mT, x_p) \in V(\epsilon)$ for $m_1(p) \leq m \leq m(p)$ and $\nu(p) = m(p) - m_1(p) \to \infty$ as $p \to \infty$ (in addition, we can take $m_1(p) > m_0$). Denote $z_p = \Xi(m_1(p)T, x_p)$. Take a vector $v_0 \in V(z_p)$ and let $v_k = F^k(z_p)v_0$. For $0 \leq k < \nu(p)$ we have the estimates

$$|v_{k+1}^s| \leq 2\mu|v_k^s| + \epsilon_1|v_k^u|$$

and

$$|v_{k+1}^u| \geq R|v_k^s| - \delta|v_k^u|,$$

hence for the inclinations we have

$$\alpha_1 \leq \rho\Delta + \frac{\epsilon_1}{R - \delta},$$

$$\cdots$$

$$\alpha_k \leq \rho^k\Delta + \frac{\epsilon_1}{(R - \delta)(1 - \rho)},$$

and it follows that $\alpha_k < \epsilon$ for large k. Since $\nu(p) \to \infty$ as $p \to \infty$, we see that, for large p, the space $V(y_p)$ is ϵ-close to $V(y) = U^0(y)$. This proves the desired continuity of V.

If the basic set Ω_i is a rest point, the proof is easier, since for the corresponding operator F on Ω_i we have the estimates

$$||F_{ss}||, ||(F_{uu})^{-1}|| \leq \lambda_0^T.$$

We leave the details to the reader. $\square$

Lemma 2.2.8 [Robb]. *The families $\{S\}, \{U\}$ are lower semicontinuous.*

Proof. We prove our lemma for the family $\{U\}$, for $\{S\}$ the proof is analogous.

Let us apply induction in i to show that for a basic set Ω_i there exists a neighborhood Z_i of Ω_i and a continuous subbundle $\mathcal{G}_i$ of $TM|_{O(Z_i)}$ such that

$$\mathcal{G}_i(x) \subset U^0(x) \text{ for } x \in O(Z_i) \tag{2.26}$$

and

$$\mathcal{G}_i(x) = U^0(x) \text{ for } x \in W^u(\Omega_i). \tag{2.27}$$

It follows from formula (2.10) and Definition 2.6 that in this case the family $\{U\}$ is lower semicontinuous in a neigborhood of $W^u(\Omega_i)$.

To establish the induction base, we consider minimal (with respect to the order introduced before Lemma 2.2.4) basic sets. A set Ω_i is minimal if and only if

$$\mathcal{W}^s(\Omega_i) \cap \mathcal{W}^u(\Omega_j) = \emptyset \text{ for } i \neq j,$$

it is easy to see that in this case $\mathcal{W}^s(\Omega_i) = \Omega_i$, and $\mathcal{W}^u(\Omega_i)$ is a neighborhood of Ω_i. If Ω_i is a minimal basic set, then the subbundle $\mathcal{G}_i(x) = U^0(x)$ for $x \in \mathcal{W}^u(\Omega_i)$ has the desired properties.

Thus, we assume that, for $j < i$, the corresponding neigborhoods Z_j and subbundles $\mathcal{G}_j$ are constructed. Fix $d \in (0,1)$ and consider the set

$$\mathcal{D} = G^{-1}(i - d) \cap \mathcal{W}^s(\Omega_i).$$

Take a point $x \in \mathcal{D}$, there exists an index $j < i$ such that $x \in \mathcal{W}^u(\Omega_j)$. It follows from the geometric STC that

$$\mathcal{U}(x) + \mathcal{S}(x) = T_x M.$$

Hence, there is a subspace $\mathcal{G}'$ of $U^0(x)$ such that

$$\mathcal{G}' \oplus \mathcal{S}(x) = T_x M.$$

By the induction hypothesis, there exists a neighborhood $Z(x)$ of x and a continuous subbundle $\mathcal{G}'_j \subset \mathcal{G}_j$ of $TM|_{Z(x)}$ such that

$$\mathcal{G}'_j(y) \subset U^0(y) \text{ for } y \in Z(x) \tag{2.28}$$

and

$$\mathcal{G}'_j(x) = \mathcal{G}'.$$

Decreasing $Z(x)$, if necessary, we may assume that $Z(x) \subset V_i$ and that

$$\mathcal{G}'_j(y) \oplus \mathcal{S}'(y) = T_y M \text{ for } y \in Z(x) \tag{2.29}$$

(recall that $\mathcal{S}'$ and U' are chosen to introduce "coordinates" in $TM|_{V_i}$). It follows from (2.18) and from (2.29) that $\mathcal{G}'_j$ can be written as the graph of a continuous bundle mapping

$$g^{(x)} : U'|_{Z(x)} \rightarrow \mathcal{S}'|_{Z(x)}.$$

Let B be a closed neighborhood of $\mathcal{D}$ in $G^{-1}(i - d)$ such that

$$B \subset \bigcup_{x \in \mathcal{D}} Z(x).$$

Take a smooth partition of unity $\{\beta(x) : x \in \mathcal{D}\}$ subordinate to the covering $\{Z(x) : x \in \mathcal{D}\}$. Since the set B is compact, all but a finite number of the $\beta(x)$ are identically zero.

Define $g^B : U'|_B \rightarrow \mathcal{S}'|_B$ by

$$g^B = \sum_{x \in \mathcal{D}} \beta(x) g^{(x)}$$

and let $\mathcal{G}_i \subset TM|_B$ be the graph of g^B. By construction, $\mathcal{G}_i$ is a continuous subbundle of $TM|_B$.

A fiber $\mathcal{G}_i(y)$ is a convex combination of $\mathcal{G}'_j(y)$, hence it follows from (2.28) that

$$\mathcal{G}_i(y) \subset U^0(y) \text{ for } y \in B. \tag{2.30}$$

Now we define $\mathcal{G}_i$ on $O(B)$ as follows:

$$\mathcal{G}_i(x) = D\Xi(t, y)\mathcal{G}_i(y) \text{ for } t \in \mathbb{R}, \ y \in B$$

(this definition is reasonable, since any trajectory in $O(B)$ intersects B at a unique point);

$$\mathcal{G}_i(x) = U^0(x) \text{ for } x \in W^u(\Omega_i).$$

Since the family U^0 is $D\Xi$-invariant (see item (1b) of Theorem 2.2.1), we deduce from (2.30) that $\mathcal{G}_i(x) \subset U^0(x)$ for $x \in O(B)$.

It follows from Lemma 2.2.7 that if the chosen neighborhood B is small enough, then the constructed subbundle $\mathcal{G}_i$ is continuous on $O(B) \cup W^u(\Omega_i)$, and by Lemma 2.2.5 this last set contains a neighborhood of Ω_i (and hence of $W^u(\Omega_i)$). Denote this neighborhoood by Z_i to complete the induction step. $\square$

Now we are able to give a proof of the following statement from [Robi1].

Lemma 2.2.9 [Robi1]. *There exist disjoint neighborhoods V_i of the basic sets Ω_i, continuous subbundles S_i, U_i of $TM|_{\overline{V}_i \cup O(V_i)}$, and a number $\lambda_1 \in (0,1)$ such that*

(1) S_i, U_i are $D\Xi$-invariant;
(2) $W_i|_{\Omega_i} = W^0|_{\Omega_i}, W = S, U$;
(3) $S_j(x) \subset S_k(x), \ U_k(x) \subset U_j(x) \text{ for } x \in O^+(V_j) \cap O^-(V_k)$;
(4)

$$S_i(x) \oplus U_i(x) = T_xM \text{ if } \Omega_i \subset \Omega^0;$$

$$S_i(x) \oplus U_i(x) \oplus <X(x)> \; = T_xM \text{ if } \Omega_i \subset \Omega';$$

(5) if $x \in V_i, v^s \in S_i(x)$, and $v^u \in U_i(x)$, then

$$|D\Xi(t, x)v^s| \leq \lambda_1^t |v^s|, \ 0 \leq t \leq 1;$$

$$|D\Xi(-t, x)v^u| \leq \lambda_1^t |v^u|, \ 0 \leq t \leq 1.$$

Proof. We show how to construct the subbundles U_i, for S_i the construction is similar. We proceed by induction assuming that U_j and V_j have been defined for $1 \leq j < i$, and that they satisfy analogs of statements (1)-(5) (of course, we have in mind only the statements about U_j). In addition, we assume that one more induction hypothesis is satisfied,

$$U_j(x) + S(x) = T_xM \tag{2.31}$$

for $x \in O(V_j), 1 \leq j < i$.

We also assume that the constructed neighborhoods V_j are closed and that

$$V_j \subset G^{-1}[j - \frac{1}{4}, j + \frac{1}{4}],$$

where G is a Lyapunov function given by Lemma 2.2.4.

We take closed neighboghoods V_j' of Ω_j such that $V_j' \subset \mathrm{Int}\, V_j$ for $1 \le j < i$, and construct U_i on V_i so that properties (1)-(5) hold with the neighborghoods $V_1', \ldots, V_{i-1}', V_i$.

By Lemma 2.2.3, any point $x \in M$ belongs to a stable manifold $\mathcal{W}^s(\gamma)$, where $\gamma \subset \Omega(\Xi)$. By statement (3) of Theorem 2.2.1, there is a point $p \in \Omega(\Xi)$ such that $x \in W^s(p)$ (obviously, $W^s(x) = W^s(p)$ in this case).

Choose a neighborhood V_i^0 of Ω_i such that

$$V_i^0 \subset G^{-1}[i - \frac{1}{4}, i + \frac{1}{4}],$$

and fix continuous subbundles U' and S' on $TM|_V$ such that

$$U'(x) = U^0(x) \text{ for } x \in \mathcal{W}^u(\Omega_i) \cap V$$

and

$$S'(x) = S(x) \text{ for } x \in \mathcal{W}^s(\Omega_i) \cap V.$$

We take the neighborhood V_i^0 so small that (2.18) holds for $x \in V_i^0$.

Apply Lemma 2.2.6 to find $d > 0$ such that the sets

$$G^{-1}(i - d') \cap \mathcal{W}^s(\Omega_i)$$

are subsets of V_i^0 for $d' \in (0, d]$. Denote

$$\mathcal{D} = G^{-1}(i - d) \cap \mathcal{W}^s(\Omega_i).$$

It follows from Lemma 2.2.2, from our choice of indices for the basic sets, and from the induction hypothesis that

$$\mathcal{D} \subset \mathcal{W}^s(\Omega_i) \subset \bigcup_{1 \le j < i} \mathcal{W}^u(\Omega_j) \subset \bigcup_{1 \le j < i} O^+(\mathrm{Int}\, V_j').$$

For $q \ge 0$ and for a set V we denote

$$V^q = \bigcup_{0 \le t \le q} \Xi(t, V).$$

Since the set $\mathcal{D}$ is compact, there exists $q > 0$ such that

$$\mathcal{D} \subset \bigcup_{1 \le j < i} V_j'^q.$$

Take a point $x \in \mathcal{D}$ and define $j = j(x)$ by the conditions

$$x \in V_j^q \tag{2.32}$$

and
$$x \notin V_k'^q \text{ for } j < k < i. \tag{2.33}$$

Since the neighborghoods V_j' are closed, we can find a neighborghood $Z(x)$ of x such that
$$Z(x) \subset V_j^q \text{ and } Z(x) \cap V_k'^q = \emptyset, \ j < k < i.$$

We claim that
$$Z(x) \cap O^+(V_k') = \emptyset \text{ for } j < k < i. \tag{2.34}$$

Indeed, assume that there is a point $y \in Z(x) \subset V_j^q$ such that
$$y \in O^+(V_k') \setminus V_k'^q \text{ for some } j < k < i.$$

Then $y \in \Xi(s, V_k')$ with some $s > q$, hence
$$y' = \Xi(-s, y) \in V_k', \text{ and } G(y') \geq k - \frac{1}{4} > j + \frac{1}{4}.$$

The Lyapunov function G is increasing along trajectories, this gives
$$G(\Xi(t, y)) > j + \frac{1}{4} \text{ for } -s \leq t \leq 0.$$

Therefore,
$$\Xi(t, y) \notin V_j \text{ for } -q \leq t \leq 0,$$

$y \notin V_j^q$. The obtained contradiction proves (2.34).

It follows from (2.31) that there exists a continuous subbundle B' of $TM|_{Z(x)}$ such that
$$B'(x) \subset U_j(x)$$
and
$$T_x M = B'(x) \oplus S'(x)$$

(note that both conditions above are imposed only on the space $B'(x)$).

Let
$$\pi : TM|_{Z(x)} \to U_j|_{Z(x)}$$

be the orthogonal projection. Set $B^{(x)} = \pi B'$. If the neighborhood $Z(x)$ is small enough, this gives $B^{(x)}$ with the properties
$$B^{(x)}(y) \subset U_j(y) \text{ and } T_y M = B^{(x)}(y) \oplus S'(y) \text{ for } y \in Z(x). \tag{2.35}$$

It follows from (2.18) and (2.35) that $B^{(x)}$ can be written as the graph of a bundle mapping
$$b^{(x)} : U'|_{Z(x)} \to S'|_{Z(x)}.$$

Let Q be a closed neighborghood of $\mathcal{D}$ in $G^{-1}(i - d)$ such that
$$Q \subset \bigcup_{x \in \mathcal{D}} Z(x).$$

Take a smooth partition of unity $\{\beta(x) : x \in \mathcal{D}\}$ subordinate to the covering $\{Z(x) : x \in \mathcal{D}\}$. Since the set Q is compact, all but a finite number of the $\beta(x)$ are identically zero.

Define $b^{\mathcal{D}} : U'|_Q \to \mathcal{S}'|_Q$ by

$$b^{\mathcal{D}} = \sum_{x \in \mathcal{D}} \beta(x) b^{(x)}$$

and let $B^{\mathcal{D}} \subset TM|_Q$ be the graph of $b^{\mathcal{D}}$. By construction, $B^{\mathcal{D}}$ is a continuous subbundle of $TM|_Q$.

Now we want to show that

$$B^{\mathcal{D}}(y) \subset U_k(y) \text{ for } y \in Q \cap O^+(V_k') \text{ with } 1 \leq k < i. \tag{2.36}$$

A fiber $B^{\mathcal{D}}(y)$ is a convex combination of $B^{(x)}(y)$, where $y \in Z(x)$, thus it is sufficient to show that

$$B^{(x)}(y) \subset U_k(y) \text{ for } y \in Z(x) \cap O^+(V_k'), \ x \in \mathcal{D}.$$

Let $j = j(x)$ satisfy (2.32) and (2.33). By (2.34), we have $k \leq j$. Then it follows from (2.35) and from the induction hypothesis that

$$B^{(x)}(y) \subset U_j(y) \subset U_k(y).$$

This proves (2.36).

Now we define U_i on $O(Q)$ as follows. For $y = \Xi(t,x), x \in Q$, we set $U_i(y) = D\Xi(t,x)B^{\mathcal{D}}(x)$. On $\mathcal{W}^u(\Omega_i)$, we let $U_i = U^0$.

By Lemma 2.2.5, the set $O^+(Q) \cup \mathcal{W}^u(\Omega_i)$ contains a neighborghood $V_i \subset V_i^0$ of the basic set Ω_i. By construction, U_i is $D\Xi$-invariant, and it coincides with U^0 on $\mathcal{W}^u(\Omega_i)$, this proves statements (1) and (2) of our lemma. Since G is increasing along trajectories, we have $O^+(V_i) \cap V_j = \emptyset$ for $j < i$, now (2.36) implies statement (3).

By Lemma 2.2.7, U_i is continuous on $O(V_i)$ if Q is small enough.

Since for $x \in \Omega_i$ analogs of the inequalities in statement (5) of our lemma hold with λ_0, we can take any $\lambda_1 \in (\lambda_0, 1)$ and take a smaller neighborghood V_i (if necessary) to establish (5).

To complete the induction step, it remains to check (2.31). Since for $x \in \mathcal{W}^u(\Omega_i) \cap V_i^0$ we have

$$U_i(x) + \mathcal{S}(x) = U^0(x) + \mathcal{S}(x) = T_x M$$

due to the strong transversality condition, since the family U_i is continuous, and since the family $\mathcal{S}$ is lower semicontinuous (see Lemma 2.2.8), it follows from the remark after Definition 2.6 that we can reduce V_i to satisfy (2.31).

It remains to establish the induction base. Take a basic set Ω_i such that $\mathcal{W}^s(\Omega_i) \cap \mathcal{W}^u(\Omega_j) = \emptyset$ for $i \neq j$. Obviously, we can apply the same construction as above taking $B^{\mathcal{D}}$ equal U' on Q. $\qquad \square$

If V is an arbitrary neighborhood of $\Omega(\Xi)$, then there exists [Bi] a number $T_0 > 0$ (usually called the *Birkhoff constant*) such that if for $x \in M$

$$\Xi(t, x) \notin V \text{ for } t \in \Theta,$$

then $\text{mes}\Theta \leq T_0$ (here mes is the Lebesgue measure).

Fix a Birkhoff constant T_0 for $V = V_1 \cup \ldots V_m$ (where V_i are given by Lemma 2.2.9). Now we construct a family of linear subspaces $S(p), U(p) \subset T_p M, p \in M$, with special properties.

Take $p \in M$. There exists $t_0 \in [0, T_0]$ such that $q = \Xi(t_0, p) \in V_l$ for some l, and

$$\Xi(t, p) \notin V_k \text{ with } l \neq k \text{ for } t \in [0, t_0]. \tag{2.37}$$

Set

$$W(p) = D\Xi(-t_0, q)W_l(q), \ W = S, U. \tag{2.38}$$

We say in this case that the point p takes the subspaces (t.s. below) from V_l.

Take points $p, q = \Xi(t, p)$, a neighborhood W of q belonging to a chart of M, and introduce local coordinates in $TM|_W$ as follows. Let

$$\beta_1 : W \to W_0 \subset \mathbb{R}^n$$

be a coordinate mapping. We consider a coordinate mapping

$$\beta : TM|_W \to W_0 \times \mathbb{R}^n$$

of the form

$$\beta(x, v) = (\beta_1(x), B(x)v)$$

for $(x, v) \in TM|_W$.

Lemma 2.2.10. *Let $q = \Xi(t, p)$. Then*
(1)

$$D\Xi(t, p)S(p) \subset S(q), \ t \geq 0;$$

$$D\Xi(t, p)U(p) \subset U(q), \ t \leq 0;$$

(2) if $r_m \to q$ for $m \to \infty$, then for large m there exist linear isomorphisms $\Pi_m : \mathbb{R}^n \to \mathbb{R}^n$ with the following properties:

$$||\Pi_m - I|| \to 0 \text{ for } m \to \infty$$

and

$$\Pi_m(B(q)D\Xi(t, p)S(p)) \subset B(r_m)S(r_m) \text{ for } t \geq T_0;$$

$$\Pi_m(B(q)D\Xi(t, p)U(p)) \subset B(r_m)U(r_m) \text{ for } t \leq 0.$$

Proof. The first statement follows immediately from the definitions of $S(p), U(p)$, and from statements (1), (3) of Lemma 2.2.9.

Before we prove (2), note that all the limits below in the proof exist due to the continuity of the subbundles S_i, U_i in V_i. First we consider the case $t \geq T_0$.

There exists $t_0 \in [0, T_0]$ such that $p' = \Xi(t_0, p) \in V_l$ with some l, and (2.37) holds, so that p t.s. from V_l. Set $\kappa = t - t_0$, then

$$r'_m = \Xi(-\kappa, r_m) \to p' \text{ as } m \to \infty,$$

hence $r'_m \in V_l$ for large m. The continuity of $S_l(x)$ in V_l implies the relation

$$S_l(r'_m) \to S_l(p').$$

Consequently,

$$D\Xi(t, p)S(p) = D\Xi(\kappa, p')S_l(p') =$$

$$= \lim_{m \to \infty} D\Xi(\kappa, r'_m)S_l(r'_m) = \lim_{m \to \infty} D\Xi(\kappa, r'_m)S(r'_m).$$

By the first statement,

$$D\Xi(\kappa, r'_m)S(r'_m) \subset S(r_m),$$

hence for large m there exist isomorphisms Π_m such that

$$\|\Pi_m - I\| \to 0 \text{ and } \Pi_m(B(q)S(q)) \subset B(r_m)S(r_m).$$

Let us prove the second statement in (2). Assume that the point p t.s. from V_i, that is, there exists a point

$$p' = \Xi(l, p) \in V_i, \ l \geq 0,$$

such that

$$U(p) = D\Xi(-l, p')U_i(p').$$

Since

$$r'_m = \Xi(-t + l, r_m) \to p',$$

we have

$$\lim_{m \to \infty} U(r'_m) = U(p').$$

It follows from statement (1) that

$$D\Xi(t - l, r'_m)U(r'_m) \subset U(r_m),$$

hence

$$D\Xi(t, p)U(p) = D\Xi(t - l, p')U(p') \subset \lim_{m \to \infty} U(r_m),$$

and this implies the existence of Π_m. $\square$

Obviously, there exists a constant C_0 such that

$$\|D\Xi(t, x)\| \leq \exp(C_0|t|)$$

for all $t \in \mathbb{R}, x \in M$ (since $D\Xi(t,x)$ satisfies a variational system analogous to (1.207), one can take

$$C_0 = \max_{p \in M} \|DX(p)\|).$$

Set

$$C = \exp(C_0 T_0)\lambda_1^{-T_0}.$$

Take a point $p \in M$. Let us show that for $v \in S(p)$ we have

$$|D\Xi(t,p)v| \le C\lambda_1^t|v| \text{ for } t \ge 0. \tag{2.39}$$

Since $D\Xi(0,p) = I$, for $t = 0$ our estimate is obvious. Consider $t > 0$, let Θ be the following subset of $[0,t]$:

$$\Theta = \{\tau \in [0,t] : \Xi(\tau,p) \notin V\}.$$

It follows from the definition of the Birkhoff constant that $\mathrm{mes}\Theta \le T_0$. For arbitrary $\epsilon > 0$ there exist open segments $(a_i, b_i), i = 1, \dots, N$, such that

$$(b_1 - a_1) + \dots + (b_N - a_N) \le T_0 + \epsilon \tag{2.40}$$

and

$$\Theta \subset \bigcup_{i=1}^{N}[a_i, b_i].$$

Take $v \in S(p)$ and set

$$v_i^- = D\Xi(a_i,p)v, \; v_i^+ = D\Xi(b_i,p)v, \; v' = D\Xi(t,p)v.$$

Since $\Xi(\tau,p) \in V$ for

$$\tau \in (0,a_1) \cup (b_1,a_2) \cup \dots \cup (b_N,t),$$

statement (5) of Lemma 2.2.2 implies the inequalities

$$|v_1^-| \le \lambda_1^{a_1}|v|, \; |v_2^-| \le \lambda_1^{a_2-b_1}|v_1^+|, \; \dots, \; |v'| \le \lambda_1^{t-b_N}|v_N^+|.$$

On the other hand, it follows from the definition of C_0 that

$$|v_i^+| \le \exp(C_0(b_i - a_i))|v_i^-|.$$

Hence, we deduce from (2.40) that

$$|v'| \le \exp(C_0\{(b_1 - a_1) + \dots + (b_N - a_N)\})\lambda_1^{(a_1+a_2-b_1+\dots+t-b_N)}|v| \le$$

$$\le \exp(C_0(T_0 + \epsilon))\lambda_1^{t-(T_0+\epsilon)}|v| = A(\epsilon)\exp(C_0T_0)\lambda_1^{-T_0}\lambda_1^t|v|$$

with

$$A(\epsilon) \to 1 \text{ as } \epsilon \to 0.$$

Obviously, this proves (2.39). Similarly one shows that for $v \in U(p)$ we have

$$|D\Xi(t,p)v| \leq C\lambda_1^{-t}|v| \text{ for } t \leq 0. \tag{2.41}$$

Now we consider a $(d,1)$-pseudotrajectory Ψ of system (2.7). Let us fix a Lipschitz constant $\mathcal{L}$ for the vector field X on M. Fix a natural number T. The same reasons as in the proof of Lemma 1.1.3 show that Ψ is a (d_T,T)-pseudotrajectory of (2.7) with

$$d_T = d(1 + \exp(\mathcal{L}) + \ldots + \exp((T-1)\mathcal{L})).$$

It is easy to understand that if we prove an analog of Theorem 2.2.3 for (d,T)-pseudotrajectories, then the same result holds for $(d,1)$-pseudotrajectories (with other constants d_0, L).

Take $\mu \in (0,1)$ and find a natural T such that

$$T \geq T_0 + 1, \ C\lambda_1^{T-1} \leq \mu. \tag{2.42}$$

For $p \in M$ we set

$$Z(p) = S(p) + U(p).$$

Note that it follows from our constructions that

$$Z(p) = S(p) \oplus U(p).$$

Estimates (2.39), (2.41), the second inequality in (2.42), and the uniform continuity of the families S_i, U_i on $\overline{V}_i$ imply the following statement.

Lemma 2.2.11. *The subspaces $S(p), U(p)$ have the properties*
(1) there exists $N > 0$ such that

$$\|P(p)\|, \|Q(p)\| \leq N, \ p \in M,$$

where $P(p), Q(p)$ are the projectors in $Z(p)$ onto $S(p)$ parallel to $U(p)$, and onto $U(p)$ parallel to $S(p)$, respectively;
(2)

$$|D\Xi(t,p)v| \leq \mu|v| \text{ for } v \in S(p), \ t \geq T - 1,$$
$$|D\Xi(-t,p)v| \leq \mu|v| \text{ for } v \in U(p), \ t \geq T - 1.$$

Take $K \geq N$ such that

$$|X(p)| \leq K \text{ for } p \in M \text{ and } \|D\Xi(t,p)\| \leq K \text{ for } |t| \leq T + 1. \tag{2.43}$$

Now we fix local coordinates in TM as follows. For $p \in M$ let

$$\exp_p : T_pM \to M$$

be the standard exponential mapping generated by our metric r. For $c > 0, p \in M$ we set

$$E_c(p) = \{v \in T_pM : |v| \leq c\}.$$

It is well known that, for some $c > 0$, any $\exp_p$ is a diffeomorphism of the ball $E_{2c}(p)$ onto its image, the first derivatives of $\exp_p, \exp_p^{-1}$ are uniformly (in p) bounded on $E_{2c}(p)$ and on $\exp_p(E_{2c}(p))$, and

$$D\exp_p(0) = I. \tag{2.44}$$

We fix this c below.

It is easy to see that a family of linear subspaces $W(p) \subset T_pM, p \in M$, is continuous at $z \in M$ if and only if

$$q \to z \text{ implies } D\exp_z^{-1}(q)W(q) \to W(z).$$

Now we formulate the uniformity of the "weak upper semicontinuity" described in statement (2) of Lemma 2.2.10. The subbundles S_i, U_i are uniformly continuous on $\overline{V}_i$. A point $p \in M$ t.s. from V_i according to formula (2.38) with $t_0 \in [0, T_0]$. We easily deduce from the proof of Lemma 2.2.10, from the compactness of M, and from the uniform boundedeness of derivatives of $\exp_z^{-1}$ that the following statement holds.

Lemma 2.2.12. *Given $\beta > 0$ there exists $\alpha > 0$ such that if $z, p \in M$, $t_1, t_2 \in \mathbb{R}$, $q = \Xi(t_1, p), y = \Xi(t_2, z)$, and the inequalities*

$$r(z,q) < \alpha, \ r(y,p) < \alpha, \ |t_1 - T| \le 1, \ |t_2 + T| \le 1$$

hold, then there is a linear isomorphism $\Pi(p, z) : T_zM \to T_zM$ such that

$$\|\Pi - I\| \le \beta, \ \Pi(p,z)(D\exp_z^{-1}(q)D\Xi(t_1,p)S(p)) \subset S(z) \tag{2.45}$$

and a linear isomorphism $\Theta(p, z) : T_pM \to T_pM$ such that

$$\|\Theta - I\| \le \beta, \ \Theta(p,z)(D\exp_p^{-1}(y)D\Xi(t_2,z)U(z)) \subset U(p). \tag{2.46}$$

For $p \in M$ and $a \in (0, c)$ denote

$$Z_a(p) = Z(p) \cap E_a(p), \ \Sigma(p) = \exp_p(Z_c(p)), \ \Sigma_a(p) = \exp_p(Z_a(p)).$$

Obviously, $\Sigma(p)$ is a smooth disk (with boundary) transverse to $X(p)$ at p.

We say below that a point $p \in M$ is of type (R) if $Z(p) = T_pM$ (in this case $\dim\Sigma_p = \dim M$), and of type (O) if $Z(p) \neq T_pM$ (in this case $\dim\Sigma_p = \dim M - 1$).

Let p be a point of type (O). The subspaces $S(p), U(p)$ are defined by (2.38). Since S_l, U_l are continuous, their sum is transverse to X in $\overline{V}_l$, and $t_0 \le T_0$, we see that the angle between $Z(p)$ and $X(p)$ and the norm $|X(p)|$ are separated from zero (by a constant independent of p). Hence, the following two statements hold.

Lemma 2.2.13. *There exist constants $d_1 > 0, \epsilon_0 < 1, K_1 \ge 1$ such that if z is of type (O) and $r(z,q) \le d_1$, then there is a unique scalar function $f(x)$ in the d_1-neighborhood of q such that $|f(x)| \le \epsilon_0$ and*

$$\Xi(f(x), x) \in \Sigma(z).$$

This function f is of class C^1 and

$$|f(x)| \le K_1 \text{dist}(x, \Sigma(z)), \ r(z, \Xi(f(x), x)) \le K_1 r(x, z).$$

Remark. This lemma is analogous to Lemma 1.5.2, we do not prove it here. The following statement is proved similarly, we leave details to the reader.

Lemma 2.2.14. *There exist positive constants $d_2 \le d_1, b \le c$ with the following properties.*

If z is of type (O), $r(\Xi(T, p), z) \le d_2$, then there is a function $\tau(x)$ of class C^1 on $\Sigma_b(p)$ such that

$$\phi(x) := \Xi(\tau(x), x) \in \Sigma(z), \ |\tau(x) - T| \le 1.$$

If p is of type (O), $r(\Xi(-T, z), p) \le d_2$, then there is a function $\tau_1(x)$ of class C^1 on $\Sigma_b(z)$ such that

$$\phi_1(x) := \Xi(\tau_1(x), x) \in \Sigma(p), \ |\tau_1(x) + T| \le 1.$$

Remark. Obviously, the norms of $D\phi, D\phi_1$ are uniformly (in p, z) bounded, below we assume that

$$\|D\phi(x)\|, \|D\phi_1(x)\| \le K.$$

In addition, $D\phi, D\phi_1$ are uniformly continuous in the following sense: given $\beta > 0$ there is $\alpha > 0$ (depending only on β) such that if $x_1, x_2 \in \Sigma_b(p)$, $x_i^* = \exp_p^{-1}(x_i)$, $y_i = \phi(x_i)$, and $r(x_1, x_2) \le \alpha$, then

$$\|D\exp_z^{-1}(y_1)D\phi(x_1)D\exp_p(x_1^*) - D\exp_z^{-1}(y_2)D\phi(x_2)D\exp_p(x_2^*)\| \le \beta$$

(and the same holds for $D\phi_1$).

Let us prove this uniform continuity for $D\phi$. In this proof, the existence of a uniform estimate for some value a means the following: there exists a number b such that for any point $p \in M$ we have the estimate of the type $|a| \le C(b)$ in the b-neighborghood of p, where the function C depends only on such global properties of system (2.7) as

$$\max |X|, \ \max \|DX(x)\|,$$

and so on.

Fix local coordinates in neighborghoods W_p, W_z of the points p, z, respectively. We may assume that z is the origin of W_z, and that in local coordinates y of W_z the disk $\Sigma(z)$ belongs to the hyperplane $y_1 = 0$. Without loss of generality, we may assume that $\Sigma(z)$ in W_z is given by the linear relation

$$y = Ls, \quad s \in \mathbb{R}^{n-1},$$

where elements l_{ij} of the matrix L have the form $l_{ij} = \delta(i, j+1)$ for $1 \le i \le n, 1 \le j \le n-1$ (here $\delta(i,j)$ is the Kronecker symbol, $\delta(i,j) = 1$ for $i = j$, $\delta(i,j) = 0$ for $i \ne j$).

Let us preserve the notation Ξ for the flow in the chosen local coordinates. Denote $p' = \phi(p) \in \Sigma_z$, and let $p' = \Xi(t_0, p)$. Consider, for $x \in W_p, t \in \mathbb{R}$, and $s \in \mathbb{R}^{n-1}$, the function

$$F(x, t, s) = \Xi(t, x) - Ls.$$

Obviously, a point $y = Ls \in \Sigma_z$ is the image of a point $x \in W_p$ under ϕ if and only if there exists a number t close to t_0 and such that $F(x, t, s) = 0$. Assume that $p' = Ls'$. Then $F(p, t_0, s') = 0$.

Consider the matrix

$$A = \frac{\partial F}{\partial(t, s)}(p, t_0, s') = [X(p'), -L]$$

(we preserve the notation X for the vector field). The vector $X(p')$ has nonzero angle with the hyperplane $\Sigma(z)$, hence $\det A \ne 0$. By the implicit function theorem, in a neighborghood W' of p there exist functions $\tau(x), \sigma(x)$ of class C^1 such that $\tau(p) = t_0, \sigma(p) = s'$, and

$$F(x, \tau(x), \sigma(x)) \equiv 0 \tag{2.47}$$

in this neighboghood. Obviously, $\tau(x)$ is the function given by Lemma 2.2.14, and $\phi(x) = L\sigma(x)$ for $x \in W'$.

For $x \in W'$ denote

$$A(x) = [X(\phi(x)), -L].$$

Denote by X_1 the first component of the vector X in the local coordinates of W_z. If a basic set Ω_i is a rest point of Ξ, then it follows from the construction of the families S, U before Lemma 2.2.10 that any point in the corresponding neighborhood V_i of Ω_i is of type (R). Hence, there exist constants $c_1, c_2 > 0$ such that for any y in a c_1-neighborghood of a point z of type (O) we have the estimate

$$|X(y)| \ge c_2.$$

The angle between the vector $X(\phi(x))$ (for x in a neighborhood of p) and the hyperplane $y_1 = 0$ (containing $\Sigma(z)$) is uniformly separated from 0, hence there is $c_3 > 0$ such that for $X_1(\phi(x))$, the projection of $X(\phi(x))$ to the y_1 axis, the inequality

$$|X_1(\phi(x))| \ge c_3|X(\phi(x))| \ge c_4 = c_2 c_3$$

holds. By the structure of the matrix L, $|\det A(x)| = |X_1(\phi(x))| \ge c_4$. Hence, the values of

$$\|A^{-1}(x)\|$$

are uniformly bounded. The second derivatives of components of F with respect to t, s_i are also uniformly bounded. It follows from the standard estimates in the implicit function theorem that the radiuses of W' can be chosen uniformly separated from 0. Set $B(x) = A^{-1}(x)$. Differentiate (2.47) to obtain

$$\frac{\partial F}{\partial x} + \frac{\partial F}{\partial t}\frac{\partial \tau}{\partial x} + \frac{\partial F}{\partial s}\frac{\partial \sigma}{\partial x} \equiv 0.$$

This gives the equality

$$-\frac{\partial F}{\partial x} = A(x)\left[\frac{\partial \tau}{\partial x}, \frac{\partial \sigma}{\partial x}\right]^*$$

($*$ means transposition). Thus, we can express

$$\frac{\partial \sigma}{\partial x} = -B_0(x)\frac{\partial F}{\partial x},$$

where B_0 is a submatrix of B. It follows that

$$D\phi(x) = -L\frac{\partial \sigma}{\partial x} = -LB_0(x)\frac{\partial F}{\partial x}.$$

The matrix

$$G(x) = \frac{\partial F}{\partial x}$$

coincides with $D\Xi(\tau(x), x)$. Since the times $\tau(x)$ are uniformly bounded, the matrix $G(x)$ is uniformly continuous in the following sense: given $\beta > 0$ there exists $\alpha > 0$ such that if $r(x_1, x_2) < \alpha$, then

$$\|G(x_1) - G(x_2)\| < \beta.$$

Standard calculation shows that

$$\frac{\partial B}{\partial x} = -A^{-1}\frac{\partial A}{\partial x}A^{-1}(x).$$

Since the value

$$\frac{\partial A}{\partial x} = \left[\frac{\partial X}{\partial x}X, 0\right]$$

is uniformly bounded, this obviously proves our remark (since the derivatives of $\exp, \exp^{-1}$ are also uniformly bounded).

Lemma 2.2.15. *Given $\beta > 0$ there exists $\alpha > 0$ such that, for any $z, p \in M$, the following holds for $\tau, \tau_1, \phi, \phi_1$ defined in Lemma 2.2.14.*

If z is of type (O) and $r(\Xi(T, p), z) \leq \alpha$, then for any $v \in S(p)$ with $|v| = 1$ we have

$$|(D\Xi(\tau(p), p) - D\phi(p))v| \leq \beta.$$

If p is of type (O) and $r(\Xi(-T,z),p) \leq \alpha$, then for any $v \in U(z)$ with $|v| = 1$ we have

$$|(D\Xi(\tau_1(z),z) - D\phi_1(z))v| \leq \beta.$$

Proof. We prove only the first statement, for the second one the proof is similar.

Take $v \in S(p)$ with $|v| = 1$. Let $\gamma(s)$, $s \in [0,1]$, be a smooth curve such that $\gamma(0) = p, \dot{\gamma}(0) = v$. Denote $\tau_0 = \tau(p)$,

$$\gamma_1(s) = \Xi(\tau_0, \gamma(s)), \quad \gamma_2(s) = \phi(\gamma(s)) = \Xi(\tau(\gamma(s)), \gamma(s)).$$

There exists $s_0 > 0$ such that if $\alpha \leq d_1$ (see Lemma 2.2.13), then, for $s \in [0, s_0]$, the function $\tau^*(s) = f(\gamma_1(s))$ is defined.

Denote

$$v_1 = D\Xi(\tau_0, p)v, \quad v_2 = D\phi(p)v.$$

Obviously,

$$v_1 = \frac{d\gamma_1}{ds}(0) = \frac{d}{ds}\Xi(\tau_0, \gamma(s))\mid_{s=0}, \tag{2.48}$$

$$v_2 = \frac{d\gamma_2}{ds}(0) = \frac{d}{ds}\Xi(\tau_0 + \tau^*(s), \gamma(s))\mid_{s=0} = X(q)\frac{d\tau^*}{ds}(0) + v_1, \tag{2.49}$$

where $q = \Xi(\tau_0, p)$.

Set $v' = D\exp_z^{-1}(q)v_1$. By Lemma 2.2.12, given $\beta_1 > 0$ there exists $\alpha_1 > 0$ such that if $\alpha \leq \alpha_1$, then there is an isomorphism $\Pi(p,z) : T_zM \to T_zM$ with

$$\|\Pi(p,z) - I\| \leq \beta_1, \quad \Pi(p,z)v' \in S(z) \subset Z(z).$$

Hence, given $\beta_2 > 0$ there exists $\alpha_2 > 0$ such that if $\alpha \leq \alpha_2$, then the angle between v' and $Z(z)$ is less than β_2. Note that these α_i depend only on the corresponding β_i.

Since derivatives of $\exp_p$ are uniformly bounded, given $\beta_3 > 0$ there exist numbers $\alpha_3 > 0$ and $s_1 > 0$ such that for $\alpha \leq \alpha_3$ we have

$$\text{dist}(\gamma_1(s), \Sigma(z)) \leq \beta_3 s, \quad s \in [0, s_1],$$

hence (by Lemma 2.2.13)

$$|\tau^*(s)| \leq K_1\beta_3 s, \ s \in [0, s_1], \quad \text{and} \quad \left|\frac{\partial\tau^*}{ds}(0)\right| \leq K_1\beta_3.$$

It follows from (2.43) that

$$\left|X(q)\frac{\partial\tau^*}{ds}(0)\right| \leq KK_1\beta_3.$$

Since β_3 can be taken arbitrary, it remains to compare (2.48) with (2.49) to complete the proof of our lemma. □

Take $\nu_0 \in (0,1)$ such that $\lambda = (1 + \nu_0)^2\mu < 1$. For this λ and N (from Lemma 2.2.11) take the corresponding N_1 (Theorem 1.3.1) and find $\kappa > 0$ such that (1.65) holds. Now we find $\nu \in (0, \nu_0)$ such that

$$4K(8K^3 + 6K^2 + 9K + 1)\nu < \kappa/2. \tag{2.50}$$

Below we denote by d' positive constants that depend only on ν, κ, K, N. At each step of the proof, we consider (d, T)-pseudotrajectories such that d does not exceed the minimal d' previously chosen. Since we choose d' finitely many times, no generality is lost. We take $d' \leq \min(b, d_2)$ (b and d_2 are given by Lemma 2.2.14).

We fix d' such that, for any points $x, y \in M$ with $r(x,y) \leq d'$ and $y' = \exp_x^{-1}(y)$, the inequalities

$$\frac{|D\exp_x^{-1}(y)v|}{|v|}, \frac{|D\exp_x(y')v|}{|v|} \leq 1 + \nu \tag{2.51}$$

hold.

We fix d' such that if

$$r(z, q) < d', \ r(y, p) < d',$$

then for the linear isomorphisms $\Theta(p, z), \Pi(p, z)$ (see Lemma 2.2.12) we have

$$||\Pi - I||, ||\Theta - I||, ||\Theta^{-1} - I|| \leq \nu. \tag{2.52}$$

Now we prove Theorem 2.2.3 for (d, T)-pseudotrajectories. Let Ψ be a (d, T)-pseudotrajectory, denote $x_k = \Psi(kT), k \in \mathbf{Z}$. Set $H_k = Z(x_k)$.

Fix a pair x_k, x_{k+1}. Denote

$$p = x_k, \ z = x_{k+1}.$$

We consider three possible cases.

Case 1. z is of type (R). In this case, define a diffeomorphism ξ by $\xi(x) = \Xi(T, x)$ and let

$$\phi_k = \exp_z^{-1} \circ \xi \circ \exp_p.$$

Obviously, we can take d' such that ϕ_k is defined on $Z_{d'}(p)$ and $\xi^{-1}(z) \in \exp_p(E_a(p))$. It follows from (2.44) that

$$D\phi_k(0) = D\exp_z^{-1}(q)D\xi(p) : H_k \to H_{k+1},$$

where $q = \xi(p)$. Denote $J = D\phi_k(0)$.

We can write

$$\phi_k(v) = Jv + \chi(v)$$

with $\chi(0) = \phi_k(0), D\chi(0) = 0$. Note that

$$|\phi_k(0)| = |\exp_z^{-1}(\xi(p))| = r(\xi(p), z) < d.$$

Since the derivatives of $\xi, \exp_p$ are uniformly continuous, there exists d' such that

$$|\chi(v) - \chi(v')| \leq \frac{\kappa}{2}|v - v'| \text{ for } v, v' \in Z_{d'}(p). \tag{2.53}$$

Set

$$A_k^s = \Pi(p, z)JP(p), \ A_k^u = F\Theta^{-1}(p, z)Q(p),$$

where

$$F = D\xi(y)D\exp_p(y^*) : H_k \to H_{k+1},$$

and $y = \xi^{-1}(z), y^* = \exp_p^{-1}(y)$.

Below for $\Pi(p, z), \Theta(p, z)$ given by Lemma 2.2.12 we write Π instead of $\Pi(p, z)$ etc.

We represent $\phi_k(v)$ in form (1.51), where

$$A_k = A_k^s + A_k^u, \ w_{k+1}(v) = (J - A_k)v + \chi(v).$$

Take $v^s \in S(p)$. By Lemmas 2.2.10 and 2.2.11, $v_1^s = D\xi(p)v^s \in S(q)$ and $|v_1^s| \leq \mu|v^s|$. By (2.51), for $v_2^s = D\exp_z^{-1}(q)v_1^s$ we have the estimate $|v_2^s| \leq (1 + \nu)|v_1^s|$. Since $\Pi v_2^s \in S(z)$ and (2.52) holds, we see that $v_3^s = \Pi v_2^s = A_k^s v^s \in S(z)$ and $|v_3^s| \leq (1 + \nu)|v_2^s| \leq \lambda|v^s|$. Hence,

$$A_k^s S(p) \subset S(z), \ \|A_k^s|_{S(p)}\| \leq \lambda. \tag{2.54}$$

Now take $v^u \in U(z)$ and let

$$U'(z) = D\xi^{-1}(z)U(z) \text{ and } U''(z) = D\exp_p^{-1}(y)U'(z).$$

By Lemmas 2.2.10 and 2.2.11, $v_1^u = D\xi^{-1}(z)v^u \in U'(y) \subset U(y)$ and $|v_1^u| \leq \mu|v^u|$. For $v_2^u = D\exp_p^{-1}(y)v_1^u \in U''(z)$ we have $|v_2^u| \leq (1 + \nu)|v_1^u|$ (see (2.51)). Now (2.52) implies that for $v_3^u = \Theta v_2^u \in \Theta U''(z) \subset U(p)$ we have $|v_3^u| \leq \lambda|v^u|$. Set $B_k = \Theta G$, where

$$G = D\exp_p^{-1}(y)D\xi^{-1}(z) : H_{k+1} \to H_k.$$

Since $v_3^u = B_k v^u$, we see that

$$B_k U(z) \subset U(p), \ \|B_k|_{U(z)}\| \leq \lambda. \tag{2.55}$$

Take $w = v_3^u \in \Theta U''(z)$, then $Qw = w, \Theta^{-1}w = v_2^u$, and

$$F\Theta^{-1}w = Fv_2^u = D\xi(y)D\exp_p(y^*)v_2^u = D\xi(y)v_1^u = v^u,$$

hence

$$A_k^u B_k|_{U(z)} = I. \tag{2.56}$$

Since

$$A_k|_{S(p)} = A_k^s, \ A_k B_k|_{U(z)} = A_k^u B_k|_{U(z)},$$

it follows from (2.54)-(2.56) that A_k, B_k satisfy conditions (1.63) and (1.64) of Theorem 1.3.1.

Obviously, there exists d' such that if $r(q, z) \leq d'$, then

$$\|D\xi(y)Dexp_p(y^*) - Dexp_z^{-1}(q)D\xi(p)\| \leq \nu.$$

For $d \leq d'$ let us estimate

$$\|J - A_k\| = \|J(P + Q) - A_k\| \leq \|JP - A_k^s\| + \|JQ - A_k^u\| =$$

$$= \|JP - \Pi JP\| + \|JQ - F\Theta^{-1}Q\|. \tag{2.57}$$

Since $\|P\| \leq K, \|J\| \leq (1 + \nu)K \leq 2K$ (see (2.51)), and $\|\Pi - I\| \leq \nu$ (see (2.52)), the first summand in (2.57) does not exceed $2K^2\nu$. Let us estimate the second summand,

$$\|JQ - F\Theta^{-1}Q\| \leq K\|J - F\Theta^{-1}\| = K\|F(\Theta^{-1} - I) + F - J\| \leq$$

$$\leq 2K^2\nu + K\|D\xi(y)Dexp_p(y^*) - Dexp_z^{-1}(q)D\xi(p)\| \leq \nu K(2K + 1).$$

This gives

$$\|J - A_k\| \leq 2K(2K + 1)\nu < \kappa/2. \tag{2.58}$$

Now we deduce from (2.53) and (2.58) that if $d \leq d'$, then ϕ_k satisfies conditions (1.63)-(1.65) of Theorem 1.3.1 with $\Delta = d'$ (of course, we identify P_k with $P(p)$ etc).

Case 2. p and z are of type (O). By Lemma 2.2.14, the functions τ, τ_1 and mappings ϕ, ϕ_1 are defined on $\Sigma_b(p), \Sigma_b(z)$, respectively. Denote

$$q = \phi(p), \; y = \phi_1(z), \; y^* = exp_p^{-1}(y), \; q^* = exp_z^{-1}(q), \; q' = \Xi(T, p).$$

Set

$$\phi_k = exp_z^{-1} \circ \phi \circ exp_p.$$

Then

$$J_0 = D\phi_k(0) = Dexp_z^{-1}(q)D\phi(p),$$

denote

$$J = Dexp_z^{-1}(q)D\Xi(\tau(p), p).$$

By Lemma 2.2.13, the inequality $r(z, q') \leq d \leq d'$ implies $r(z, q) \leq K_1 d$, hence $|\phi_k(0)| = |z - q^*| \leq K_1 d$.

It follows from the properties of ϕ (see the remark after Lemma 2.2.14) that we can take d' such that

$$\phi_k(v) = J_0 v + \chi(v),$$

and for $d \leq d'$ (2.53) holds.

Set

$$F = D\Xi(\tau(y), y)Dexp_p(y^*), \; G = D exp_p^{-1}(y)D\Xi(\tau_1(z), z)$$

and note that $\tau(y) = -\tau_1(z)$, hence

$$FG = I. \tag{2.59}$$

We again take $A_k^s = \Pi(p, z)JP(p)$ ($\Theta(p, z), \Pi(p, z)$ are given by Lemma 2.2.12 for the points q, y fixed above). The same reasons as in case 1 show that (2.54) holds (note that $D\Xi(\tau(p), p)S(p) \subset S(q)$).

Set $B_k = \Theta G$ and $U' = B_k U(z) \subset U(p)$. Repeat the proof of case 1 to show that (2.55) holds. Let U^* be the orthogonal complement of U' in $U(p)$. Represent $Q(p) = Q' + Q^*$, where Q', Q^* are the corresponding projectors. Obviously,

$$||Q'||, ||Q^*|| \le K.$$

Set

$$A_k^u = F\Theta^{-1}Q' + J_0 Q^*.$$

For $v \in U(z)$ we have $v' = B_k v = \Theta G v \in U'$, hence $Q^* v' = 0$ and

$$A_k^u B_k v = F\Theta^{-1}\Theta G v = v$$

(we apply (2.59) here). This proves (2.56). Now we take $A_k = A_k^s + A_k^u$ and represent $\phi_k(v)$ in form (1.51) with $w_{k+1}(v) = (J_0 - A_k)v + \chi(v)$.

Apply Lemma 2.2.15 to find $d_3 = \alpha$ corresponding to $\beta = \nu$. Take $d_4 > 0$ so small that if $d \le d_4$, then

$$||D\phi_1(q)D\exp_z(q^*) - D\exp_p^{-1}(y)D\phi_1(z)| \le \nu,$$

(see the remark after Lemma 2.2.14).

For $d \le d' \le \min(d_3, d_4)$ let us estimate

$$||J_0 - A_k|| \le ||A_k^s - J_0 P|| + ||A_k^u - J_0 Q||.$$

Consider the first summand. By Lemma 2.2.15,

$$||JP - J_0 P|| \le K(1 + \nu) \max_{v \in S(p), |v|=1} |(D\Xi(\tau(p), p) - D\phi(p))v| \le$$

$$\le K(1 + \nu)\nu \le 2K\nu,$$

hence

$$||A_k^s - J_0 P|| = ||\Pi JP - J_0 P|| \le ||\Pi - I|| \cdot ||JP|| + ||JP - J_0 P|| \le$$

$$\le 2K(K + 1)\nu. \tag{2.60}$$

Now consider

$$||A_k^u - J_0 Q|| = ||F\Theta^{-1}Q' + J_0 Q^* - J_0(Q' + Q^*)|| = ||F\Theta^{-1}Q' - J_0 Q'|| \le$$

$$\le ||F(\Theta^{-1} - I)Q'|| + ||FQ' - J_0 Q'||. \tag{2.61}$$

The first summand in (2.61) does not exceed $2K^2\nu$. Let us estimate the second one. Take $v_0 \in Z(p), |v_0| = 1$, let $w = Q'v_0$, we have $|w| \le K|v_0| = K$. There exists $v \in U(z)$ such that $w = B_k v$. Since $v = F\Theta^{-1}w$, we have $|v| \leq 4K|w| \leq 4K^2$. We obtain

$$FQ'v_0 - J_0 Q'v_0 = F\Theta Gv - J_0 \Theta Gv =$$
$$= F(\Theta - I)Gv - J_0(\Theta - I)Gv + FGv - J_0 Gv.$$

Obviously,

$$|F(\Theta - I)Gv| \leq 4K^2\nu|v| \leq 16K^4\nu, |J_0(\Theta - I)Gv| \leq 16K^4\nu.$$

Since $FGv = v$, we come to the estimates

$$|FGv - J_0 Gv| = |v - J_0 Gv| = |J_0(J_0^{-1}v - Gv)| \leq$$

$$\leq \|J_0\|(|(J_0^{-1} - D\exp_p^{-1}(y)D\phi_1(z))v| + |(D\exp_p^{-1}(y)D\phi_1(z) - G)v|) \leq$$

$$\leq 2K|(D\phi_1(q)D\exp_z(q^*) - D\exp_p^{-1}(y)D\phi_1(z))v| + \qquad (2.62)$$

$$+2K|(D\exp_p^{-1}(y)D\phi_1(z) - D\exp_p^{-1}(y)D\Xi(\tau_1(z), z))v| \leq \qquad (2.63)$$

$$\leq 6K\nu|v| \leq 24K^3\nu.$$

Note that we apply the choice of d_4 to estimate (2.62) and the choice of d_3 to estimate (2.63).

It follows from our estimates that

$$\|A_k - J_0\| \leq 4K(8K^3 + 6K^2 + K + 1)\nu < \kappa/2.$$

Thus, for $d \leq d'$, ϕ_k satisfies conditions (1.63)-(1.65) of Theorem 1.3.1.

Case 3. p is of type (R) and z is of type (O). Consider τ and ϕ defined on $\Sigma_b(p)$ (see Lemma 2.2.14). Take q, q^*, ϕ_k, J, J_0 the same as in case 2. Represent $\phi_k(v) = J_0 v + \chi(v)$, we assume that for $d \leq d'$ (2.53) holds. Note that similarly to case 2 we have $|\phi_k(0)| \leq K_1 d$.

Denote $\tau_0 = \tau(p)$ and define a diffeomorphism ξ by $\xi(x) = \Xi(\tau_0, x)$. Denote

$$\Sigma' = \xi^{-1}(\Sigma_b(z)), \ y = \xi^{-1}(z), \ y^* = \exp_p^{-1}(y).$$

Obviously, $p \in \Sigma'$, and for $x \in \Sigma'$ we have $\xi(x) \in \Sigma_z$. Hence, $\phi(x) = \xi(x)$ on Σ', and

$$D\xi(p)w = D\phi(p)w \text{ for } w \in T_p\Sigma'. \qquad (2.64)$$

Set

$$F = D\xi(y)D\exp_p(y^*), \ G = D\exp_p^{-1}(y)D\xi^{-1}(z).$$

Consider $\Theta(p, z), \Pi(p, z)$ given by Lemma 2.2.12 for p, z, q, y. Define A_k^s, B_k, U', U^*, Q', Q^*, A_k^u, A_k as in case 2. The same reasons as in case 2 show that (2.59),(2.54)-(2.56) hold.

Let us estimate $\|A_k^u - J_0 Q\|$. Since derivatives of $\exp_p$ are uniformly bounded and $|\tau_0 - T| \leq 1$ (see Lemma 2.2.14), we can find d' such that if $d \leq d'$, then the following two statements are true.

If $v \in T_y \Sigma'$ (or $v \in T_p \Sigma'$), then $w = D \exp_p{}^{-1}(y)v$ (correspondingly, $w = D \exp_p(y^*)v$) can be represented as $w = w' + w''$, where $w' \in T_p \Sigma'$ (correspondingly, $w' \in T_y \Sigma'$) and $|w''| \leq \nu |w|$.

The inequality

$$\|D \exp_z^{-1}(q) D \xi(p) - D \xi(y) D \exp_p(y^*)\| \leq \nu \qquad (2.65)$$

holds.

Take $d \leq d'$. The first summand in (2.61) is estimated similarly to case 2. Now we estimate $\|FQ' - J_0 Q'\|$.

Since $D\xi^{-1}(z)U(z) \subset U(y)$ (Lemma 2.2.10) and $U(z) \subset Z(z) = T_z \Sigma_z$, for $v \in U(z)$ we have

$$v_1 = D\xi^{-1}(z)v \in U(y) \subset T_y \Sigma'.$$

By our choice of d',

$$v_2 = Gv = D\exp_p{}^{-1}(y)v_1 = v_2' + v_2'' \text{ with } v_2' \in T_p \Sigma', |v_2''| \leq \nu |v_2|.$$

Since (2.52) holds,

$$v_3 = \Theta(p, z)v_2 = B_k v = v_2 + v_3', \ |v_3'| \leq \nu |v_2|.$$

Obviously, $\kappa \leq 1$, hence it follows from (2.50) that $\nu < 1/8$. Since

$$|v_2|(1 - \nu) \leq |v_2'|,$$

we have $|v_2| \leq 2|v_2'|$. From the inequalities

$$|v_3'| \leq \nu |v_2| \leq 2\nu |v_2'| \text{ and } |v_2'' + v_3'| \leq 4\nu |v_2'|$$

we deduce that $B_k v = v_2' + v_2'''$ with $v_2' \in T_p \Sigma'$ and $|v_2'''| \leq 4\nu |v_2'|$.

Hence, any $w \in U'$ we can represent in the form $w = w' + w''$ with $w' \in T_p \Sigma'$ and $|w''| \leq 4\nu |w'|$.

Take $v_0 \in Z(p)$ with $|v_0| = 1$, let $w = Q'v_0$. We can write $w = w' + w''$ with w', w'' as above. Since $|w| \leq K$ and $|w'|(1-4\nu) \leq |w|$, we obtain the inequalities $|w'| \leq 2K$, $|w''| \leq 8K\nu$.

Let us write

$$FQ'v_0 = D\xi(y)D\exp_p(y^*)w =$$

$$= D\xi(y)D\exp_p(y^*)w' + w_1'', \ |w_1''| \leq 2K|w''| \leq 16K^2\nu; \qquad (2.66)$$

$$J_0 Q'v_0 = D\exp_z^{-1}(q)D\phi(p)w =$$

$$= D\exp_z^{-1}(q)D\phi(p)w' + w_2'', \ |w_2''| \leq 16K^2\nu. \qquad (2.67)$$

Since $D\phi(p)w' = D\xi(p)w'$ (see (2.64)), it follows from (2.65) that

$$|D\xi(y)D\exp_p(y^*)w' - D\exp_z^{-1}(q)D\phi(p)w'| =$$

$$= |D\xi(y)D\exp_p(y^*)w' - D\exp_z^{-1}(q)D\xi(p)w'| \leq \nu |w'|.$$

Now we deduce from (2.66) and (2.67) that

$$|FQ'v_0 - J_0Q'v_0| \leq \nu|w'| + 32K^2\nu \leq 2K(16K+1)\nu.$$

Since the first summand in (2.61) does not exceed $2K^2\nu$, we come to the estimate $||A_k^u - J_0Q|| \leq 2K(17K+1)\nu$. This gives

$$||A_k - J_0|| \leq 4K(9K+1)\nu < \kappa/2.$$

Hence, in case 3 for $d \leq d'$, ϕ_k also satisfies conditions (1.63)-(1.65) of Theorem 1.3.1.

Apply Theorem 1.3.1 to ϕ_k and find d'_0, L' (corresponding to $\lambda, N, \Delta = d'$). Take $d_0 = d'_0/K_1$. Then for $d \leq d_0$ we have

$$|\phi_k(0)| \leq K_1 d \leq d'_0.$$

It follows from Theorem 1.3.1 that there exist vectors $v_k \in H_k$ such that $|v_k| \leq L''d$ (with $L'' = K_1L'$) and $\phi_k(v_k) = v_{k+1}$. The definitions of ϕ_k imply that the points $p_k = \exp_{x_k}(v_k)$ have the following property: any p_{k+1} belongs to the trajectory of the flow Ξ through p_k. Hence, these points belong to one trajectory of Ξ. Note that the inequalities

$$r(p_k, x_k) \leq L''d$$

hold. Define numbers t_k by $p_{k+1} = \Xi(t_k, p_k)$. In case 1, we have $t_k = T$. In cases 2, 3, the inequalities

$$r(\Xi(T, p_k), z) \leq r(\Xi(T, p), z) + r(\Xi(T, p_k), \Xi(T, p))$$

hold. The first summand does not exceed d, the second one does not exceed $\exp(\mathcal{L}T)L''d$ (recall that $\mathcal{L}$ is the Lipschitz constant of X we fixed above). It follows from Lemma 2.2.13 that $p_{k+1} = \Xi(t_k, p_k)$ with $|t_k - T| \leq L_1 d$, where $L_1 = K_1(1 + \exp(\mathcal{L}T)L'')$.

Set $\tau_0 = 0, \tau_k = t_1 + \ldots + t_k$ for $k > 0$, and $\tau_k = t_k + \ldots + t_{-1}$ for $k < 0$.

Now we define $\alpha : \mathbb{R} \to \mathbb{R}$, for $k \in \mathbb{Z}$ set $\alpha(Tk) = \tau_k$, and for $t \in [Tk, T(k+1)]$ set

$$\alpha(t) = \tau_k + (t - Tk)\frac{t_{k+1}}{T}.$$

The same reasons as in the proof of Theorem 1.5.1 show that $\alpha \in \text{Rep}(L_2 t)$, where $L_2 = L_1/T$, and that

$$|\alpha(t) - \tau_k - t'| \leq L_1 d \qquad (2.68)$$

for $t \in [Tk, T(k+1)]$ and $t' = t - Tk$.

Set $p = p_0$. Obviously, $\Xi(\alpha(Tk), p) = \Xi(\tau_k, p) = p_k$. For $t \in [Tk, T(k+1)]$ we have

$$r(\Xi(\alpha(t), p), \Psi(t)) \leq r(\Xi(\alpha(t) - \tau_k, p_k), \Xi(t', p_k)) +$$

$$+ r(\Xi(t', p_k), \Xi(t', x_k)) + r(\Psi(Tk + t'), \Xi(t', \Psi(Tk)))$$

(note that $\Xi(\alpha(t), p) = \Xi(\alpha(t) - \tau_k, p_k)$, $x_k = \Psi(Tk)$, $\Psi(t) = \Psi(Tk + t')$).

The first term on the right does not exceed $KL_1 d$ (we apply (2.68) and the inequality $|X| \leq K$). The second term does not exceed

$$\exp(\mathcal{L}T)r(p_k, x_k) \leq L'' \exp(\mathcal{L}T)d,$$

and the third term is not more than d (since Ψ is a (d, T)-pseudotrajectory). This leads to the inequality

$$r(\Xi(\alpha(t), p), \Psi(t)| \leq L_3 d,$$

where $L_3 = KL_1 + L'' \exp(\mathcal{L}T) + 1$.

Take $L = \max(L_2, L_3)$ to complete the proof of Theorem 2.2.3.

Remark. If $\Omega' = \emptyset$, then all x_k are of type (R), hence $t_k = T$ for all k, and we can take $\alpha(t) = t$.

2.2.2 The Case of a Diffeomorphism

To introduce the notion of structural stability, we define the C^1-topology on the set of diffeomorphisms of M as follows.

We again fix an embedding

$$f : M \to \mathbb{R}^{2n+1}$$

(see the previous subsection), and consider M as a smooth submanifold of $\mathbb{R}^{2n+1}$. We preserve notation $|.|$ for the norm in $T_p M, p \in M$, generated by the Euclidean norm of $\mathbb{R}^{2n+1}$.

For a diffeomorphism ϕ of M and for a point $p \in M$, the derivative $D\phi(p)$ is a linear mapping of a linear subspace $T_p M$ of $\mathbb{R}^{2n+1}$ into $\mathbb{R}^{2n+1}$. Thus, for two diffeomorphisms ϕ and ψ, the value

$$||D\phi(p) - D\psi(p)|| = \max_{v \in T_p M, |v|=1} |D\phi(p)v - D\psi(p)v|$$

is defined.

Now we define the C^1-distance ρ_1 between two C^1 diffeomorphisms ϕ, ψ of M by the formula

$$\rho_1(\phi, \psi) = \rho_0(\phi, \psi) + \max_{p \in M} ||D\phi(p) - D\psi(p)||$$

(the C^0-metric ρ_0 was defined in the previous section). We denote by $\text{Diff}^1(M)$ the space of diffeomorphisms of class C^1 of M with the topology induced by ρ_1.

Definition 2.7 *We say that $\phi \in Diff^1(M)$ is "structurally stable" if given $\epsilon > 0$ there is $\delta > 0$ such that for any $\psi \in Diff^1(M)$ with $\rho_1(\phi, \psi) < \delta$ there is a homeomorphism $h : M \to M$ with the following properties:*

(a) $h \circ \phi = \psi \circ h$;

(b) $r(x, h(x)) < \epsilon$ for $x \in M$.

Remark. One can give another definition of structural stability. Let us say that $\phi \in \mathrm{Diff}^1(M)$ is structurally stable if there exists $\delta > 0$ such that for any $\psi \in \mathrm{Diff}^1(M)$ with $\rho_1(\phi, \psi) < \delta$ there is a homeomorphism $h : M \to M$ with property (a) of Definition 2.7.

It follows from the main results of [Ma] and [Robi1] that these two definitions are equivalent.

Let ϕ be a structurally stable diffeomorphism, and let $\delta > 0$ have the property formulated above. Take ψ such that the inequality $\rho_1(\phi, \psi) < \delta$ holds and let h be a homeomorphism with property (a). There exists $\delta_1 > 0$ such that for any diffeomorphism χ with $\rho_1(\chi, \psi) < \delta_1$ we have $\rho_1(\phi, \chi) < \delta$, hence there is a homeomorphism h_1 such that $h_1 \circ \phi = \chi \circ h_1$. Then we have $h' \circ \psi = \chi \circ h'$, where $h' = h_1 \circ h^{-1}$. This shows that the set of structurally stable diffeomorphisms is an open subset of $\mathrm{Diff}^1(M)$.

Let $\Omega(\phi)$ be the set of nonwandering points for a diffeomorphism ϕ (see Definition 1.10). It is easy to show that $\Omega(\phi)$ is a compact ϕ-invariant set.

Smale introduced in [Sm2] the following property of ϕ.

Axiom A.

(a) $\Omega(\phi)$ *is a hyperbolic set;*

(b) *the set of periodic points of ϕ is dense in $\Omega(\phi)$.*

For a point $p \in \Omega(\phi)$ we define the *stable manifold*

$$W^s(p) = \{x \in M : r(\phi^k(x), \phi^k(p)) \to 0 \text{ as } k \to +\infty\}$$

and the *unstable manifold*

$$W^u(p) = \{x \in M : r(\phi^k(x), \phi^k(p)) \to 0 \text{ as } k \to -\infty\}.$$

One easily shows that if ϕ satisfies Axiom A and $p \in \Omega(\phi)$, then $W^s(p)$ and $W^u(p)$ are related to the local stable and unstable manifolds described in Theorem 1.2.1 as follows:

$$W^s(p) = \{x \in M : \phi^k(x) \in W^s_\Delta(p) \text{ for some } k \in \mathbb{Z}\},$$

$$W^u(p) = \{x \in M : \phi^k(x) \in W^u_\Delta(p) \text{ for some } k \in \mathbb{Z}\}.$$

In this case, the sets $W^s(p), W^u(p)$ are the images of Euclidean spaces $\mathbb{R}^k$ under immersions $\mathbb{R}^k \to M$ of class C^1 (see [Pi1]).

Now we describe one more condition on ϕ introduced by Smale [Sm2].

Definition 2.8 *We say that a diffeomorphism ϕ satisfies the "geometric STC" (the "geometric strong transversality condition") if, for any $p, q \in \Omega(\phi)$, the stable and unstable manifolds $W^s(p)$ and $W^u(q)$ are transverse.*

Necessary and sufficient conditions for structural stability are given by the following statement.

Theorem 2.2.4 [Robb, Robi2, Ma]. *A diffeomorphism ϕ is structurally stable if and only if ϕ satisfies Axiom A and the geometric STC.*

Remark. Robbin showed in [Robb] that the conditions of Theorem 2.2.4 are sufficient for the structural stability of a diffeomorphism ϕ of class C^2, Robinson proved the analogous statement for diffeomorphisms of class C^1. Mañé established the necessity part.

Nitecki showed in [Ni2] that the same conditions imply topological stability.

Theorem 2.2.5 [Ni2]. *If a diffeomorphism ϕ satisfies Axiom A and the geometric STC, then ϕ is topologically stable.*

The following statement is a corollary of Theorems 2.1.1, 2.2.4, and 2.2.5.

Theorem 2.2.6. *If a diffeomorphism ϕ is structurally stable, then ϕ has the POTP.*

Remark. This result was established independently by Robinson [Robi3], Sawada [Saw], and Morimoto [Mori2] (to be exact, they showed that Axiom A combined with the geometric STC implies the POTP).

Indeed, a stronger statement holds.

Theorem 2.2.7. *If a diffeomorphism ϕ is structurally stable, then ϕ has the LpSP.*

Remark. A proof of Theorem 2.2.7 was given in [Pi2], Appendix A. Here we describe another proof, close to the proof of Theorem 2.2.3 given in the previous subsection (note that the problem for flows is really more complicated than the one for diffeomorphisms). We omit all the auxiliary statements and formulate only the final result we need (Lemma 2.2.16), the details are left to the reader (note that an analog of Lemma 2.2.9 for a diffeomorphism was proved in [Robb]).

Thus, we consider a structurally stable diffeomorphism ϕ of M. There exist families of subspaces $S(p), U(p)$ of T_pM for $p \in M$ with the properties described in the lemma below (the corresponding statements are proved similarly to Lemmas 2.2.10, 2.2.11, 2.2.12).

Lemma 2.2.16. *One can construct subspaces $S(p), U(p)$ of $T_pM, p \in M$, such that*

(1)
$$S(p) \oplus U(p) = T_p M;$$

(2)
$$D\phi(p)S(p) \subset S(\phi(p)), \ D\phi^{-1}(p)U(p) \subset U(\phi^{-1}(p));$$

(3) there exist constants $C > 0, \lambda_1 \in (0,1)$ such that

$$|D\phi^k(p)v| \le C\lambda_1^k|v| \text{ for } v \in S(p), \ k \ge 0,$$

$$|D\phi^{-k}(p)v| \le C\lambda_1^k|v| \text{ for } v \in U(p), \ k \ge 0;$$

(4) there exists $N > 0$ such that if $P(p), Q(p)$ are the complementary projectors corresponding to $S(p), U(p)$, then

$$\|P(p)\|, \|Q(p)\| \le N;$$

(5) given $\beta > 0$ and a natural T there exists $\alpha > 0$ such that if $z, p \in M$, $q = \phi^T(p), y = \phi^{-T}(z)$, and $r(z,q) < \alpha$, then there is a linear isomorphism $\Pi(p,z) : T_z M \to T_z M$ such that

$$\|\Pi - I\| \le \beta, \ \Pi(p,z)(D\exp_z^{-1}(q)D\phi^T(p)S(p)) \subset S(z),$$

and a linear isomorphism $\Theta(p,z) : T_p M \to T_p M$ such that

$$\|\Theta - I\| \le \beta, \ \Theta(p,z)(D\exp_p^{-1}(y)D\phi^{-T}(z)U(z)) \subset U(p).$$

Now we find $T > 0$ such that

$$\mu = C\lambda_1^T < 1. \tag{2.69}$$

It follows from Lemma 1.1.3 that if $\phi' = \phi^T$ has the LpSP, then the same holds for ϕ. To simplify notation, below we write ϕ insread of ϕ'. We will apply Lemma 2.2.16 to this new ϕ with $C = 1, \lambda_1 = \mu$ in (2), and with $T = 1$ in (5).

Take $\nu_0 \in (0,1)$ such that $\lambda = (1 + \nu_0)^2\mu < 1$. For this λ and N (from statement (4) of Lemma 2.2.16) take the corresponding N_1 (Theorem 1.3.1) and find $\kappa > 0$ such that (1.65) holds. Now we fix $K \ge N$ such that

$$\|D\phi(p)\|, \|D\phi^{-1}(p)\| \le K \text{ for } p \in M.$$

Find a number $\nu \in (0, \nu_0)$ with the property

$$2K(2K + 1)\nu < \kappa/2.$$

In the previous subsection, a number c was fixed such that, for any $p \in M$, the mapping $\exp_p$ is a diffeomorphism of the ball $E_{2c}(p) \subset T_p M$ onto its image. We take a number $d' < c$ such that, for any points $x, y \in M$ with $r(x,y) \le d'$ and for $y' = \exp_x^{-1}(y)$, inequalities (2.51) hold.

We assume that d' has also the following property: if for $p, z \in M$ and $q = \phi(p)$ we have $r(z, q) < d'$, then for the corresponding linear isomorphisms $\Theta(p, z), \Pi(p, z)$ (see statement (5) of Lemma 2.2.16) inequalities (2.52) are true.

Now we take a sequence $\xi = \{x_k\}$ such that

$$r(x_{k+1}, \phi(x_k)) \leq d$$

with $d \leq d'$, and consider the spaces $H_k = T_{x_k}M$. We fix two points x_k, x_{k+1}, and denote

$$p = x_k, \ z = x_{k+1}, \ q = \phi(p), \ y = \phi^{-1}(z), \ y^* = \exp_p^{-1}(y).$$

Define $\phi_k : H_k \to H_{k+1}$ by the formula

$$\phi_k = \exp_z^{-1} \circ \phi \circ \exp_p.$$

It follows from our choice of d' that ϕ_k is defined on $E_{d'}(p)$ and that

$$|\phi_k(0)| \leq d.$$

Set $J = D\phi_k(0)$, we have

$$J = D\exp_z^{-1}(q)D\phi(p).$$

Define an operator $F = D\phi(y)D\exp_p(y^*)$, now we set

$$A_k^s = \Pi(p, z)JP(p), \ A_k^u = F\Theta^{-1}(p, z)Q(p),$$

and

$$A_k = A_k^s + A_k^u.$$

Let us write

$$\phi_k(v) = Jv + \chi(v)$$

with $\chi(0) = \phi_k(0), D\chi(0) = 0$. Since the derivatives of $\phi, \exp_p$ are uniformly continuous, we can choose d' so that (2.53) holds.

Now we represent $\phi_k(v)$ in form (1.51), where $w_{k+1}(v) = (J - A_k)v + \chi(v)$. The same reasons as in the proof of Theorem 2.2.3 (case 1) show that relations (2.54)-(2.56) hold. To estimate the value $\|A_k - J\|$, let us note that

$$\|J\| \leq (1 + \nu)K \leq 2K, \ \|F\| \leq 2K, \ \|P\|, \|Q\| \leq K.$$

Let us take d' such that the inequality

$$r(q, z) = r(\phi(x_k), x_{k+1}) \leq d'$$

implies the inequality

$$\|D\exp_z^{-1}(q)D\phi(p) - D\phi(y)D\exp_p(y^*)\| < \nu$$

(obviously, we can find d' depending only on ν and ϕ). Similarly to the proof of Theorem 2.2.3, we show that if $d \leq d'$, then

$$\|A_k - J\| \leq 2K(2K+1)\nu < \frac{\kappa}{2}. \tag{2.70}$$

Now we fix $\Delta = d'$ and apply Theorem 1.3.1 to find numbers d_0, L (depending only on ϕ) such that if $|\phi_k(0)| \leq d \leq d_0$, then there exists a sequence $v_k \in H_k$ with the properties

$$|v_k| \leq Ld \text{ and } \phi_k(v_k) = v_{k+1}.$$

Set $x = \exp_{x_0}(v_0)$. Then

$$\phi^k(x) = \exp_{x_k}(v_k),$$

and it follows that

$$r(x_k, \phi^k(x)) = |v_k| \leq Ld.$$

This proves Theorem 2.2.7.

Since the set of structurally stable diffeomorphisms is open in the C^1-topology (see the remark after Definition 2.7), it follows from Theorem 2.2.7 that if a diffeomorphism ϕ is structurally stable, then there exists a neighborhood W of ϕ in $\mathrm{Diff}^1(M)$ such that any diffeomorphism ψ in W has the LpSP. Below we show that one can take a neighborhood W such that the LpSP is uniform, i.e., the numbers d_0 and L are the same for all $\psi \in W$ [Beg]. It was first noticed by Sakai [Sak1] that the POTP is uniform in a neighborhood of a structurally stable diffeomorphism ϕ (see statement (4) of Theorem 2.2.8 below). Later Sakai showed [Sak2] that the C^1 interior of the set of diffeomorphisms with the POTP consists of structurally stable diffeomorphisms (note that for $\dim M \leq 2$ this result was proved by Moriaysu [Moriy]). Let us sum up the information related to this problem.

Theorem 2.2.8. *The following statements are equivalent:*

(1) a diffeomorphism ϕ is structurally stable;

(2) there is a neighborhood W of ϕ such that any $\psi \in W$ has the POTP;

(3) there is a neighborhood W of ϕ such that any $\psi \in W$ has the LpSP;

(4) there is a neighborhood W of ϕ such that for any $\epsilon > 0$ one can find $d > 0$ with the following property: any d-pseudotrajectory of a $\psi \in W$ is (ϵ, ψ)-shadowed by a point of M;

(5) there is a neighborhood W of ϕ and numbers $L, d_0 > 0$ such that any $\psi \in W$ has the LpSP with constants L, d_0.

Proof. Obviously, the implications

$$(5) \Rightarrow (4) \Rightarrow (2) \text{ and } (5) \Rightarrow (3) \Rightarrow (2)$$

hold. We prove here that $(1) \Rightarrow (5)$ and that $(5) \Rightarrow (1)$ (this establishes the equivalence of (1) and (5)), for the proof of the remaining part (i.e., of the implication $(2) \Rightarrow (1)$), we refer the reader to the original paper [Sak2].

First we prove the implication (1) $\Rightarrow$ (5). Let ϕ be a structurally stable diffeomorphism. We fix families $\{S(p), U(p)\}$ having the properties described in Lemma 2.2.16 (for ϕ). Take T such that (2.69) holds. There exists a neighborhood W_0 of ϕ in $\text{Diff}^1(M)$ such that all diffeomorphisms ψ in W have the same Lipschitz constant. Therefore, it follows from Lemma 1.3.1 that if all diffeomorhpisms ψ^T, where $\psi \in W_0$, have the property described in statement (5) of our theorem, then all $\psi \in W_0$ have a similar property.

We fix a number $d' > 0$ satisfying all the conditions imposed on Δ in the proof of Theorem 2.2.7. In addition, we assume that for a neighborhood $W_1 \subset W_0$ of ϕ and for d' the following holds. If $\psi \in W_1$, and for points $p, z \in M$ the inequality $r(\psi(p), z) \leq d'$ is satisfied, then the mapping

$$\exp_z^{-1} \circ \psi \circ \exp_p \qquad (2.71)$$

is a diffeomorphism from $E_{d'}(p)$ into $T_z M$ (obviously, one can find W_1 and d' with these properties).

Now let us take $\psi \in W_1$ and a sequence $\xi = \{x_k\}$ such that

$$r(x_{k+1}, \psi(x_k)) \leq d$$

with $d \leq d'$. Set $H_k = T_{x_k} M$, fix a pair of points $p = x_k, z = x_{k+1}$, and let $\phi_k : H_k \to H_{k+1}$ be defined by (2.71). For $\psi \in W_1$ and for $u \in H_k$ with $|u| \leq d'$ we set

$$X(\psi, u) = D(\exp_z^{-1} \circ \psi \circ \exp_p)(u),$$

so that

$$J' = D\phi_k(0) = X(\psi, 0) \text{ and } J = D\phi'(0) = X(\phi, 0), \qquad (2.72)$$

where $\phi' = \exp_z^{-1} \circ \phi \circ \exp_p$.

Represent

$$\phi_k(v) = J'v + \chi(\psi, v).$$

Take $v, v' \in H_k$ with $|v|, |v'| \leq d'$ and let $\theta(s), s \in [0, 1]$, be a linear parametrization of the segment joining v and v'. Then we can apply the standard formula

$$\chi(\psi, v) - \chi(\psi, v') = \int_0^1 X(\psi, \theta(s)) \, ds \, (v - v') - J'(v - v').$$

The uniform continuity of $X(\phi, u)$ and the second equality in (2.72) imply the existence of a number $\Delta \leq d'$ such that

$$\left| \int_0^1 X(\phi, \theta(s)) \, ds \, (v - v') - J(v - v') \right| \leq \frac{\kappa}{8} |v - v'|$$

for $|v|, |v'| \leq \Delta$. Hence, there exists a neighborhood $W \subset W_1$ of ϕ such that if $\psi \in W$, then

$$\|J - J'\| \leq \frac{\kappa}{4}$$

and

$$|\chi(\psi, v) - \chi(\psi, v')| =$$

$$= \left| \int_0^1 X(\psi, \theta(s)) \, ds \, (v - v') - J'(v - v') \right| \leq \frac{\kappa}{4} |v - v'|$$

for $|v|, |v'| \leq \Delta$.

Now we represent

$$\phi_k(v) = A_k v + w_{k+1}(v),$$

where A_k is constructed for ϕ, p, and z by the same formulas as in the proof of Theorem 2.2.7, and

$$w_{k+1}(v) = (J - A_k)v + (J' - J)v + \chi(\psi, v).$$

Since the mappings A_k satisfy all the conditions of Theorem 1.3.1, and

$$|w_{k+1}(v) - w_{k+1}(v')| \leq \kappa |v - v'|$$

for $|v|, |v'| \leq \Delta$ (here we refer to the estimates above and to (2.70)), we can apply Theorem 1.3.1 (in the same way as in the proof of Theorem 2.2.7) to show that that there exist constants d_0, L with the desired properties, and these constants do not depend on ψ. This proves the implication $(1) \Rightarrow (5)$.

To prove the implication $(5) \Rightarrow (1)$, we first formulate some known results. Let us denote by $\mathcal{G}^1(M)$ the set of diffeomorphisms ϕ with the following property: ϕ has a neighborhood W in $\text{Diff}^1(M)$ such that every periodic point of a diffeomorphism $\psi \in W$ is hyperbolic.

Theorem 2.2.9 [Hay1]. *If $\phi \in \mathcal{G}^1(M)$, then ϕ satisfies Axiom A.*

Now we formulate some properties of diffeomorphisms satisfying Axiom A. The following three statements are analogs of Lemmas 2.2.1 - 2.2.3 for the case of a diffeomorphism ϕ satisfying Axiom A.

Lemma 2.2.17 [Sm2]. *There exists a unique decomposition of the nonwandering set $\Omega(\phi)$,*

$$\Omega(\phi) = \Omega_1 \cup \ldots \cup \Omega_m,$$

where Ω_i are compact disjoint invariant sets, and each Ω_i contains a dense trajectory.

The sets Ω_i are called *basic sets*. For a basic set Ω_i we define the stable and unstable sets

$$W^s(\Omega_i) = \{x \in M : \text{dist}(\phi^k(x), \Omega_i) \to 0 \text{ as } k \to +\infty\}$$

and

$$W^u(\Omega_i) = \{x \in M : \text{dist}(\phi^k(x), \Omega_i) \to 0 \text{ as } k \to -\infty\}.$$

Lemma 2.2.18 [Sm2].

$$M = \bigcup_{1 \leq i \leq m} W^s(\Omega_i) = \bigcup_{1 \leq i \leq m} W^u(\Omega_i).$$

Lemma 2.2.19 [Hirs1]. *For a basic set Ω_i we have*

$$W^s(\Omega_i) = \bigcup_{p \in \Omega_i} W^s(p), \ W^u(\Omega_i) = \bigcup_{p \in \Omega_i} W^u(p).$$

It follows from Lemmas 2.2.18 and 2.2.19 that if a diffeomorphism ϕ satisfies Axiom A, then any point $x \in M$ belongs to the intersection

$$W^s(p) \cap W^u(q),$$

where $p, q \in \Omega(\phi)$. Below we will apply to hyperbolic sets for a diffeomorphism of a manifold an analog of Theorem 1.2.1 (we will simply refer to Theorem 1.2.1 in this case; the needed modifications are obvious). Let us note that if $p \in \Omega(\phi)$, then we have

$$W^s(p) = \bigcup_{k \geq 0} \phi^{-k}\left(W^s_\Delta(\phi^k(p))\right), \ W^u(p) = \bigcup_{k \geq 0} \phi^k\left(W^u_\Delta(\phi^{-k}(p))\right),$$

where Δ is given by Theorem 1.2.1 for the hyperbolic set $\Omega(\phi)$ (these equalities are easy consequences of Theorem 1.2.1, we leave their proof to the reader).

Take a point $x \in W^s(p)$ and find $k_0 > 0$ such that

$$x' = \phi^{k_0}(x) \in \mathrm{Int}W^s_\Delta(\phi^{k_0}(p)),$$

where the interior is taken with respect to the inner topology of the C^1 disk $D = W^s_\Delta(\phi^{k_0}(p))$. Take an open (with respect to the inner topology of D) C^1 disk C_0 in D containing x' and consider the C^1 disk

$$C = \phi^{-k_0}(C_0) \subset W^s(p).$$

Let us prove the following auxiliary statement (Sakai applied an analog of this statement for two-dimensional diffeomorphisms in [Sak3]).

Lemma 2.2.20. *For any point $x \in W^s(p)$ and sets C_0 and C as above there exists $\epsilon = \epsilon(x, C)$ such that if for a point y we have*

$$r(\phi^k(x), \phi^k(y)) < \epsilon \tag{2.73}$$

for all $k \geq 0$, then $y \in C$.

Proof. Take $\epsilon < \Delta$, then it follows from (2.73) and from the inclusion $x \in W^s(p)$ that

$$r(\phi^k(y), \phi^k(p)) < \Delta$$

for large k, hence there exists $k_1 \geq k_0$ such that

$$\phi^{k_1}(y) \subset W_\Delta^s(\phi^{k_1}(p)).$$

The set

$$C' = \phi^{k_1}(C) = \phi^{k_1-k_0}(C_0)$$

contains a neighborhood of $x'' = \phi^{k_1}(x)$ in $W_\Delta^s(\phi^{k_1}(p))$. Since $W_\Delta^s(\phi^{k_1}(p))$ is a C^1 disk, there exists $\epsilon_1 > 0$ such that any point $z \in W_\Delta^s(\phi^{k_1}(p))$ with $r(z, x'') < \epsilon_1$ belongs to C'. Take $\epsilon = \epsilon_1$, then $\phi^{k_1}(y) \in C'$, and it follows that this ϵ has the required property.   □

Remark. A similar statement is true for a point $x \in W^u(q)$, below for $x \in W^u(q)$ we also refer to Lemma 2.2.20.

Lemma 2.2.21.   *If for a diffeomorphism ϕ statement (2) of Theorem 2.2.8 holds, then $\phi \in \mathcal{G}^1(M)$.*

Proof. Take a neighborhood W of ϕ such that any diffeomorphism $\psi \in W$ has the POTP, we claim that, for any $\psi \in W$, all periodic points are hyperbolic.

To obtain a contradiction, assume that a diffeomorphism $\psi' \in W$ has a nonhyperbolic periodic point p (to simplify notation, we assume that p is a fixed point, other cases are treated similarly). Hence, the linear mapping

$$L' = D\psi'(p) : T_p M \to T_p M$$

has an eigenvalue μ_1 with $|\mu_1| = 1$. We consider the case $\mu_1 = 1$ (it is useful for the reader to consider other possible cases). There exists a perturbation $\psi'' \in W$ of ψ' such that p is a fixed point of ψ'', and the eigenvalues $\mu_1, \ldots, \mu_n$ of the linear mapping

$$L = D\psi''(p) : T_p M \to T_p M$$

satisfy the conditions

$$\mu_1 = 1; \ |\mu_j| \neq 1, \ j \neq 1.$$

Obviously, we can find a neighborhood V of p, a diffeomorphism $\psi \in W$, and local coordinates $(y_1, \ldots, y_n)$ in V such that $p = 0$, and ψ coincides with L in V with respect to these coordinates. Let Q be the one-dimensional invariant subspace of L corresponding to μ_1, we assume that Q is given by

$$y_2 = \ldots = y_n = 0.$$

Then any point of the set

$$Q' = \{y_i = 0 : i = 2, \ldots, n\} \cap V$$

is a fixed point of ψ, while for any point $z \in V \setminus Q$ its trajectory $\{\psi^k(z) : k \in \mathbb{Z}\}$ cannot be a subset of V.

Let us show that ψ does not have the POTP. Take $\epsilon > 0$ such that the set

$$Q_\epsilon = Q' \cap \{|y_1| \le \epsilon\}$$

is a subset of V, and $\mathrm{dist}(x, Q_\epsilon) \ge \epsilon$ for $x \in M \setminus V$.

Take an arbitrary $d > 0$ and find a natural number N such that $\epsilon < Nd$. Consider a sequence $\xi = \{x_k : k \in \mathbb{Z}\}$ defined as follows. Set $x_0 = (\epsilon, 0, \ldots, 0)$. Represent $k \in \mathbb{Z}$ in the form $k = 2mN + l$ with $0 \le l \le 2N - 1$, and set $x_k = (z_k, 0, \ldots, 0)$, where

$$z_{k+1} = \begin{cases} z_k + \frac{\epsilon}{N} & \text{for } m \text{ odd}, \\ z_k - \frac{\epsilon}{N} & \text{for } m \text{ even}. \end{cases}$$

By construction, $\xi \subset Q_\epsilon$. Since any point of Q_ϵ is a fixed point of ψ, it follows from the choice of N that ξ is a d-pseudotrajectory of ψ. For any point x of Q' there is an index k such that $r(x, x_k) \ge \epsilon$. For any point $x \notin Q'$, the trajectory $\psi^k(x)$ contains a point $x' \in M \setminus V$, hence $\mathrm{dist}(x', \xi) \ge \epsilon$. This shows that ψ does not have the POTP. $\qquad\square$

Since $(5) \Rightarrow (2)$, it follows from Lemma 2.2.21 and from Theorem 2.2.9 that if for a diffeomorphism ϕ condition (5) holds, then ϕ satisfies Axiom A. By Theorem 2.2.4, to establish the implication $(5) \Rightarrow (1)$, it remains to prove that ϕ satisfies the strong transversality condition. Fix a number $\mathcal{L}$ such that any $\psi \in W$ has a Lipschitz constant not exceeding $\mathcal{L}$.

Assume that v is a point of nontransverse intersection for $W^s(p), W^u(q)$, where $p, q \in \Omega(\phi)$. Since the set $\Omega(\phi)$ is hyperbolic, we see that $v \notin \Omega(\phi)$. Let us take a small neighborhood V of v, denote $v' = \phi(v)$ and $V' = \phi(V)$. We can take V so small that

$$\phi^k(V) \cap V = \emptyset \text{ for } k \ne 0. \tag{2.74}$$

Find numbers k_+, k_- such that

$$\phi^{k_+}(v') \in \mathrm{Int}\, W_\Delta^s(\phi^{k_+}(p)), \quad \phi^{k_-}(v) \in \mathrm{Int}\, W_\Delta^u(\phi^{k_-}(q)).$$

We consider open disks

$$C_0^+ \subset \mathrm{Int}\, W_\Delta^s(\phi^{k_+}(p)), \quad C_0^- \subset \mathrm{Int}\, W_\Delta^u(\phi^{k_-}(q)),$$

and the corresponding disks

$$C^+ = \phi^{-k_+}(C_0^+), \quad C^- = \phi^{-k_-}(C_0^-),$$

containing v', v, respectively, and so small that

$$C^- \subset V, \, C^+ \subset V',$$

then by (2.74) we have

$$\phi^k(C^+) \cap V = \emptyset, k \ge 0. \tag{2.75}$$

Apply Lemma 2.2.20 to find $\epsilon > 0$ such that if

$$r(\phi^k(v'), \phi^k(x)) < \epsilon, \ k \geq 0,$$

then $x \in C^+$, and if

$$r(\phi^k(v), \phi^k(x)) < \epsilon, \ k \leq 0,$$

then $x \in C^-$.

Since v is a point of nontransverse intersection of $W^s(p)$ and $W^u(q)$, we have the inequality

$$T_v W^s(p) + T_v W^u(q) \neq T_v M.$$

Assume that the stable manifold $W^s(p)$ is m-dimensional, and the unstable manifold $W^u(q)$ is l-dimensional (note that $m, l < n$). We consider two cases.

Case 1. $l + m \geq n$. In this case, we introduce local coordinates $y = (y_1, \ldots, y_n)$ in V so that $v = 0$,

$$T_v W^s(p) + T_v W^u(q) \subset \{y_n = 0\},$$

and

$$C^- \subset \{y_n = 0\}. \tag{2.76}$$

Let $C = \phi^{-1}(C^+) \subset V$. We choose coordinates so that

$$T_v C = \{y_{m+1} = \ldots = y_n = 0\}.$$

Assume that in a neighborhood of the point v the disk C is given by the equation $\zeta = g(\eta)$, where

$$\eta = (y_1, \ldots, y_m), \ \zeta = (y_{m+1}, \ldots, y_n),$$

with $g \in C^1$, $g(0) = 0$, and

$$\frac{\partial g}{\partial \eta}(0) = 0.$$

The spaces $T_v C = T_v W^s(p)$ and $T_v C^- = T_v W^u(q)$ have dimensions m and l with $m + l \geq n$ and belong to the $(n-1)$-dimensional space $\{y_n = 0\}$. Hence, the dimension of their intersection is at least one. We assume that

$$\{y_2 = \ldots = y_n = 0\} \subset T_v C \cap T_v C^-. \tag{2.77}$$

We can perturb ϕ on V so that for the perturbed diffeomorphism ψ the following statements hold:

(1) $\psi \in W$, where W is the neighborhood from statement (5) of Theorem 2.2.8;

(2) there exists a C^1 disk C' containing v such that

$$\psi(C') = C^+,$$

and C' is given by the equation $\zeta = G(\eta)$, where $G \in C^1$, $G(0) = 0$,

$$\frac{\partial G}{\partial \eta}(0) = 0,$$

and

$$G_n(\eta) \neq 0 \text{ for } 0 < |\eta| < b \tag{2.78}$$

with some $b > 0$ (here G_n is the nth component of G).

Let us note some properties of the disk C'. Relations (2.76) and (2.78) imply that if a point $w \neq v$ belongs to $C' \cap C^-$, then $r(v, w) \geq b$.

It follows from (2.75) that

$$\psi^{k+1}(C') = \phi^k(C^+) \text{ for } k \geq 0. \tag{2.79}$$

For small $t > 0$ we consider two smooth curves,

$$h(t) = (t, 0, \ldots, 0, G_{m+1}(t, 0, \ldots, 0), \ldots, G_n(t, 0, \ldots, 0)) \subset C',$$

and

$$c(t) = (c_1(t), \ldots, c_n(t)) \subset C^-$$

such that

$$c(0) = 0, \quad \frac{d}{dt}c(0) = (1, 0, \ldots, 0).$$

For small $t > 0$ we define a sequence of points $\xi(t) = \{x_k(t) : k \in \mathbb{Z}\}$ as follows:

$$x_0(t) = c(t), \quad x_1(t) = \psi(h(t)); \quad x_k(t) = \phi^k(x_0(t)), k < 0;$$
$$x_k(t) = \phi^{k-1}(x_1(t)), \quad k > 1.$$

The diffeomorphisms ϕ and ψ coincide outside V, hence

$$r(x_{k+1}(t), \psi(x_k(t))) \leq d(t) := \mathcal{L}r(h(t), c(t)).$$

Since $d(t) \to 0$ as $t \to 0$, we have

$$d(t) < 2\nu t < d_0$$

for small $t > 0$. By our assumption, there is a point $x(t)$ such that

$$r(x_k(t), \psi^k(x(t))) \leq Ld(t) \tag{2.80}$$

(d_0 and L above are from statement (5) of our theorem).

By our choice of ϵ and the construction of $\xi(t)$, for small t the point $x(t)$ belongs to $C' \cap C^-$.

As was noted, for any point $w \in C' \cap C^-$ such that $w \neq v$, we have $r(v, w) \geq b$. It follows from (2.80) that

$$r(x_0(t), x(t)) \to 0 \text{ for } t \to 0.$$

Since $x_0(t) \to 0$ as $t \to 0$, we see that $x(t) = v$ for small $t > 0$.

Since

$$\frac{d}{dt}c_1(0) = 1; \ \frac{d}{dt}c_i(0) = 0, \ i \neq 1,$$

we have $r(x(t), c(t)) \geq t + o(t)$ for small $t > 0$. On the other hand,

$$\frac{d}{dt}c(0) = \frac{d}{dt}h(0),$$

hence $d(t) = o(t)$, and it follows that

$$r(x(t), c(t)) = r(x(t), x_0(t)) = o(t).$$

The obtained contradiction shows that case 1 is impossible.

Case 2. $l + m < n$. In this case, we can apply the transversality theorem [Hirs2] to obtain a diffeomorphism $\psi \in W$ that coincides with ϕ on $M \setminus V$, and a C^1 disk $C' \subset V$ such that

$$\psi(C') = C^+, \ C^- \cap C' = \emptyset.$$

Consider the sequence $\xi = \{x_k : k \in \mathbb{Z}\}$ defined as follows:

$$x_k = \phi^k(v), \ k \leq 0; \ x_k = \phi^{k-1}(v'), \ k \geq 1.$$

Since ψ coincides with ϕ outside V, (2.74) holds, and $\psi(v) \neq v'$, we have

$$r(x_{k+1}, \psi(x_k)) \leq d,$$

where $d = r(\psi(v), v')$. We can take ψ such that

$$r(\psi(v), v') < \min(d_0, \frac{\epsilon}{L}),$$

where d_0, L are from condition (5) of our theorem, and ϵ is given by Lemma 2.2.20 for the disks C^-, C^+.

If for a point x we have

$$r(\psi^k(x), x_k) \leq Ld,$$

then it follows from Lemma 2.2.20 that $x \in C^- \cap C'$. The obtained contradiction shows that case 2 is also impossible. This proves the implication (5) $\Rightarrow$ (1). $\square$

2.3 Shadowing in Two-Dimensional Diffeomorphisms

It was shown in the previous sections of Chap. 2 that topological (structural) stability implies the POTP (the LpSP, correspondingly). Without additional assumptions on the dynamical system, no necessary conditions for the POTP are known.

In this section, we assume that $\dim M = 2$ and that ϕ is a diffeomorphism of class C^1 satisfying Axiom A (see Subsect. 2.2.2). Below we call ϕ an (A,2)-*diffeomorphism*.

We show that in this case necessary conditions under which ϕ has the POTP (or the LpSP) have very natural geometric structure. For an (A,2)-diffeomorphism ϕ, the POTP is equivalent to a sort of topological transversality condition [Sak3], and the LpSP is equivalent to the strong transversality condition (and hence to structural stability).

We consider in this subsection also the so-called weak shadowing property (WSP) introduced in [Cor2]. Plamenevskaya [Pla2] studied this property for an (A,2)-diffeomorphism ϕ with finite nonwandering set and with a nontransverse heteroclinic trajectory joining two saddle fixed points. In this case, necessary and sufficient conditions for the WSP are very delicate, they are connected with arithmetic properties of eigenvalues of $D\phi$ at the saddle points (Theorem 2.3.4).

First we describe a result of Sakai [Sak3] giving necessary and sufficient conditions under which an (A,2)-diffeomorphism ϕ has the POTP. Let us begin with some notation and definitions.

We assume that an analog of Theorem 1.2.1 holds for the hyperbolic set $\Omega(\phi)$, take an arbitrary neighborhood U of this set, and fix the corresponding constants Δ and ν.

Take points $x \in M$ and $p \in \Omega(\phi)$ such that $x \in W^\sigma(p)$, where $\sigma \in \{s, u\}$, and $\dim W^\sigma(p) = 1$. We can find $k \in \mathbb{Z}$ such that

$$\phi^k(x) \in \operatorname{Int} W_\Delta^\sigma(\phi^k(p)).$$

Since the set

$$W_\Delta^\sigma(\phi^k(p))$$

is an embedded segment of class C^1, for a small $a > 0$, the connected component of the intersection

$$N_a(x) \cap \phi^{-k}(W_\Delta^\sigma(\phi^k(p)))$$

containing x is an embedded open segment of class C^1, we denote this set by $C_a^\sigma(x)$.

If $x \in W^s(p)$, and $a > 0$ is small enough, we denote by $B_{i,a}(x), i = 1, 2$, the open components of the set

$$N_a(x) \setminus C_a^s(x).$$

Definition 2.9 *We say that ϕ satisfies the "C^0 transversality condition" if for any point $x \in M$ such that*

$$x \in W^s(p) \cap W^u(q), \quad \text{where } p, q \in \Omega(\phi)$$

with $\dim W^s(p) = \dim W^u(q) = 1$, and for any small positive a we have

$$C_a^u(x) \cap B_{i,a}(x) \neq \emptyset, \quad i = 1, 2.$$

It is easy to see that if the condition above holds for some point x, then it holds for any point of the trajectory of x.

The main result obtained by Sakai in [Sak3] can be stated as follows.

Theorem 2.3.1. *An (A,2)-diffeomorphism ϕ has the POTP if and only if ϕ satisfies the C^0 transversality condition.*

We will prove here the "only if" part of Theorem 2.3.1, the reader is referred to [Sak3] for the remaining part of the proof.

Proof. To obtain a contradiction, let us assume that ϕ has the POTP, but the C^0 transversality condition is not satisfied.

Hence, there exists a point $z \in W^s(p) \cap W^u(q)$ such that $p, q \in \Omega(\phi)$, $\dim W^s(p) = \dim W^u(q) = 1$, and for small $a > 0$ we have

$$C_a^u(z) \cap B_{1,a}(z) = \emptyset.$$

Denote $p_k = \phi^k(p), k \in \mathbb{Z}$.

The stable and unstable Δ-disks

$$W_\Delta^s(x) \text{ and } W_\Delta^u(x)$$

through points $x \in \Omega(\phi)$ described in Theorem 1.2.1 are constructed in the case of a manifold as follows. Let $\{S(p), U(p)\}$ be the hyperbolic structure on $\Omega(\phi)$. There exist mappings

$$f_x^s : \{v \in S(x) : |v| \leq \Delta\} \to U(x)$$

and

$$f_x^u : \{v \in U(x) : |v| \leq \Delta\} \to S(x)$$

of class C^1 such that

$$W_\Delta^s(x) = \exp_x(\{(v, f_x^s(v)) : v \in S(x), |v| \leq \Delta\})$$

and

$$W_\Delta^u(x) = \exp_x(\{(v, f_x^u(v)) : v \in U(x), |v| \leq \Delta\}).$$

It follows from the proof of Theorem 1.2.1 (in the case of a compact manifold) that the norms $\|Df_x^s\|, \|Df_x^u\|$ are uniformly (in $x \in \Omega(\phi)$) bounded, and the angles between the spaces $S(x), U(x)$ are uniformly separated from 0.

Hence, there exist positive numbers $b \geq \epsilon_0$ with the following properties. If $N_k = N_{2b}(p_k)$, then

- the set

$$(W_\Delta^u(p_k) \cap N_k) \setminus p_k$$

is the union of two smooth open segments (they are denoted $W_{k,1}^u$ and $W_{k,2}^u$);

- the set

$$N_k \setminus W_\Delta^s(p_k)$$

is the union of two open two-dimensional disks (they are denoted D_k^1 and D_k^2 so that $W_{k,i}^u \subset D_k^i, i = 1, 2$);

- if $v \in W_{k,2}^u$ and $r(v, p_k) = b$, then

$$\text{dist}(v, D_k^1) \geq \epsilon_0.$$

Denote

$$W_b^\sigma(p_k) = W_\Delta^\sigma(p_k) \cap N_b(p_k), \quad \sigma = s, u.$$

We choose the indices of the segments $W_{k,i}^u$ so that $W_{k+1,i}^u \subset \phi(W_{k,i}^u)$, it is easy to understand that in this case for any point $y \in D_k^i, i = 1, 2$, we have either $\phi(y) \in D_{k+1}^i$ or $\phi(y) \notin N_{k+1}$.

Obviously, the point z fixed above can be taken so that $z \in W_b^s(p)$. It follows from our assumption that there exists an open segment $C \subset C_a^u(z)$ containing z and such that one of the sets $C \cap D_0^i, i = 1, 2$, is empty. Assume that $C \cap D_0^2 = \emptyset$. By Lemma 2.2.20, there exists $\epsilon_1 > 0$ such that if

$$r(\phi^k(x), \phi^k(z)) \leq \epsilon_1 \text{ for } k \leq 0,$$

then $x \in C$. Take $\epsilon < \min(\epsilon_0, \epsilon_1)$. Since ϕ is assumed to have the POTP, for this ϵ we can find the corresponding d.

There exist natural numbers l, m with the following properties:

$$r(\phi^l(z), p_l) < \frac{d}{2},$$

and for the point $w \in W_{l+m,2}^u$ with $r(w, p_{l+m}) = b$ the inequality

$$r(\phi^{-m}(w), p_l) < \frac{d}{2}$$

holds.

Construct a sequence $\xi = \{x_k : k \in \mathbb{Z}\}$ as follows:

$$x_k = \phi^k(z), \ k < l; \ x_k = \phi^{k-l-m}(w), \ k \geq l.$$

Since $\phi(x_{l-1}) = \phi^l(z)$ and $x_l = \phi^{-m}(w)$, our choice of l and m implies that ξ is a d-pseudotrajectory for ϕ. Assume that ξ is (ϵ, ϕ)-shadowed by a point x. Since $\epsilon < \epsilon_1$, we see that $x \in C$, and hence $x \in \overline{D_0^1}$.

If $x \in W^s(p)$ or $\phi^k(x) \in D_k^1$ for all $k \in [0, l + m]$, we have

$$r(x_{l+m}, \phi^{l+m}(x)) = r(w, \phi^{l+m}(x) \geq \epsilon_0.$$

If there exists $k \in [0, l+m]$ such that $\phi^{k-1}(x) \in D_{k-1}^1$ but $\phi^k(x) \notin D_k^1$, then $\phi^k(x) \notin N_k$, hence

$$r(x_k, \phi^k(x)) \geq b \geq \epsilon_0$$

since by construction

$$x_k \in W_b^s(p_k) \cup W_b^u(p_k) \subset N_b(p_k)$$

for $0 \leq k \leq l + m$. The obtained contradiction completes the proof.   □

Now we give conditions under which an (A,2)-diffeomorphism has the LpSP.

Theorem 2.3.2. *An (A,2)-diffeomorphism ϕ has the LpSP if and only if ϕ is structurally stable.*

Proof. By Theorem 2.2.7, the structural stability of ϕ implies the LpSP. It follows from Theorem 2.2.4 that to prove our theorem it remains to show that an (A,2)-diffeomorphism having the LpSP satisfies the geometric strong transversality condition.

To obtain a contradiction, assume that an (A,2)-diffeomorphism ϕ has the LpSP (with constants (L, d_0)) but does not satisfy the geometric STC.

Hence, there exist points $p, q \in \Omega(\phi)$ having a point z of nontransverse intersection of $W^s(p)$ and $W^u(q)$. Obviously, in this case we have

$$\dim W^s(p) = \dim W^u(q) = 1.$$

Apply Lemma 2.2.20 to find open segments $C^+ \subset W^s(p)$ and $C^- \subset W^u(q)$ containing z and a number ϵ such that if

$$r(\phi^k(z), \phi^k(x)) < \epsilon, \ k \geq 0,$$

then $x \in C^+$, and if

$$r(\phi^k(z), \phi^k(x)) < \epsilon, \ k \leq 0,$$

then $x \in C^-$.

Decreasing the segments C^+, C^-, if necessary, we can introduce coordinates (y, v) in a neighborhood of z so that z is the origin, the inclusion

$$C^- \subset \{y = 0\}$$

holds, and

$$C^+ = \{(y, g(y)) : |y| < b\}$$

with some $b > 0$, where

$$g \in C^1, \ g(0) = 0, \ \text{and} \ \frac{dg}{dy}(0) = 0.$$

If there exists $b_1 > 0$ such that

$$g(y) \equiv 0 \text{ for } |y| < b_1, \tag{2.81}$$

then ϕ does not satisfy the C^0 transversality condition, and by Theorem 2.3.1 ϕ does not have the POTP. Hence, in this case ϕ does not have the LpSP, and our theorem is proved.

If there exists no b_1 such that relation (2.81) is satisfied, then there is a sequence y'_m such that $y'_m \to 0$ and $g(y'_m) \neq 0$. For definiteness, we assume that $y'_m > 0$. Obviuosly, in this case there is a sequence y_m such that

$$g(y_m) = 0 \text{ and } g(y) \neq 0 \text{ for } y \in (y_m, y'_m)$$

(of course, it is possible that $y_m \equiv 0$). For the numbers

$$h_m = \left| \frac{dg}{dy}(y_m) \right|$$

we have $\lim h_m = 0$ as $m \to \infty$.

Let $\mathcal{L}$ be a Lipschitz constant of ϕ. Take $h > 0$ such that

$$2hL\mathcal{L} < 1 \tag{2.82}$$

and find m such that $h_m \leq h$. There exists $y_0 \in (y_m, y'_m)$ such that

$$|y_0 - y_m| < |y_0 - y'_m| \text{ and } |g(y_0)| < 2h|y_0 - y_m|.$$

Fix points $x_0 = (y_0, 0)$, $x' = (y_0, g(y_0))$, and $x_1 = \phi(x')$. Consider the sequence $\xi = \{x_k : k \in \mathbb{Z}\}$ defined as follows:

$$x_k = \phi^k(x_0), \ k \leq 0; \quad x_k = \phi^{k-1}(x_1), \ k \geq 1.$$

Since $r(x_0, x') = |g(y_0)|$, we have

$$r(x_1, \phi(x_0)) \leq \mathcal{L}|g(y_0)|,$$

it follows that

$$r(x_{k+1}, \phi(x_k)) \leq d := \mathcal{L}|g(y_0)|.$$

We can take m and y_0 such that

$$|g(y_0)| < \frac{d_0}{\mathcal{L}}$$

and $Ld < \epsilon$, then there is a point x such that

$$r(\phi^k(x), x_k) \leq Ld$$

and $x \in C^+ \cap C^-$. Since $x \in C^-$, we have $x = (y^*, 0)$, and it follows from $x \in C^+$ that $y^* \notin (y_m, y'_m)$.

Now we see that

$$|y_0 - y_m| \leq |y_0 - y^*| = r(x_0, x) \leq Ld = L\mathcal{L}|g(y_0)| <$$

$$< 2hL\mathcal{L}|y_0 - y_m|,$$

and this is impossible, since (2.82) holds. This completes the proof.   □

Let us mention a result by Sakai related to Theorems 2.3.1 and 2.3.2 (see [Sak5] for the details).

Theorem 2.3.3. *An (A,2)-diffeomorphism ϕ belongs to the C^2 interior of the set of diffeomorphisms having the POTP if and only if ϕ is structurally stable.*

The following shadowing property was introduced in [Cor2] in connection with the problem of genericity of shadowing (see the next section). To define it, take a system $\phi \in Z(M)$ and let $\xi = \{x_k : k \in \mathbb{Z}\}$ be a d-pseudotrajectory of ϕ.

Definition 2.10 *We say that ξ is "weakly (ϵ)-shadowed" by a trajectory $O(p)$ if*

$$\xi \subset N_\epsilon(O(p)). \tag{2.83}$$

Definition 2.11 *A system $\phi \in Z(M)$ has the "WSP" (the "weak shadowing property") if given $\epsilon > 0$ there exists $d > 0$ such that any d-pseudotrajectory of ϕ is weakly (ϵ)-shadowed by a trajectory of ϕ.*

It is easy to give an example of a system which does not have the WSP.

Example 2.12 Take $M = S^1$ (with coordinate $\alpha \in [0,1)$), and consider $\phi \in Z(M)$ such that $\phi(\alpha) \equiv \alpha$. Fix an arbitrary $d > 0$, and define a system $\psi \in Z(M)$ by

$$\psi(\alpha) = \alpha + \frac{\delta}{2} \pmod 1.$$

Obviously, trajectories of ψ are d-pseudotrajectories of ϕ, and for any $d \in (0, 1/2)$, $x, y \in M$ we have

$$\{\psi^k(y) : k \in \mathbb{Z}\} \not\subset N_{1/4}(x).$$

Since $\{x\}$ is the trajectory of x in ϕ, this proves that ϕ does not have the WSP.

It is also easy to show that there exist systems which have the WSP but do not have the POTP.

Example 2.13 Take again $M = S^1$ with coordinate $\alpha \in [0,1)$, and consider the system $\phi \in Z(M)$ generated by the mapping

$$f(\alpha) = \alpha + \beta \pmod 1,$$

where β is irrational. Every trajectory of ϕ is dense in S^1, this obviously implies the WSP.

Assume that ϕ has the POTP, take $\epsilon = 1/4$, and find the corresponding d (see Definition 1.3). There exists a rational $\gamma = l/m$ such that $|\gamma - \beta| < d$. Consider the system ψ generated by

$$g(\alpha) = \alpha + \gamma \pmod 1,$$

obviously, trajectories of ψ are d-pseudotrajectories of ϕ. It follows from the structure of γ that any $x \in S^1$ is a fixed point of ψ^m. Since for $p \in S^1$ the sets $\{\phi^{km}(p)\}$ are dense in S^1, we see that for any pair x, p there is $k \in \mathbb{Z}$ such that

$$r(\phi^{km}(p), \psi^{km}(x)) \geq \frac{1}{4}.$$

Hence, ϕ does not have the POTP.

Plamenevskaya studied conditions under which an $(A,2)$-diffeomorphism has the WSP [Pla2]. She considered this problem for an Axiom A diffeomorphism of the two-dimensional torus T^2 with finite nonwandering set. She showed that, in contrast with the simple geometric condition for the POTP given in Theorem 2.3.1, necessary and sufficient conditions for the WSP are very delicate. Let us describe the example of Plamenevskaya.

Represent T^2 as the square $[-2, 2] \times [-2, 2]$ with identified opposite sides. Consider the metric r generated by the supnorm.

We study a diffeomorphism ϕ of T^2 with the following properties. The nonwandering set $\Omega(\phi)$ is the union of 4 hyperbolic fixed points,

$$\Omega(\phi) = \{o, p_1, p_2, s\},$$

where the point s is asymptotically stable, the point o is completely unstable (i. e., it is asymptotically stable for ϕ^{-1}), and p_1, p_2 are saddles.

It is assumed that with respect to coordinates $(u, v) \in [-2, 2] \times [-2, 2]$ the following conditions hold:

(c1)
$$o = (1, 2), \ p_1 = (-1, 0), \ p_2 = (1, 0), \ s = (-1, 2);$$

(c2)
$$W^s(p_1) = W^u(p_2) = [-2, 2] \times \{0\};$$
$$W^u(p_1) = \{-1\} \times (-2, 2), \ W^s(p_2) = \{1\} \times (-2, 2);$$

(c3) there exist neighborhoods O_1, O_2 of p_1, p_2 such that

$$\phi(x) = p_i + D\phi(p_i)(x - p_i) \text{ in } O_i, \ i = 1, 2;$$

(c4) there exists a neighborhood O of the point $z = (0, 0)$ such that

$$\phi(O) \subset O_1, \ \phi^{-1}(O) \subset O_2,$$

and ϕ^{-1} is affine on $\phi(O)$;

(c5) the eigenvalues of $D\phi(p_1)$ are $-\mu, \nu$ with $\mu > 1, \nu \in (0, 1)$, the eigenvalues of $D\phi(p_2)$ are $-\lambda, \kappa$ with $\lambda \in (0, 1), \kappa > 1$.

It follows from conditions (c2), (c3), and (c5) that

$$\phi(u, v) = (\nu(u + 1) - 1, -\mu v) \text{ in } O_1$$

and

$$\phi(u, v) = (\kappa(u - 1) + 1, -\lambda v) \text{ in } O_2.$$

Note that by Theorem 2.3.1 the diffeomorphism ϕ does not have the POTP since it does not satisfy the C^0 transversality condition.

Theorem 2.3.4. *The diffeomorphism ϕ has the WSP if and only if the number*

$$\frac{\log \lambda}{\log \mu} \tag{2.84}$$

is irrational.

Remark. It is worth noting that the value (2.84) appeared in the qualitative theory of dynamical systems as a functional modulus of local topological conjugacy in a neighborhood of a heteroclinic curve joining two saddles [DM].

Proof. First we show that if the value (2.84) is irrational, then ϕ has the WSP. Note that in this part of the proof we do not apply condition (c4). Below $\xi = \{x_k : k \in \mathbb{Z}\}$ is a d-pseudotrajectory of ϕ (we do not repeat this when we impose conditions on d).

Fix $\epsilon > 0$. To simplify notation, we will find d corresponding to 2ϵ instead of ϵ in Definition 2.11. We take ϵ so small that

$$N_\epsilon(p_i) \subset O_i, \ i = 1, 2.$$

Consider the sets

$$V_1 = (-1 - \epsilon, -1 + \epsilon) \times \left(\frac{\mu^{-2}\epsilon}{2}, \epsilon\right)$$

and

$$V_2 = (1 - \epsilon, 1 + \epsilon) \times \left(\frac{\lambda^2 \epsilon}{2}, \epsilon\right).$$

It is easy to see that there exists a number $d_1 \in (0, \epsilon)$ and neighborhoods $V(s), V(o)$ of the points s, o such that

$$V(s) \subset N_\epsilon(s), \ V(o) \subset N_\epsilon(o),$$

and

$$N_{d_1}(\phi(V(s))) \subset V(s), \ N_{d_1}(\phi^{-1}(V(o))) \subset V(o).$$

It follows that if $d < d_1$ and $x_k \in V(s)$, then $x_l \in V(s)$ for $l > k$ (similarly, if $x_k \in V(o)$, then $x_l \in V(o)$ for $l < k$).

The set $\overline{V_1}$ belongs to the basin of attraction of the point s, and the set $\overline{V_2}$ belongs to the basin of attraction of o for ϕ^{-1}. There exist numbers $T_1 > 0$ and $d_2 \in (0, d_1)$ such that

$$\phi^{T_1}(N_{d_2}(V_1)) \subset V(s), \ \phi^{-T_1}(N_{d_2}(V_2)) \subset V(s),$$

and if $d < d_2$ and $r(x_0, y) < d_2$, then $r(x_k, \phi^k(y)) < d_1$ for $|k| \leq T_1$.

We see that if $d < d_2$, $x_k \in V_1$, and $r(y, x_k) < d_2$, then

$$\{x_l : l \geq k\} \subset N_{2\epsilon}(O^+(y))$$

(we recall that $O^+(X)$ and $O^-(X)$ are the positive and the negative ϕ-trajectories of a set X). Similarly, if $d < d_2$, $x_k \in V_2$, and $r(y, x_k) < d_2$, then

$$\{x_l : l \leq k\} \subset N_{2\epsilon}(O^-(y)).$$

In addition, we take d_2 such that

$$d_2 < \min \left(\frac{\mu^{-2}\epsilon}{2}, \frac{\lambda^2 \epsilon}{2} \right). \tag{2.85}$$

Now we fix the following subsets of V_1 and V_2:

$$W_1 = (-1 - d_2, -1 + d_2) \times \left(\frac{\mu^{-2}\epsilon}{2}, \epsilon \right)$$

and

$$W_2 = (1 - d_2, 1 + d_2) \times \left(\frac{\lambda^2 \epsilon}{2}, \epsilon \right).$$

Take $d_3 > 0$ such that

$$\nu d_2 + d_3 < d_2 \text{ and } 2d_3(\mu + 1) < \epsilon. \tag{2.86}$$

We claim that this d_3 has the following property: if $d < d_3$, $x_{k_0} \in N_{d_2}(p_1)$, and $x_{k_1} \notin N_\epsilon(p_1)$ for some $k_1 > k_0$, then there exists $k \in (k_0, k_1)$ such that $x_k \in W_1$. Indeed, let $x_k = (u_k, v_k)$. We can consider k_1 such that $x_k \in N_\epsilon(p_1)$ for $k < k_1$, it follows from the choice of ϵ that for $k_0 \leq k < k_1$ we have

$$\phi(u_k, v_k) = (\nu(u_k + 1) - 1, -\mu v_k).$$

Hence, if

$$u_k \in (-1 - d_2, -1 + d_2) \text{ and } r(\phi(x_k), x_{k+1}) < d_3,$$

then

$$|u_{k+1} + 1| < \nu d_2 + d_3 < d_2.$$

Thus, if x_k belongs to the region $|u + 1| < d_2$, then either x_{k+1} belongs to the same region or x_{k+1} does not belong to $N_\epsilon(p_1)$. If

$$|v_k| < \frac{\mu^{-2}\epsilon}{2},$$

then

$$|v_{k+1}| < \mu \frac{\mu^{-2}\epsilon}{2} + d_3 = \frac{\mu^{-1}\epsilon}{2} + d_3,$$

and we deduce from (2.86) that

$$|v_{k+2}| < \frac{\epsilon}{2} + \mu d_3 + d_3 < \epsilon.$$

In addition, if $x_k \in N_\epsilon(p_1)$ and $|v_k| > d_3$, then the signs of v_k and v_{k+1} are opposite. Since (2.85) holds, this establishes the property of d_3 formulated above.

We assume, in addition, that d_3 satisfies the conditions

$$\kappa^{-1}d_2 + d_3 < d_2 \text{ and } d_3(\lambda^{-1} + 1) < \frac{\epsilon}{2}$$

similar to (2.86). The same reasons as above show that if $d < d_3$, $x_{k_0} \in N_{d_2}(p_2)$, and $x_{k_1} \notin N_\epsilon(p_2)$ for some $k_1 < k_0$, then there exists $k \in (k_1, k_0)$ such that $x_k \in W_1$.

Consider the set

$$F = \left[\left(1 - \epsilon, 1 - \frac{\kappa^{-2}\epsilon}{2}\right) \cup \left(1 + \frac{\kappa^{-2}\epsilon}{2}, 1 + \epsilon\right)\right] \times \{0\}.$$

Since $\overline{F} \subset W^s(p_1)$, there exists $d_4 \in (0, d_3)$ and a number $T_2 > 0$ such that for the set

$$W = F \times (-d_4, d_4)$$

the inclusion

$$\phi^{T_2}(\overline{F}) \subset N_{d_2}(p_1)$$

holds. In addition, we take d_4 so small that, for $d < d_4$, the inclusion $x_0 \in W$ implies the inclusion $x_{T_2} \in N_{d_2}(p_1)$, and the inequality $r(x_0, y) < 2d_4$ implies the inequalities $r(x_k, \phi^k(y)) < \epsilon$ for $|k| \leq T_2$.

Repeating the arguments applied to find d_3, we can find $d_5 \in (0, d_4)$ such that if $d < d_5$, $x_{k_0} \in N_{d_4}(p_2)$, and $x_{k_1} \notin N_\epsilon(p_2)$ for some $k_1 > k_0$, then there exists $k \in (k_0, k_1)$ such that $x_k \in W$.

Set $Q_1 = N_{d_2}(p_1)$ and $Q_2 = N_{d_4}(p_2)$. We claim that ϕ has the LpSP on each of the sets $M_i = T^2 \setminus Q_i, i = 1, 2$. Consider the sets

$$M_1' = T^2 \setminus N_{d_2/2}(p_1), \; M_2' = T^2 \setminus N_{d_4/2}(p_2).$$

It is easy to see (the details are left to the reader) that an analog of Lemma 2.2.16 holds for each of the sets $M_i', i = 1, 2$. This can be done so that $C = 1$ in statement (3), and all the statements are true for points $p \in M_i$ such that $\phi(p), \phi^{-1}(p) \in M_i'$. Of course, we can find $d > 0$ with the following property: if ξ is a d-pseudotrajectory of ϕ such that $\xi \subset M_i$, then $\phi(x_k), \phi^{-1}(x_k) \in M_i'$ for any k. Now it remains to repeat the proof of Theorem 2.2.7 to establish our claim. Let d_0, L be the corresponding constants (for both sets M_i). Take $d \in (0, \min(d_0, d_5))$ and such that $dL < 2\epsilon$.

We want to show that for any d-pseudotrajectory ξ of ϕ there is a point y such that $\xi \subset N_{2\epsilon}(O(y))$. It follows from our previous considerations that it is enough to consider a d-pseudotrajectory ξ such that

$$\xi \cap Q_i \neq \emptyset, \; i = 1, 2.$$

We assume that the d-pseudotrajectory ξ we work with is such that $x_k \in N_\epsilon(o)$ for negative k with large $|k|$, and $x_k \in N_\epsilon(s)$ for large positive k (other possible cases ase treated similarly). Then it follows from the choice of d that ξ intersects the sets W, W_1, W_2. Set

$$k_1 = \max\{k : x_k \in W_2\}, \quad k_2 = \max\{k : x_k \in W\}, \quad k_3 = \max\{k : x_k \in W_1\},$$

then we have $k_1 < k_2 < k_3$.

By the choice of T_2, the inclusion $x_{k_2+T_2} \in Q_2$ holds. Our condition (c3) on ϕ implies that

$$\{x_k : k_1 \leq k \leq k_2\} \subset N_\epsilon(p_2)$$

and

$$\{x_k : k_2 + T_2 \leq k \leq k_3\} \subset N_\epsilon(p_1). \tag{2.87}$$

Let $x_{k_i} = (u_i, v_i), i = 1, 2, 3$. We can find $d' > 0$ such that, for the sets

$$X_1 = (-1 - d_2, -1 + d_2) \times (v_3 - d', v_3 + d')$$

and

$$X_2 = (1 - d_2, 1 + d_2) \times (v_1 - d', v_1 + d'),$$

the inclusions $X_i \subset W_i, i = 1, 2$, hold.

By the choice of d_2, for any point

$$x \in N_\epsilon(p_2) \cap \left[\bigcup_{j \geq 0} \phi^j(X_2)\right]$$

we have

$$\{x_k : k \leq k_1\} \subset N_{2\epsilon}(O(x)).$$

Set

$$S_2 = \left[\bigcup_{j \geq 0} \phi^j(X_2)\right] \cap [(u_2 - d_4, u_2 + d_4) \times (-d_4, d_4)].$$

The inclusions

$$S_2 \subset W \subset N_\epsilon(p_2)$$

hold, and it follows from the choice of d_4 that for any point $x \in S_2$ we have

$$\{x_k : k \leq k_3\} \subset N_{2\epsilon}(O(x))$$

(here we take (2.87) into account). Thus, to prove our statement it remains to find a point $x \in S_2$ such that its trajectory intersects the set X_1, since in this case we have

$$\{x_k : k \geq k_3\} \subset N_{2\epsilon}(O(x)).$$

Set

$$S_1 = Q_1 \cap \left[\bigcup_{j \geq 0} \phi^{-j}(X_1)\right].$$

If we show that

$$S_1 \cap \phi^{T_2}(S_2) \neq \emptyset, \tag{2.88}$$

this will complete the proof of the first part of Theorem 2.3.4.

Take the points $y_j = (u_2, (-1)^j \lambda^j v_1), j \geq 0$. Property (c3) implies that $y_j \in S_2$ for large j. Set $z_j = \phi^{T_2}(y_j)$, let $z_j = (u'_j, v'_j)$. By condition (c2), the linear mapping $D\phi^{T_2}(u_2, 0)$ has an eigenvector $(1,0)$, let the matrix of this mapping be

$$\begin{pmatrix} a & b \\ 0 & c \end{pmatrix}.$$

Then we can write

$$v'_j = c(-1)^j \lambda^j v_1 + \alpha(j)\lambda^j, \tag{2.89}$$

where $c \neq 0$ and $\alpha(j) \to 0$ as $j \to \infty$.

Obviously, for $\Delta > 0$ small enough there exists m_0 such that

$$S_1 \cap [(-1 - d_2, -1 + d_2) \times (-\Delta, \Delta)] =$$

$$= \bigcup_{m \geq m_0} (-1 - d_2, -1 + d_2) \times ((-1)^m \mu^{-m} v_3 - \mu^{-m} d', (-1)^m \mu^{-m} v_3 + \mu^{-m} d').$$

Since there exists j_0 such that

$$z_j \in (-1 - d_2, -1 + d_2) \times (-\Delta, \Delta)$$

for $j \geq j_0$, it is enough to find $j \geq j_0$ and $m \geq m_0$ such that

$$|v'_j - (-1)^m \mu^{-m} v_3| < \mu^{-m} d'$$

or

$$|(-1)^m \mu^m v'_j - v_3| < d'. \tag{2.90}$$

Since the numbers $\log \lambda$ and $\log \mu$ are incommensurable and have different signs, the set

$$\{(-1)^{m+j} \mu^m \lambda^j : m, j > 0\}$$

is dense in $\mathbb{R}$. Now relation (2.89) implies that there exists a solution of (2.90). This completes the proof of the first part of Theorem 2.3.4.

Now we show that if the value (2.84) is rational, then ϕ does not have the WSP.

Assume that

$$\frac{\log \lambda}{\log \mu} = -\frac{r}{s}$$

for some natural numbers r, s, i.e.,

$$\lambda = \gamma^r, \ \mu = \gamma^{-s} \text{ with } \gamma \in (0,1).$$

Take $a > 0$ such that $N_{2a}(p_1) \subset O_1$ and $N_{2a}(p_2) \subset O_2$. Let $\epsilon > 0$ be so small that

$$N_\epsilon(z) \subset O \tag{2.91}$$

(recall that $z = (0,0)$) and

$$\epsilon < \frac{a}{2} \gamma^{1/2}(1 - \gamma^{1/2}). \tag{2.92}$$

To obtain a contradiction, we assume that there exists $d > 0$ such that any d-pseudotrajectory of ϕ is weakly ϵ-shadowed. Let us construct a d-pseudotrajectory as follows. Fix the points $z_1 = (-1, \gamma^{1/2}a)$ and $z_2 = (1, a)$. We have

$$\phi^{-k}(z_1) \to p_1, \quad \phi^k(z_2) \to p_2, \quad \phi^k(z) \to p_1, \quad \text{and} \quad \phi^{-k}(z) \to p_2$$

as $k \to \infty$. Hence, there exist numbers $k_i, i = 1, 2, 3, 4$, such that the set

$$\xi = \{\phi^k(z_1) : k \geq k_1\} \cup \{\phi^k(z) : k_2 \geq k \geq k_3\} \cup \{\phi^k(z_2) : k \leq k_4\}$$

is a d-pseudotrajectory of ϕ.

Assume that ξ is weakly (ϵ)-shadowed by $O(p)$. Then $O(p) \cap N_\epsilon(z) \neq \emptyset$, we assume that $p \in N_\epsilon(z)$. In addition, $O(p) \cap N_\epsilon(z_2) \neq \emptyset$.

It follows from property (c5) of ϕ and from (2.91) that the coordinate u decreases along the trajectory of p, hence $O^-(p) \cap N_\epsilon(z_2) \neq \emptyset$. Then

$$p \in N_\epsilon(z) \cap \left[\bigcup_{j>0} \phi^j(N_\epsilon(z_2))\right].$$

By (c3) and (c4), for $x \in N_\epsilon(z_2) \cap \phi^{-j}(N_\epsilon(z))$ with $j > 0$ we have

$$\phi^j(x) = p_2 + (D\phi(p_2))^j(x),$$

hence

$$N_\epsilon(z) \cap \left[\bigcup_{j>0} \phi^j(N_\epsilon(z_2))\right] \subset$$

$$\subset \bigcup_{j>0}(-\epsilon, \epsilon) \times ((-1)^j a\gamma^{rj} - \gamma^{rj}\epsilon, (-1)^j a\gamma^{rj} + \gamma^{rj}\epsilon).$$

It follows that

$$p \in \bigcup_{j>0}(-\epsilon, \epsilon) \times ((-1)^j a\gamma^{rj} - \gamma^{rj}\epsilon, (-1)^j a\gamma^{rj} + \gamma^{rj}\epsilon).$$

Similarly,

$$p \in \bigcup_{m>0}(-\epsilon, \epsilon) \times ((-1)^m a\gamma^{sm+1/2} - \gamma^{sm}\epsilon, (-1)^j a\gamma^{sm+1/2} + \gamma^{sm}\epsilon).$$

Hence, for some $j, m > 0$ the inequality

$$|a\gamma^{rj} - a\gamma^{sm+1/2}| < (\gamma^{rj} + \gamma^{sm})\epsilon$$

holds.

We deduce from (2.92) that the last inequality is equivalent to the inequality

$$1 - \gamma^{sm-rj+1/2} < \gamma^{1/2}(1 - \gamma^{1/2})$$

if $sm \geq rj$, and to the inequality

$$\gamma^{1/2}(1 - \gamma^{rj-sm-1/2}) < \gamma^{1/2}(1 - \gamma^{1/2})$$

if $sm < rj$. None of these two inequalities is true. The obtained contradiction completes the proof of Theorem 2.3.4. $\qquad\square$

2.4 C^0-Genericity of Shadowing for Homeomorphisms

Let X be a topological space. A subset Y of X is called *residual* if Y contains a countable intersection of open and dense subsets of X. If P is a property of elements of X, we say that this property is *generic* if the set

$$\{x \in X : x \text{ satisfies } P\}$$

is residual. Sometimes in this case we say that a generic element of X satisfies P.

The space X is called a *Baire space* if every its residual subset is dense in X. The classical theorem of Baire says that every complete metric space is a Baire space.

We consider the space $Z(M)$ of discrete dynamical systems with the metric ρ_0 introduced in Sect. 2.1. It is an easy exercise for the reader to show that the space $Z(M)$ is complete (and hence it is a Baire space).

The main result of this section is the following statement [Pi5].

Theorem 2.4.1. *A generic system in $Z(M)$ has the POTP.*

Remark. The genericity of the POTP for $M = S^1$ was proved by Yano in [Y2]. Odani [Od] established the genericity of the POTP in the case $n =\dim M \leq 3$. His proof was based on the possibility of approximation of an arbitrary system $\phi \in Z(M)$ by a diffeomorphism [Mu, Wh] and on the theorem of Shub [Shu1] on the C^0-density of structurally stable diffeomorphisms. Unfortunately, in the case $\dim M > 3$ not every homeomorphism is C^0-approximated by diffeomorphisms [Mu], thus this method is not applicable if the dimension of M is arbitrary.

Note also that the genericity of the weak shadowing property (see the previous section) in $Z(M)$ for any $\dim M$ was established in [Cor2].

The proof of Theorem 2.4.1 is based on the theory of topological transversality [Ki, Q]. We do not give it in this book.

3. Systems with Special Structure

3.1 One-Dimensional Systems

Consider the circle S^1 with coordinate $x \in [0,1)$, we denote by r the distance on S^1 induced by the usual distance on the real line.

We fix a homeomorphism ϕ_0 of S^1 and consider the family of homeomorphisms

$$[\phi_0] = \{\phi_0^m : m \in \mathbb{N}\}.$$

It was mentioned that a homeomorphism ϕ has the POTP if and only if the homeomorphism ϕ^m with some natural m has this property (see Sect. 1.1). Thus, if one homeomorphism of the family $[\phi_0]$ has the POTP, then all homeomorphisms of this family have it. Obviously, we may restrict our consideration to homeomorphisms preserving orientation.

For a homeomorphism ϕ we denote by $\mathrm{Fix}(\phi)$ the set of fixed points of ϕ. This set is closed.

Let $P : \mathbb{R} \to S^1$ be the mapping defined by the relations

$$P(x) \in [0,1), \ P(x) \equiv x \pmod 1,$$

with respect to the considered coordinates on S^1.

To study the dynamical system generated by a homeomorphism ϕ preserving orientation, we introduce the so-called *lift* Φ of ϕ [Ni1], i.e., a continuous increasing function

$$\Phi : \mathbb{R} \to \mathbb{R}$$

such that
  (a) the function $\Phi(t) - t$ is 1-periodic;
  (b) $P \circ \Phi = \phi \circ P$.

We fix a lift Φ of ϕ such that $\Phi(0) \in [0,1)$. It is well known (see Chap. 1 of the book [Ni1]) that for any point $x \in \mathbb{R}$ there exists the limit

$$\lim_{n \to \infty} \frac{\Phi^n(x)}{n},$$

and this limit does not depend on x. This number is called the *rotation number* of ϕ, we denote it $\mu(\phi)$. The main property of the rotation number is the following one: ϕ has a periodic point if and only if the number $\mu(\phi)$ is rational [Ni1].

For two points $a, b \in S^1$ (we identify them with the corresponding points of $[0,1)$), we denote by (a, b) the open arc of S^1 corresponding to the set $(a, b) \subset [0, 1)$ if $a < b$, and to the set $(a, 1) \cup [0, b) \subset [0, 1)$ if $b < a$. Similar notation is applied for closed arcs.

Let a, b be two fixed points of a homeomorphism ϕ preserving orientation. Assume that $(a, b) \cap \mathrm{Fix}(\phi) = \emptyset$. We say that (a, b) is an *r-interval* if

$$\Phi(t) - t > 0 \text{ for } t \in (a, b) \text{ or for } (a, 1) \cup [0, b), \text{ correspondingly.}$$

Otherwise, we say that (a, b) is an *l-interval*. Obviously, if (a, b) is an r-interval, then for $x \in (a, b)$ we have

$$\phi^k(x) \to a, \ k \to -\infty; \ \phi^k(x) \to b, \ k \to \infty,$$

and an l-interval has a similar property.

Now we state necessary and sufficient conditions under which a homeomorphism of S^1 has the POTP [Pla1]. The same problem was solved earlier by Yano [Y2], but we prefer to follow [Pla1] here, since this allows us to apply similar methods to treat both shadowing and limit shadowing for homeomorphisms of S^1.

Theorem 3.1.1. *A homeomorphism ϕ_0 of S^1 has the POTP if and only if the family $[\phi_0]$ contains a homeomorphism ϕ such that*

(a) ϕ preserves orientation;

(b) the set $\mathrm{Fix}(\phi)$ is nowhere dense and contains at least two points;

(c) for any two r-intervals (l-intervals) (a, b) and (c, d) there exist l-intervals (correspondingly, r-intervals) (p, q) and (s, t) such that

$$(p, q) \subset (b, c) \text{ and } (s, t) \subset (d, a).$$

We begin the proof of Theorem 3.1.1 by some auxiliary lemmas.

One easily proves the following statement (compare with examples 1.4 and 1.18).

Lemma 3.1.1. *If for a homeomorphism ϕ the set $\mathrm{Fix}(\phi)$ contains a nondegenerate arc (i.e., an arc that is not a point), then ϕ has neither the POTP nor the LmSP.*

Below, in Lemmas 3.1.2 and 3.1.3, we assume that ϕ has the POTP. By Lemma 3.1.1, it is enough to consider homeomorphisms ϕ such that the set $\mathrm{Fix}(\phi)$ contains no arcs (i.e., this set is nowhere dense).

Lemma 3.1.2. *If a homeomorphism ϕ of S^1 preserves orientation and has the POTP, then ϕ has periodic points.*

Proof. To obtain a contradiction, assume that ϕ has the POTP but its set of periodic points is empty. Then $\mathrm{Fix}(\phi) = \emptyset$, and we can find $\epsilon > 0$ such that

$$r(x, \phi(x)) > 3\epsilon \text{ for } x \in S^1. \tag{3.1}$$

Take a number d given for this ϵ by Definition 1.3 and such that $d < \epsilon$. Let us construct a d-pseudotrajectory ξ of ϕ as follows. Take a point $x \in S^1$ and consider the set $O(x)$, the trajectory of x. By our assumption, $\phi^k(x) \neq \phi^l(x)$ for $k \neq l$, hence the infinite set $O(x)$ contains two points $\phi^m(x)$ and $\phi^n(x)$ such that $r(\phi^m(x), \phi^n(x)) < d$. Assume that $m < n$ and denote $x_0 = \phi^m(x)$, then $\phi^n(x) = \phi^N(x_0)$, where $N = n - m > 0$. It follows that

$$r(x_0, \phi^N(x_0)) < d. \tag{3.2}$$

Represent $k \in \mathbb{Z}$ in the form $k = lN + s$, where $l \in \mathbb{Z}, 0 \leq s < N$, and set $x_k = \phi^s(x_0)$. Then $\xi = \{x_k\}$ is a d-pseudotrajectory of ϕ.

Indeed, for $s < N - 1$ we have

$$\phi(x_k) = \phi(\phi^s(x_0)) = \phi^{s+1}(x_0) = x_{k+1},$$

and for $s = N - 1$ we have

$$x_k = \phi^{N-1}(x_0), \ \phi(x_k) = \phi^N(x_0), \ x_{k+1} = x_0,$$

hence it follows from (3.2) that $r(\phi(x_k), x_{k+1}) < d$ for all k.

By our assumption, there exists a point $x \in S^1$ that (ϵ)-shadows ξ. Take points $x_0', x' \in [0, 1) \subset \mathbb{R}$ such that $P(x_0') = x_0, P(x') = x$. Since $r(x, x_0) < \epsilon$, the length of one of the arcs joining x_0 with x is less than ϵ. We assume for definiteness that 0 does not belong to this arc, then

$$|x_0' - x'| < \epsilon. \tag{3.3}$$

Now we construct a sequence $\xi' = \{x_k' \in \mathbb{R} : k \in \mathbb{Z}\}$ such that $P(x_k') = x_k$ and

$$x_k' < x_{k+1}' < x_k' + 1, \ k \in \mathbb{Z}. \tag{3.4}$$

The point x_0' is fixed, it follows from $x_k \neq x_{k+1}$ that the sequence is uniquely determined.

The set Fix(ϕ) is empty, hence $\Phi(t) - t \neq 0$ for $t \in [0, 1)$. It follows that $\Phi(t) - t \notin \mathbb{Z}$ for $t \in \mathbb{R}$, and now the inequality $0 < \Phi(0) < 1$ implies that

$$0 < \Phi(t) - t < 1 \text{ for } t \in \mathbb{R}. \tag{3.5}$$

We claim that

$$|\Phi^k(x') - x_k'| < \epsilon \text{ for } k \in \mathbb{Z}. \tag{3.6}$$

We prove (3.6) by induction (and consider only the case $k \geq 0$). For $k = 0$, (3.6) coincides with (3.3). Assume that (3.6) holds for some $k \geq 0$. Then it follows from (3.5) that

$$x_k' + \epsilon + 1 > \Phi^k(x') + 1 > \Phi^{k+1}(x') > \Phi^k(x') > x_k' - \epsilon. \tag{3.7}$$

Since $P(\Phi^{k+1}(x')) = \phi^{k+1}(x)$, and the point x (ϵ)-shadows ξ, the point $\Phi^{k+1}(x')$ belongs to the ϵ-neighborhood of a point $x'_{k+1} + l, l \in \mathbb{Z}$. We have to show that $l = 0$. Assume that

$$|\Phi^{k+1}(x') - (x'_{k+1} + l)| < \epsilon \text{ for some } l \neq 0.$$

Then it follows from (3.7) that

$$x'_k + 1 - 2\epsilon > x'_{k+1} + l > x'_k - 2\epsilon.$$

We deduce from (3.4) and from

$$|x'_{k+1} + l - x'_{k+1}| = |l| \geq 1$$

that $x'_{k+1} + l \notin [x'_k, x'_k + 1]$, hence either

$$x'_k - 2\epsilon < x'_{k+1} + l < x'_k$$

or

$$x'_k + 1 < x'_{k+1} + l < x'_k + 1 + 2\epsilon.$$

This means that $r(x_k, x_{k+1}) < 2\epsilon$ on S^1.

By the choice of ϵ and d, we have

$$r(x_k, x_{k+1}) \geq r(x_k, \phi(x_k)) - r(\phi(x_k), x_{k+1}) > 3\epsilon - d > 2\epsilon.$$

The obtained contradiction establishes (3.6). It follows from (3.6) that

$$\lim_{k \to \infty} \frac{\Phi^k(x)}{k} = \lim_{k \to \infty} \frac{x'_k}{k}.$$

The equalities $P(x'_0) = P(x'_N) = x_0$ imply that $x'_N - x'_0 = m$ for some natural m, this gives $x'_{kN} = x'_0 + km$, and we see that

$$\mu(\phi) = \lim_{k \to \infty} \frac{x'_0 + km}{kN} = \frac{m}{N}.$$

Hence, the rotation number of ϕ is rational, and ϕ must have periodic points. The obtained contradiction proves our lemma.   □

Thus, if ϕ_0 has the POTP, then the family $[\phi_0]$ contains a homeomorphism preserving orientation and having periodic points. Hence, this family contains a homeomorphism ϕ such that the set $\text{Fix}(\phi)$ is not empty. For this diffeomorphism ϕ, the following statement holds.

Lemma 3.1.3. *If $a, b \in \text{Fix}(\phi)$, then either $(a, b) \cap \text{Fix}(\phi) = \emptyset$ or $\Phi(t) - t$ changes sign on (a, b).*

Proof. To obtain a contradiction, we assume that there exist points $a, b, c \in \text{Fix}(\phi)$ such that $c \in (a, b)$ and $\Phi(t) - t \geq 0$ on (a, b). Let $a < b$.

Take $\epsilon > 0$ such that

$$2\epsilon < \min(c - a, b - c).$$

Find $d > 0$ corresponding to this ϵ by Definition 1.3 (recall that ϕ has the POTP by our assumption). Let us construct a d-pseudotrajectory of ϕ as follows. Since the set $\mathrm{Fix}(\phi)$ is nowhere dense, there is a point $y_0 \in (a, b)$ such that

$$c - \epsilon > y_0 > a + \epsilon, \text{ and } y_0 \notin \mathrm{Fix}(\phi).$$

Then we can find points $a_0, b_0 \in \mathrm{Fix}(\phi)$ such that

$$y_0 \in (a_0, b_0) \subset (a, b),$$

and $\Phi(t) - t \neq 0$ on (a_0, b_0). It follows that (a_0, b_0) is an r-interval, hence

$$\phi^k(y_0) \to b_0 \text{ for } k \to \infty.$$

Take $k_1 > 0$ such that

$$r(\phi^{k_1}(y_0), b_0) < \frac{d}{2}.$$

Find $y_1 \notin \mathrm{Fix}(\phi)$ such that

$$b_0 + \frac{d}{4} < y_1 < b_0 + \frac{d}{2}.$$

There exists an r-interval (a_1, b_1) containing y_1, it follows that

$$\phi^k(y_1) \to b_1 \text{ for } k \to \infty.$$

Similarly we find a number k_2 and a point y_2, and so on. Obviously, in the course of this process we obtain a point b_n such that $b_n - c > \epsilon$.

Set $x_k = \phi^k(x_0)$ for $k \leq 0$,

$$x_k = \phi^k(y_1) \text{ for } 1 \leq k \leq k_1,$$

$$x_k = \phi^{k-k_1-1}(y_2) \text{ for } k_1 + 1 \leq k \leq k_1 + k_2 + 1,$$

and so on, and $x_{k'+k} = \phi^k(y_n)$ for $k \geq 0$, where $k' = k_1 + \ldots + k_{n-1} + n - 1$. It follows from our construction that $\xi = \{x_k\}$ is a d-pseudotrajectory of ϕ. Let x be a point that (ϵ)-shadows ξ. By the choice of x_0, $x \in (a, c)$. Then $r(\phi^k(x), b_n) > \epsilon$ for $k \geq 0$, while $x_k \to b_n$, and x cannot (ϵ)-shadow ξ. The obtained contradiction completes the proof. $\qquad\square$

Remark. The same reasons show that if ϕ is a preserving orientation homeomorphism of S^1 such that the set $\mathrm{Fix}(\phi)$ consists of one point, then ϕ does not have the POTP.

Obviously, Lemmas 3.1.1 - 3.1.3 imply the necessity of conditions of Theorem 3.1.1. Now we prove their sufficiency.

Fix $\epsilon > 0$. It follows from conditions (b) and (c) that we can find fixed points $a_1, b_1, \ldots, a_{2N}, b_{2N}$ of ϕ with the following properties:

$$0 \leq a_1 < b_1 \leq a_2 < \ldots \leq a_{2N} < b_{2N} \leq 1;$$

- the length of any segment $[b_1, a_2], \ldots, [b_{2N}, a_1]$ is less than ϵ (note that some of these segments may be points);
- $\Phi(t) - t \neq 0$ on (a_n, b_n), $1 \leq n \leq 2N$;
- if (a_n, b_n) is an r-interval, then (a_{n+1}, b_{n+1}) is an l-interval, and vice versa (for convenience, we denote by a_{2N+1} the point a_1, and so on).

For definiteness, we assume that (a_n, b_n) with n odd are r-intervals, and (a_n, b_n) with n even are l-intervals.

For $n = 1, \ldots, 2N$, we fix points a'_n, b'_n so that $a_n < a'_n < b'_n < b_n$, and the length of any $[b'_n, a'_{n+1}]$ is less than ϵ. Denote

$$I_n = [a'_n, b'_n], \quad I'_n = [b'_n, a'_{n+1}].$$

Obviously, it is possible to fix these points so that $\phi(I'_n) \cap I'_m = \emptyset$ for $m \neq n$.

Note that by construction we have

$$\phi(I'_n) \subset \text{Int} I'_n \text{ for } n \text{ odd}$$

and

$$\phi^{-1}(I'_n) \subset \text{Int} I'_n \text{ for } n \text{ even}.$$

Find $\Delta \in (0, 1)$ such that

$$r(\phi(x), x), r(\phi^{-1}(x), x) > \Delta \text{ for } x \in I_n, \ 1 \leq n \leq 2N.$$

Now we take $d_1 > 0$ such that

$$d_1 < \min(\epsilon, \frac{\Delta}{2}); \tag{3.8}$$

$$N_{d_1}(\phi(I'_n)) \cap I'_m = \emptyset \text{ for } n \neq m; \tag{3.9}$$

$$N_{d_1}(\phi(I'_n)) \subset \text{Int} I'_n \text{ for } n \text{ odd}; \tag{3.10}$$

$$N_{d_1}(\phi^{-1}(I'_n)) \subset \text{Int} I'_n \text{ for } n \text{ even}. \tag{3.11}$$

Take $d \in (0, d_1)$ and such that the inequalities

$$r(\phi^k(x_l), x_{l+k}) < d_1 \text{ for } |k| \leq \frac{2}{\Delta} \tag{3.12}$$

hold for any d-pseudotrajectory $\{x_k\}$ of ϕ.

We claim that this d has the property described in Definition 1.3.

Take a d-pseudotrajectory $\xi = \{x_k\}$ of ϕ. First, assume that $x_k \in I'_n$ for n odd. Then the inequality $r(x_{k+1}, \phi(x_k)) < d < d_1$ and inclusion (3.10) imply that $x_{k+1} \in I'_n$. This shows that if $x_k \in I'_n$ for n odd, then $x_{k+l} \in I'_n$ for $l \geq 0$.

Similarly we prove that if $x_k \in I'_n$ for n even, then $x_{k+l} \in I'_n$ for $l \le 0$. To do this, we apply inclusion (3.11) and the inequality

$$r(\phi^{-1}(x_k), x_{k-1}) < d_1 \qquad (3.13)$$

following from (3.12).

Now assume that $x_k \in I_n$ with n odd, i.e., (a_n, b_n) is an r-interval. Then

$$x_{k+1} > \Phi(x_k) - d_1 > x_k + \frac{\Delta}{2}$$

and

$$x_{k-1} < \Phi(x_k) + d_1 < x_k - \frac{\Delta}{2}.$$

It follows from these inequalities and from the properties of the corresponding segments I'_{n-1} and I'_n established above that there exist indices k_- and k_+ such that

$$k \le k_+ \le k + \frac{2}{\Delta}, \ k \ge k_- \le k - \frac{2}{\Delta},$$

and

$$x_l \in I'_n \text{ for } l \ge k_+, \ x_l \in I'_{n-1} \text{ for } l \le k_-. \qquad (3.14)$$

A segment I_n with n even has a similar property.

For the fixed d-pseudotrajectory ξ, one of the following two cases is possible.

Case 1.

$$\xi \subset \bigcup_{n=1}^{2N} I'_n.$$

It follows from (3.9) that in this case there exists a segment I'_n such that $\xi \subset I'_n$. Since the length of I'_n is less than ϵ, and this segment contains a fixed point b_n of ϕ, ξ is (ϵ)-shadowed by b_n.

Case 2. There exists a segment I_n such that $\xi \cap I_n \ne \emptyset$. Assume that $x_0 \in I_n$ and that n is odd. It was shown above that in this case there exist indices $k_+ > 0, k_- < 0$ such that $|k_+|, |k_-| \le 2/\Delta$, and relations (3.14) hold. It follows from our considerations that

$$\phi^l(x_0) \in I'_n \text{ for } l \ge k_+, \ \phi^l(x_0) \in I'_{n-1} \text{ for } l \le k_-.$$

These inclusions and inequalities (3.12) imply that ξ is (ϵ)-shadowed by the point x_0.

The case of n even is considered similarly. Thus, Theorem 3.1.1 is proved.

□

Remark. The example of Yano [Y1] mentioned in Sect. 2.1 is based on the following construction. Set $a_k = 1/k$ for $k \in \mathbb{N}$. Consider a homeomorphism ϕ of S^1 such that any interval (a_{k+1}, a_k) with k odd is an r-interval, and any interval (a_{k+1}, a_k) with k even is an l-interval. By Theorem 3.1.1, ϕ has the POTP. On the other hand, 0 is a nonisolated fixed point of ϕ, and it is easy to show that ϕ is not topologically stable (see [Pi2] for details).

Let us describe necessary and sufficient conditions of the LmSP for a class of homeomorphisms of S^1 [Pla1].

Theorem 3.1.2. *A homeomorphism ϕ preserving orientation and with nonempty set $\mathrm{Fix}(\phi)$ has the LmSP if and only if the set $\mathrm{Fix}(\phi)$ is nowhere dense, and $\Phi(t) - t$ changes sign on $[0, 1)$.*

Proof. It follows from Lemma 3.1.1 that if the set $\mathrm{Fix}(\phi)$ contains a nondegenerate arc, then ϕ does not have the LmSP. If the set $\mathrm{Fix}(\phi)$ is nowhere dense but $\Phi(t) - t > 0$ (or $\Phi(t) - t < 0$) at all t such that $P(t) \notin \mathrm{Fix}(\phi)$, then one can apply the construction from the proof of Lemma 3.1.2 to obtain a sequence $\xi = \{x_k\}$ such that

$$r(x_{k+1}, \phi(x_k)) \to 0, \ k \to \infty, \tag{3.15}$$

and ξ "rotates" around S^1 infinitely many times. Obviously, in this case the relation

$$r(x_k, \phi^k(p)) \to 0, \ k \to \infty, \tag{3.16}$$

cannot hold for a point $p \in S^1$. This proves the necessity of the conditions of our theorem.

Let us prove their sufficiency. Take a sequence ξ such that (3.15) holds. We claim that this sequence converges to a fixed point p of ϕ.

Fix a sequence

$$\epsilon_n > 0, \ \epsilon_n \to 0 \text{ as } n \to \infty.$$

Let us show that for any n there exists a number $\kappa(n)$ and a segment J_n of length less than ϵ_n such that $x_k \in J_n$ for $k \geq \kappa(n)$. Then the family $\{J_n\}$ has a unique common point p. Obviously, in this case we have $\phi(p) = p$, and $x_k \to p$.

We apply a construction analogous to the one used in the proof of Theorem 3.1.1. Fix $\epsilon = \epsilon_n$ (below we omit indices n). Find a system of r-intervals and l-intervals

$$(a_1, b_1), \ldots, (a_N, b_N)$$

such that it contains both r- and l-intervals, and the length of any segment $[b_m, a_{m+1}]$ is less than ϵ (as previously, $a_{N+1} = a_1$ etc). Now we choose a'_m, b'_m such that $a_m < a'_m < b'_m < b_m$, and the length of any $[b'_m, a'_{m+1}]$ is also less than ϵ. There exists d having with respect to the system of segments

$$I_m = [a'_m, b'_m], \ I'_m = [b'_m, a'_{m+1}],$$

the same properties as in the proof of Theorem 3.1.1. Find k_0 such that $r(\phi(x_k), x_{k+1}) < d$ for $k \geq k_0$. If there exists m such that $x_k \in I'_m$ for $k \geq k_0$, we take $\kappa(n) = k_0$. Otherwise there exists $k_1 > k_0$ and an index m such that $x_{k_1} \in I_m$. We consider the case when I_m lies in an r-interval. Then there exists $k'_1 > k_1$ such that $x_{k'_1} \in I'_m \cup I_{m+1}$. If I_{m+1} lies in an l-interval, then $x_k \in I'_m$ for $k \geq k'_1$, thus we can take $\kappa(n) = k'_1$. If I_{m+1} lies in an r-interval, there are two possibilities. Either $x_k \in I'_m$ for $k \geq k'_1$ (and again we take $\kappa(n) = k'_1$) or there exists $k_2 > k_1$ such that $x_{k_2} \in I_{m+1}$. One of the segments $I_1, \ldots, I_N$

lies in an l-interval, hence there exist $\kappa = \kappa(n)$ and q such that $x_\kappa \in I_q'$, I_q is an r-interval, and I_{q+1} is an l-interval. Obviously, this $\kappa(n)$ has the desired property. The theorem is proved. $\square$

Now we show that for one-dimensional dynamical systems the POTP implies the LmSP [Pla1].

Theorem 3.1.3. *If a homeomorphism ϕ of S^1 has the POTP, then ϕ has the LmSP.*

Proof. Assume that a homeomorphism ϕ of S^1 has the POTP. By Theorem 3.1.1, there exists a natural m such that $\psi = \phi^m$ satisfies the conditions of Theorem 3.1.2.

Consider a sequence $\{x_k\}$ such that (3.15) holds. It follows from the inequality

$$r(x_{k+m}, \psi(x_k)) \le \sum_{n=0}^{m-1} r(\phi^{m-n-1}(x_{k+n+1}), \phi^{m-n}(x_{k+n}))$$

and from (3.15) that

$$r(x_{k+m}, \psi(x_k)) \to 0 \text{ as } k \to \infty.$$

Hence, for $y_k = x_{km}$ we have

$$r(y_{k+1}, \psi(y_k)) \to 0 \text{ as } k \to \infty,$$

and it follows from Theorem 3.1.2 that there is a point x such that

$$r(\psi^k(x), y_k) = r(\phi^{km}(x), x_{km}) \to 0 \text{ as } k \to \infty.$$

We claim that for any $l, 0 \le l \le m - 1$, we have

$$r(\phi^{km+l}(x), x_{km+l}) \to 0 \text{ as } k \to \infty,$$

this will prove our theorem.

Our claim follows from the inequalities

$$r(\phi^{km+l}(x), x_{km+l}) \le r(\phi^{km+l}(x), \phi^l(x_{km})) + r(\phi^l(x_{km}), \phi^{l-1}(x_{km+1})) +$$

$$+r(\phi^{l-1}(x_{km+1}), \phi^{l-2}(x_{km+2})) + \ldots + r(\phi(x_{km+l-1}), x_{km+l}),$$

since every term on the right tends to zero as k tends to infinity. $\square$

Now let us mention some results on shadowing for semi-dynamical systems generated by continuous (but not necessarily invertible) mappings $f : [0,1] \to [0,1]$. In this case, it is natural to study an analog of the POTP$_+$ on $[0,1]$ (we also call it POTP$_+$).

Let f be a continuous function on $[0,1]$. Denote by F the set of all interior fixed points of f. Define the set

$$H = \{x \in F : \text{ for any } \epsilon > 0 \text{ there exist } y, z \in (x - \epsilon, x + \epsilon) \text{ such that}$$

$$f(y) < y \text{ and } f(z) > z\}.$$

Pennings and Van Eeuwen [Pe] proved the following statement.

Theorem 3.1.4. *Let f be a nondecreasing continuous function mapping $[0, 1]$ to itself. The semi-dynamical system on $[0, 1]$ generated by f has the $POTP_+$ if and only if $F = H$.*

Remark. It follows from Theorem 3.1.4 that if a function f has no interior fixed points (for example, it behaves like $f(x) = \sqrt{x}$), then the corresponding semi-dynamical system has the $POTP_+$. Note that the dynamical system on S^1 generated by the homeomorphism corresponding to $\phi(x) = \sqrt{x}$ does not have the POTP on S^1 (see the remark after Lemma 3.1.3).

Let us mention some papers devoted to one-dimensional shadowing. Coven, Can, and Yorke [Cov] studied shadowing properties for families of tent maps. Mizera [Mi] showed that $POTP_+$ is a generic property in the space of semi-dynamical systems on $[0,1]$ with the C^0-topology. The $POTP_+$ for mappings of $[0,1]$ was also studied by Gedeon and Kuchta [Ge] and Slackov [Sl].

3.2 Linear and Linearly Induced Systems

We begin with linear dynamical systems on $\mathbf{C}^n$ and $\mathbb{R}^n$. Let A be a nonsingular matrix, complex in the case of $\mathbf{C}^n$, and real in the case of $\mathbb{R}^n$. We consider the dynamical system $\phi(x) = Ax$. As usual, the matrix A is called *hyperbolic* if its spectrum does not intersect the circle $\{\lambda : |\lambda| = 1\}$.

Theorem 3.2.1. *For the system ϕ the following statements are equivalent:*
(1) ϕ has the POTP;
(2) ϕ has the LpSP;
(3) the matrix A is hyperbolic.

Remark. In the case of $\mathbb{R}^n$, the equivalence of (1) and (3) was published by Morimoto [Morim3], later the proof of the implication (1) $\Rightarrow$ (3) given in [Morim3] was refined by Kakubari [Ka]. For a linear mapping in a Banach space, an analogous statement was proved in [Om2]. We prove here Theorem 3.2.1 only in the case of $\mathbf{C}^n$.

We begin with a lemma.

Lemma 3.2.1. *Let (X, r) be a metric space. Assume that for two dynamical systems ϕ an ψ on X there exists a homeomorphism H of X such that $\phi \circ H =$*

$H \circ \psi$, and H, H^{-1} are Lipschitz. Then ϕ has the POTP (or the LpSP) if and only if ψ has the same property.

Proof. We prove the lemma only for the POTP, in the case of the LpSP the proof is similar. Assume that ϕ has the POTP. Fix arbitrary $\epsilon > 0$. Find $\epsilon_1 > 0$ such that the inequality $r(x, y) < \epsilon_1$, $x, y \in X$, implies that $r(H^{-1}(x), H^{-1}(y)) < \epsilon$. Take $\Delta > 0$ such that any Δ-pseudotrajectory $\{\xi_k\}$ for ϕ is (ϵ_1, ϕ)-shadowed. Now we find $\delta > 0$ such that $r(x, y) < \delta$ implies $r(H(x), H(y)) < \Delta$.

Consider a δ-pseudotrajectory $\{x_k\}$ for ψ. Set $\xi_k = H(x_k)$ for $k \in \mathbb{Z}$. Since $r(x_{k+1}, \psi(x_k)) < \delta$ for $k \in \mathbb{Z}$ and $\phi \circ H = H \circ \psi$, we see that

$$r(\xi_{k+1}, \phi(\xi_k)) = r(H(x_{k+1}), \phi(H(x_k))) = r(H(x_{k+1}), H(\psi(x_k))) < \Delta,$$

so that $\{\xi_k\}$ is a Δ-pseudotrajectory for ϕ. Hence, there exists ξ such that

$$r(\xi_k, \phi^k(\xi)) < \epsilon_1, \ k \in \mathbb{Z}.$$

Let $x = H^{-1}(\xi)$, then for any k we have

$$r(x_k, \psi^k(x)) = r(H^{-1}(\xi_k), \psi^k(H^{-1}(\xi))) = r(H^{-1}(\xi_k), H^{-1}(\phi^k(\xi))) < \epsilon$$

(we take into account here that $\phi \circ H = H \circ \psi$ implies $H^{-1} \circ \phi = \psi \circ H^{-1}$ and $H^{-1} \circ \phi^k = \psi^k \circ H^{-1}$ for any k). $\qquad\square$

Now we prove Theorem 3.2.1. First we prove the implication $(1) \Rightarrow (3)$. let us assume that ϕ has the POTP. To obtain a contradiction, assume that the matrix A has an eigenvalue λ such that $|\lambda| = 1$.

Find a nonsingular matrix T such that $J = T^{-1}AT$ is a Jordan form of A. Then, for the dynamical system $\psi(x) = Jx$ and for the homeomorphism $H(x) = Tx$, the equality $\phi \circ H = H \circ \psi$ holds. Since the homeomorphisms H, H^{-1} are Lipschitz in $\mathbf{C}^n$, Lemma 3.2.1 implies that ψ has the POTP.

We can choose the matrix T so that the matrix J has the form

$$\begin{pmatrix} B & 0 \\ 0 & D \end{pmatrix},$$

where $B \in GL(m, \mathbf{C})$ is of the form

$$\begin{pmatrix} \lambda & 0 & \ldots & 0 & 0 \\ 1 & \lambda & \ldots & 0 & 0 \\ \vdots & \vdots & \ddots & \vdots & \vdots \\ 0 & 0 & \ldots & 1 & \lambda \end{pmatrix}.$$

Fix $d > 0$ and consider the vectors $x_k = (x_1^k, \ldots, x_n^k), k \in \mathbb{Z}$, such that $x_1^k = k\lambda^k d$, $x_2^k, \ldots, x_m^k$ satisfy the relations

$$x_i^{k+1} = \lambda x_i^k + x_{i-1}^k \text{ for } i = 2, \ldots, m, \ k \in \mathbb{Z},$$

and $x_i^k = 0$ for $i = m+1,\ldots,n, k \in \mathbb{Z}$.

It follows from the equalities

$$x_{k+1} - Jx_k = (\lambda^{k+1}d, 0, \ldots, 0)$$

that $|x_{k+1} - Jx_k| = d$, hence $\xi = \{x_k\}$ is a $2d$-pseudotrajectory of ψ. For any vector $x = (y_1,\ldots,y_n)$ we have $\psi^k(x) = J^k x = (\lambda^k y_1,\ldots)$, and this implies the inequality

$$|\psi^k(x) - x_k| \geq |kd - y_1|.$$

Since the right-hand side of the last inequality is unbounded for any y_1, we see that ψ does not have the POTP. The obtained contradiction proves the implication (1) $\Rightarrow$ (3).

Obviously, (2) $\Rightarrow$ (1). To prove the remaining implication (3) $\Rightarrow$ (2), we assume that the matrix A is hyperbolic. We show that $\phi(x) = Ax$ has the LpSP with any finite d_0. It will be also shown that the shadowing trajectory is unique in the following sense: if for a sequence $\{x_k : k \in \mathbb{Z}\}$ we have

$$\sup_{k \in \mathbb{Z}} |Ax_k - x_{k+1}| < \infty, \tag{3.17}$$

then the inequality

$$\sup_{k \in \mathbb{Z}} |A^k x - x_k| < \infty \tag{3.18}$$

holds for not more than one x.

Denote by S the invariant subspace corresponding to the eigenvalues λ_j of A such that $|\lambda_j| < 1$, and by U the invariant subspace corresponding to the eigenvalues λ_j of A such that $|\lambda_j| > 1$. It follows from our assumption that $\mathbb{C}^n = S \oplus U$.

We can find a natural m and $\lambda \in (0,1)$ such that

$$||A^m |_S ||, ||A^{-m} |_U || \leq \lambda. \tag{3.19}$$

By Lemma 1.1.3, to prove that ϕ has the LpSP, it is enough to show that $\psi(x) = A^m x$ has this property. To simplify notation, we assume that inequalities (3.19) hold with $m = 1$ (another possibility is to introduce a norm in $\mathbb{C}^n$ equivalent to the standard one and such that inequalities (3.19) hold with $m = 1$, we leave the details to the reader).

Now we find $N > 0$ such that the projectors P and Q onto S and U with the property $P + Q = I$ satisfy the inequalities

$$||P||, ||Q|| \leq N.$$

Take a sequence $\xi = \{x_k\}$ such that (3.17) holds, and let

$$\sup_{k \in \mathbb{Z}} |Ax_k - x_{k+1}| = d.$$

We are looking for a sequence v_k such that for the point $x = x_0 + v_0$ the relations

$$A^k x = x_k + v_k, \ k \in \mathbb{Z}, \tag{3.20}$$

and inequalities (3.18) are satisfied.

We can write (3.20) in the form $x_{k+1} + v_{k+1} = A(x_k + v_k)$ or

$$v_{k+1} = A v_k + (A x_k - x_{k+1}), \ k \in \mathbb{Z}. \tag{3.21}$$

It follows from Theorem 1.3.1 (with $A_k = A$, $w_{k+1}(v) = A x_k - x_{k+1}$, $P_k = P$ etc) that there exists a sequence $\{v_k\}$ satisfying (3.21) and such that

$$\sup_{k \in \mathbb{Z}} |v_k| \le L d,$$

where $L = L(\lambda, N)$. This proves that ϕ has the LpSP. The uniqueness of a sequence v_k such that

$$\sup_{k \in \mathbb{Z}} |v_k| < \infty$$

follows from Theorem 1.3.2. Obviosly, this implies the uniqueness of a point x with property (3.18). This completes the proof of Theorem 3.2.1.   □

It is easy to see that a similar statement holds for a mapping $\phi : \mathbf{C}^n \to \mathbf{C}^n$ of the form

$$\phi(x) = A x + b, \ b \in \mathbf{C}^n$$

(and for a mapping of the same form in $\mathbb{R}^n$, see [Morim3]). An analogous shadowing problem for a linear system of differential equations,

$$\dot{x} = A x, \tag{3.22}$$

where the matrix A is constant, was studied by Ombach [Om3]. It is shown in [Om3] that system (3.22) has the Lipschitz shadowing property (analogous to the one established in Theorem 1.5.1) if and only if all eigenvalues of the matrix A have nonzero real parts.

Now we pass to shadowing for linearly induced systems. Here we treat in detail one class of linearly induced systems, the so-called spherical linear transformations [Sas], later we mention known results for some other classes.

Consider the space $\mathbb{R}^{n+1}$ with coordinates $x = (x_0, \dots, x_n)$. Let S^n be the unit sphere, $S^n = \{x : |x| = 1\}$. We denote by r the canonical distance on S^n.

Fix a real nonsingular $(n+1) \times (n+1)$ matrix A, i.e., an element of $GL(n+1, \mathbb{R})$. We define the *spherical linear transformation* ϕ corresponding to A by the formula

$$\phi(x) = \frac{A x}{|A x|}.$$

Obviously, ϕ is a diffeomorphism of S^n. Denote by $\lambda_1, \dots, \lambda_{n+1}$ the eigenvalues of the matrix A.

Theorem 3.2.2. *For the system ϕ on S^n the following statements are equivalent:*

(1) ϕ has the POTP;
(2) ϕ has the LpSP;
(3) $|\lambda_i| \neq |\lambda_j|$ for $i \neq j$.

Proof. If $A, B \in GL(n+1, \mathbb{R})$, and ϕ, ψ are the corresponding spherical linear transformations, then it is easy to see that the system $\phi \circ \psi$ is the spherical linear transformation corresponding to the product AB. Since, for any matrix $T \in GL(n+1, \mathbb{R})$, the corresponding spherical transformation is a Lipschitz homeomorphism of S^n, it follows from Lemma 3.2.1 that we may consider the matrix A in its Jordan form.

To prove the implication $(1) \Rightarrow (3)$, we first establish two auxiliary statements.

Lemma 3.2.2. *Assume that the matrix A has the form*

$$\begin{pmatrix} B & 0 \\ C & D \end{pmatrix}, \qquad (3.23)$$

where $B \in GL(m+1, \mathbb{R})$, $m < n$. Let ψ be the spherical linear transformation of S^m generated by the matrix B. If ϕ has the POTP, then ψ has the same property.

Proof. For $x = (x_0, \dots, x_n) \in \mathbb{R}^{n+1}$ we set $x' = (x_0, \dots, x_m)$ and $x'' = (x_{m+1} \dots, x_n)$. Consider

$$S^m = \{x \in S^n : x'' = 0\} \text{ and } P = \{x \in S^n : x' = 0\}.$$

We define the projection $\pi : S^n \setminus P \to S^m$ by

$$\pi(x) = \frac{x'}{|x'|}.$$

Obviuosly, for the projection π, the inequality $r(x, y) \geq r'(\pi(x), y)$ holds for $x \in S^n$ and $y \in S^m$, where r' is the distance on S^m. It follows from (3.23) that $\pi \circ \phi(x) = \psi \circ \pi(x)$ for $x \in S^n \setminus P$.

Assume that ϕ has the POTP. Take $\epsilon > 0$, we assume that $\epsilon < \text{dist}(S^m, P)$. There exists $d > 0$ such that any d-pseudotrajectory of ϕ is (ϵ, ϕ)-shadowed.

Take a d-pseudotrajectory $\xi = \{x_k\}$ of ψ. Since ξ is a d-pseudotrajectory of ϕ, it is (ϵ, ϕ)-shadowed by a point $x \in S^n$, and it follows from the choice of ϵ that $x \notin P$. The property of π mentioned above implies the inequalities

$$r'(\psi^k(\pi(x)), x_k) \leq r(\pi(\phi^k(x)), x_k) < \epsilon.$$

This proves our lemma. $\qquad \qquad \square$

Lemma 3.2.3. *Assume that the matrix A has one of the following forms:*

$$A_1 = \begin{pmatrix} \lambda & 0 \\ 0 & \lambda \end{pmatrix}, \; A_2 = \begin{pmatrix} a & -b \\ b & a \end{pmatrix},$$

or

$$A_3 = \begin{pmatrix} \lambda & 0 & \cdots & 0 & 0 \\ 1 & \lambda & \cdots & 0 & 0 \\ \vdots & \vdots & \ddots & \vdots & \vdots \\ 0 & 0 & \cdots & 1 & \lambda \end{pmatrix}.$$

Then the spherical linear transformation ϕ generated by the matrix A does not have the POTP.

Proof. The first two matrices A_1 and A_2 generate diffeomorphisms of the circle S^1. In the case of A_1, every point of S^1 is a fixed point of ϕ, hence ϕ does not have the POTP (see Theorem 3.1.1). In the case of A_2, ϕ reduces to rotation of S^1; if $\alpha \in [0, 2\pi)$ is angular coordinate on S^1, and if $a + bi = \rho \exp(i\theta)$, then $\phi(\alpha) = \alpha + \theta \pmod{2\pi}$. If the number $\mu = \theta/\pi$ is rational, then there exists a natural m such that every point of S^1 is a fixed point of ϕ^m, and the same reasons as above show that ϕ does not have the POTP. If the number μ is irrational, then the rotation number of ϕ is also irrational, and ϕ does not have the POTP by Theorem 3.1.1.

It remains to consider the case of A_3. We assume for definiteness that A_3 is an $(n+1) \times (n+1)$ matrix and that $\lambda > 0$. It is easy to see that, for $c > 0$, matrices A and cA induce the same spherical transformation. Since we can find a Jordan form of A_3 such that all nonzero off-diagonal terms equal λ, we may assume that $\lambda = 1$.

A simple calculation shows that for a vector $x = (x_0, \ldots, x_n)$ and for a natural m we have

$$\phi^m(x) = \frac{y}{|y|},$$

where $y = (y_0, \ldots, y_n)$,

$$y_0 = x_0, \; \ldots, \; y_i = \sum_{j=0}^{i} C_m^j x_{i-j}, \; \ldots, \; y_n = \sum_{j=0}^{n} C_m^j x_{n-j},$$

and C_m^j are the usual binomial coefficients,

$$C_m^j = \frac{m!}{j!(m-j)!}.$$

Denote

$$S_+^n = \{x \in S^n : x_0 \geq 0\}, \; S_-^n = \{x \in S^n : x_0 \leq 0\},$$

it follows from the formulas above that if $x_0 \geq 0$, then $\phi^m(x) \in S_+^n$, and if $x_0 \leq 0$, then $\phi^m(x) \in S_-^n$, and both inclusions hold for $m \in \mathbb{Z}$.

Take $z^+ = (1, 0, \ldots, 0)$ and denote $z^{+,m} = \phi^m(z^+)$. For natural m we have

$$z^{+,m} = \frac{y^m}{|y^m|}, \; \text{where } y^m = (1, C_m^1, \ldots, C_m^n).$$

Since each C_m^i is a polynomial in m of degree i, we see that

$$z^{+,m} \to p = (0,0,\dots,1) \text{ as } m \to \infty.$$

Take arbitrary $d > 0$. There exists $m_0 > 0$ such that $r(z^{+,m_0},p) < d$. Similarly one shows that there exists a point z^- and a natural number m_1 such that $r(z^-,p) < d$ and $\phi^{m_1}(z^-) = (-1,0,\dots,0)$. Construct a sequence $\xi = \{z_k : k \in \mathbb{Z}\}$ as follows:

$$z_0 = z^-, \; z_k = \phi^k(z_0) \text{ for } k \geq 0; \; z_{-1} = z^{+,m_0}, \; z_k = \phi^{k+1}(z_{-1}) \text{ for } k < 0.$$

Let $\mathcal{L}$ be a Lipschitz constant of ϕ. Since $\phi(p) = p$, we have

$$r(\phi(z_{-1}),p) \leq \mathcal{L}r(z_{-1},p) < \mathcal{L}d,$$

hence ξ is an $(\mathcal{L}+1)d$-pseudotrajectory of ϕ. For any trajectory $O_+ \subset S_+^n$ we have $\mathrm{dist}(O_+,(-1,0,\dots,0)) \geq \pi/2$, for any trajectory $O_- \subset S_-^n$ we have $\mathrm{dist}(O_+,(1,0,\dots,0)) \geq \pi/2$. Since ξ contains both points $(1,0,\dots,0)$ and $(-1,0,\dots,0)$, and d is arbitrary, we see that ϕ does not have the POTP. The lemma is proved.   □

Now we can establish the implication (1) $\Rightarrow$ (3). First assume that the matrix A has two real eigenvalues λ and λ' with $|\lambda| = |\lambda'|$. Since the linear spherical transformation ϕ^2 generated by A^2 has the POTP if and only if ϕ has, we may assume that $\lambda = \lambda' > 0$, then by Jordan's theorem A is conjugate to a matrix of form (3.23), where B is one of the matrices A_1 or A_3 (see Lemma 3.2.3). If A has a pair of conjugate complex eigenvalues, then in a Jordan form (3.23) of A, the matrix B coincides with A_2. In all these cases, Lemmas 3.2.2 and 3.2.3 imply that ϕ does not have the POTP.

Obviously, (2) $\Rightarrow$ (1), hence it remains to show that (3) $\Rightarrow$ (2). We may assume that

$$A = \mathrm{diag}(\lambda_0,\dots,\lambda_n), \text{ where } \lambda_0 > \lambda_1 > \dots \lambda_n > 0.$$

In this case, ϕ has fixed points $P_i^\pm$, $i = 0,\dots,n$, corresponding to the eigenvectors of A,

$$P_i^\pm = (a_0,\dots,a_n) \text{ with } a_i = \pm 1, \; a_j = 0 \text{ for } j \neq i.$$

Let us identify $T_{P_i^\pm}S^n$ with the subspace $\{x \in \mathbb{R}^{n+1} : x_i = 0\}$. Direct calculation shows that for $y = (y_1,\dots,y_{i-1},y_{i+1},\dots,y_n)$ we have

$$D\phi(P_i^\pm)y = \left(\frac{\lambda_0}{\lambda_i}y_0,\dots,\frac{\lambda_{i-1}}{\lambda_i}y_{i-1},\frac{\lambda_{i+1}}{\lambda_i}y_{i+1},\dots,\frac{\lambda_n}{\lambda_i}y_n \right),$$

hence any fixed point of ϕ is hyperbolic.

It is easily seen that, for a fixed point P_i^+, the stable and unstable manifolds are given by

$$W^s(P_i^+) = \{x \in S^n : x_0 = \dots = x_{i-1} = 0, x_i > 0\},$$

$$W^u(P_i^+) = \{x \in S^n : x_i > 0, x_{i+1} = \ldots = x_n = 0\},$$

and for a point P_i^- these manifolds are given by

$$W^s(P_i^-) = \{x \in S^n : x_0 = \ldots = x_{i-1} = 0, x_i < 0\},$$

$$W^u(P_i^-) = \{x \in S^n : x_i < 0, x_{i+1} = \ldots = x_n = 0\}.$$

It follows that every point of S^n belongs to the intersection of a stable manifold of a fixed point and an unstable manifold of a fixed point. Hence, the nonwandering set $\Omega(\phi)$ consists of fixed points of ϕ, and this means that ϕ satisfies Axiom A.

Let us show that ϕ satisfies the geometric STC. Take a point $x = (x_0, \ldots, x_n) \in S^n$, let $x \in W^s(P) \cap W^u(Q)$. If $x_0 \neq 0$, then $P = P_0^+$ or $P = P_0^-$. Since both manifolds $W^s(P_0^+)$ and $W^s(P_0^-)$ are n-dimensional, x is a point of transverse intersection of $W^s(P)$ and $W^u(Q)$. Similarly, if $x_n \neq 0$, then $Q = P_n^+$ or $Q = P_n^-$, and $W^s(P), W^u(Q)$ are transverse at x.

If $x = (0, \ldots, 0, x_l, \ldots, x_m, 0, \ldots, 0)$ with $x_l \neq 0$ and $x_m \neq 0$, then $P = P_l^+$ or $P = P_l^-$. In this case, $T_x W^s(P)$ is the intersection of the $(n-l+1)$-dimensional subspace $\{0, \ldots, 0, y_l, \ldots, y_n\}$ of $\mathbb{R}^{n+1}$ with $T_x S^n$, hence $\dim W^s(P) = n - l$. Similar reasons show that $\dim W^u(Q) = m$ and

$$\dim(T_x W^s(P) \cap T_x W^u(Q)) = m - l,$$

hence

$$\dim W^s(P) + \dim W^u(Q) - \dim(T_x W^s(P) \cap T_x W^u(Q)) = n - l + m - m + l = n,$$

and this means that $W^s(P), W^u(Q)$ are transverse at x.

Now Theorem 2.2.7 implies that ϕ has the LpSP. This proves our theorem.
□

Remark. In [Sas], the equivalence of statements (1) and (3) of Theorem 3.2.2 was established. In this paper, Sasaki showed that the same conditions on the matrix A are equivalent to the POTP for the corresponding real projective linear transformation.

In [Kato1, Kato2], Kato established similar results for Grassmann transformations and Poincaré diffeomorphisms on spheres, he also considered the corresponding flows.

3.3 Lattice Systems

We work in this book with a particular class of autonomous lattice dynamical systems defined as follows. Consider the Banach space

$$\mathcal{B} = \{u = \{u_j\} : u_j \in \mathbb{R}^k, j \in \mathbb{Z}\}$$

with the norm

$$||u|| = \sup_{j \in \mathbf{Z}} |u_j|.$$

Fix a natural number s and denote

$$\{u_j\}^s = \{u_{j-s}, \ldots, u_j, \ldots, u_{j+s}\} \in (\mathbb{R}^k)^{2s+1}.$$

Consider a smooth mapping

$$F : (\mathbb{R}^k)^{2s+1} \to \mathbb{R}^k$$

and define a corresponding operator T as follows:

$$[T(u)]_j = F(\{u_j\}^s). \tag{3.24}$$

Under appropriate conditions on the mapping F, T maps $\mathcal{B}$ into itself, hence it defines a semi-dynamical system $(T^n, \mathcal{B})$ called a *lattice dynamical system* with discrete time. A sequence

$$\{u(n) = \{u_j(n) : j \in \mathbb{Z}\} : n \geq 0\}$$

is a trajectory of this system if and only if the relations

$$u_j(n+1) = F(u_{j-s}(n) \ldots, u_{j+s}(n)), \; n \geq 0, \; j \in \mathbb{Z}, \tag{3.25}$$

hold.

Lattice dynamical systems are models for a wide class of physical phenomena in space-time (see [Cou], for example). Another source of lattice dynamical systems are discretizations of partial differential equations. Consider, for example, a parabolic equation

$$v_t = v_{xx} + f(v, v_x), \tag{3.26}$$

where $v, x \in \mathbb{R}$ (we do not fix boundary conditions in this example). Let us discretize Eq. (3.26) with space step D and time step h. Denote by $u_j(n)$ the corresponding approximate values of $v(jD, nh)$ for $n \geq 0, j \in \mathbb{Z}$. Taking standard approximations

$$v_t(jD, nh) \approx \frac{u_j(n+1) - u_j(n)}{h},$$

and

$$v_x(jD, nh) \approx \frac{u_{j+1}(n) - u_j(n)}{D}, \; v_{xx}(jD, nh) \approx \frac{u_{j+1}(n) - 2u_j(n) + u_{j-1}(n)}{D^2},$$

we obtain the relations

$$\frac{u_j(n+1) - u_j(n)}{h} = \frac{u_{j+1}(n) - 2u_j(n) + u_{j-1}(n)}{D^2} + f\left(u_j(n), \frac{u_{j+1}(n) - u_j(n)}{D}\right)$$

easily reduced to the following lattice dynamical system:

$$u_j(n+1) = \frac{h}{D^2}(u_{j+1}(n) + u_{j-1}(n)) + hf\left(u_j(n), \frac{u_{j+1}(n) - u_j(n)}{D}\right) +$$

$$+ \left(1 - \frac{2h}{D^2} \right) u_j(n)$$

(here $s = 1$). We take the explicit scheme above only to simplify presentation. These schemes are not of real practical interest, usually implicit or semi-implicit discretizations are applied (see Sect. 4.3).

In [Cho8], Chow and Van-Vleck established a finite-time shadowing result for lattice dynamical systems and applied it to discretizations of some classes of partial differential equations, such as Burger's equation,

$$v_t = a(x,t)v_{xx} + b(x,t)vv_x,$$

and the Korteweg-de Vries equation,

$$v_t = a(x,t)v_{xxx} + b(x,t)vv_x.$$

In this section, we introduce some special classes of pseudotrajectories for lattice dynamical systems and show that it is possible to reduce the shadowing problem for them to the same problem for auxiliary finite-dimensional dynamical systems [Af2].

Three types of solutions are usually studied for lattice dynamical systems.

Steady-state solutions. These solutions do not depend on time n, we denote $u_j(n) = v_j$. They satisfy the equations

$$v_j = F(v_{j-s}, \ldots, v_{j+s}), \ j \in \mathbb{Z}. \tag{3.27}$$

Travelling wave solutions. Fix integer numbers l and m and consider solutions of the form $u_j(n) = v(lj + mn)$. They are called (l, m)-travelling waves and satisfy the equations

$$v(lj+mn+m) = F(v(lj-ls+nm), \ldots, v(lj+ls+mn)), \ j \in \mathbb{Z}, \ n \geq 0. \tag{3.28}$$

Spatially-homogeneous solutions. They do not depend on the spatial coordinate j, we denote $u_j(n) = v(n)$, and they satisfy the equations

$$v(n + 1) = F(v(n), \ldots, v(n)), \ n \geq 0. \tag{3.29}$$

These types of solutions are governed by finite-dimensional dynamical systems (see below).

The definition of a pseudotrajectory for a lattice dynamical system is similar to Definition 1.1. We say that

$$\xi = \{z_j(n) : z_j \in \mathbb{R}^k, j \in \mathbb{Z}, n \geq 0\}$$

is a d-pseudotrajectory for system $(\mathcal{T}^n, \mathcal{B})$ if

$$|z_j(n + 1) - F(z_{j-s}(n), \ldots, z_{j+s}(n))| < d, \ j \in \mathbb{Z}, \ n \geq 0. \tag{3.30}$$

Now we define three types of pseudotrajectories corresponding to the three types of solutions introduced above.

d-static pseudotrajectory. A d-pseudotrajectory $\{z_j(n)\}$ is called d-static if

$$|z_j(n+1) - z_j(n)| < d \text{ for } j \in \mathbb{Z}, \ n \geq 0 \qquad (3.31)$$

(i.e., it almost does not depend on time).

d-travelling pseudotrajectory. A d-pseudotrajectory $\{z_j(n)\}$ is called a d-(m, l)-travelling wave pseudotrajectory if

$$|z_{j-m}(n+l) - z_j(n)| < d \text{ for } j \in \mathbb{Z}, \ n \geq -l. \qquad (3.32)$$

d-homogeneous pseudotrajectory. A d-pseudotrajectory $\{z_j(n)\}$ is called d-homogeneous if

$$|z_{j+1}(n) - z_j(n)| < d \text{ for } j \in \mathbb{Z}, \ n \geq 0. \qquad (3.33)$$

We begin with some standard results connected with the global inverse mapping theorem.

Let $f : \mathbb{R}^p \to \mathbb{R}^p$ be a mapping of class C^1. We say that f satisfies the *Hadamard conditions* (and we write $f \in \mathrm{HC}(p)$ in this case) if

(HC1) $\det Df(x) \neq 0$ for $x \in \mathbb{R}^p$;

(HC2) $|f(x)| \to \infty$ as $|x| \to \infty$.

Note that if $p = 1$ and there exists $k > 0$ such that $|f'(x)| \geq k$ for all x, then $f \in \mathrm{HC}(1)$.

A proof of the following statement can be found in [Z].

Theorem 3.3.1. *If $f \in \mathrm{HC}(p)$, then f is a diffeomorphism of $\mathbb{R}^p$ onto $\mathbb{R}^p$.*

Now consider a mapping $f : \mathbb{R}^q \times \mathbb{R}^p \to \mathbb{R}^p$ of class C^1, and let y, x be coordinates in $\mathbb{R}^q, \mathbb{R}^p$. We say that f satisfies the *generalized Hadamard conditions* (and we write $f \in \mathrm{GHC}(q, p)$ in this case) if

(GHC1) $\det \frac{\partial f}{\partial x}(y, x) \neq 0$ for all y, x;

(GHC2) for any fixed y, $|f(y, x)| \to \infty$ as $|x| \to \infty$.

If $p = 1$ and there is $k > 0$ such that

$$\left| \frac{\partial f}{\partial x}(y, x) \right| \geq k,$$

then $f \in \mathrm{GHC}(q, 1)$.

It follows from Theorem 3.3.1 that if $f \in \mathrm{GHC}(q, p)$, then for any fixed y, the mapping $f(y, .)$ is a diffeomorphism of $\mathbb{R}^p$ onto $\mathbb{R}^p$, hence a mapping $\Phi : \mathbb{R}^q \times \mathbb{R}^p \to \mathbb{R}^p$ is defined such that

$$f(y, \Phi(y, z)) = z.$$

By the implicit function theorem, Φ is of class C^1. Hence, the following statement is true.

Theorem 3.3.2. *If $f \in \mathrm{GHC}(q, p)$, then*

(a) for any $y \in \mathbb{R}^q, z \in \mathbb{R}^p$ there exists a unique $\Phi(y, z)$ such that

$$f(y, \Phi(y, z)) = z,$$

and the mapping $\Phi(y, z)$ is of class C^1;

(b) for any compact susbet $K \subset \mathbb{R}^q \times \mathbb{R}^p$ there exists a constant $c_0 = c_0(K)$ such that for $(y, z_1), (y, z_2) \in K$ we have

$$|\Phi(y, z_1) - \Phi(y, z_2)| \le c_0 |z_1 - z_2|.$$

Shadowing of d-static pseudotrajectories

Let us assume that for the mapping $F(u_{j-s}, \ldots, u_{j+s})$ defining the system $(T^n, \mathcal{B})$ we have $F \in \mathrm{GHC}(2sk, k)$ (here $u_{j-s}, \ldots, u_{j+s-1}$ are coordinates in $\mathbb{R}^{2sk}$, and u_{j+s} are coordinates in $\mathbb{R}^k$).

It follows from Theorem 3.3.2 that in this case there exists a mapping

$$G \in C^1(\mathbb{R}^{2sk}, \mathbb{R}^k)$$

such that for any $(y_1 \ldots, y_{2s+1}) \in \mathbb{R}^{(2s+1)k}$ (here $y_i \in \mathbb{R}^k$) the equality

$$y_{s+1} = F(y_1, \ldots, y_{2s+1})$$

implies that

$$y_{2s+1} = G(y_1, \ldots, y_{2s}).$$

Hence, it follows from Eq. (3.27) that

$$v_{j+s} = G(v_{j-s}, \ldots, v_{j+s-1}). \tag{3.34}$$

Let us introduce, for $j \in \mathbb{Z}$,

$$x_j^{(1)} = v_{j-s}, \ldots, x_j^{(2s)} = v_{j+s-1}.$$

Then Eq. (3.34) may be represented as a dynamical system (with time j) on the set of steady-state solutions,

$$x_{j+1}^{(1)} = x_j^{(2)}, \ldots, x_{j+1}^{(2s-1)} = x_j^{(2s)}, \ x_{j+1}^{(2s)} = G(x_j^{(1)}, \ldots, x_j^{(2s)}). \tag{3.35}$$

We assume, in addition, that

$$\det \left(\frac{\partial G}{\partial x_j^{(1)}} \right) \ne 0, \ |G(x_j^{(1)}, \ldots, x_j^{(2s)})| \to \infty \text{ as } |x_j^{(1)}| \to \infty.$$

Denote

$$\phi(x^{(1)}, \ldots, x^{(2s)}) = (x^{(2)}, \ldots, x^{(2s)}, G(x^{(1)}, \ldots, x^{(2s)})).$$

Since

$$D\phi = \begin{pmatrix} 0 & I & \dots & 0 \\ \vdots & \vdots & \ddots & \vdots \\ 0 & 0 & \dots & I \\ \psi & * & \dots & * \end{pmatrix},$$

where I is the unit $k \times k$ matrix, and

$$\psi = \frac{\partial G}{\partial x^{(1)}},$$

it follows from Theorem 3.3.1 that ϕ is a diffeomorphism $\mathbb{R}^{2sk} \to \mathbb{R}^{2sk}$.

Theorem 3.3.3.   *Assume that the diffeomorphism ϕ has a hyperbolic set Λ. Then there exist numbers $L, d_0 > 0$ such that if*
    (a) $\xi = \{z_j(n) : j \in \mathbb{Z}, n \geq 0\}$ is a d-static d-pseudotrajectory of $(\mathcal{T}^n, \mathcal{B})$ with $d < d_0$;
    (b)

$$\text{dist}((z_{j-s}(n), \dots, z_{j+s-1}(n)), \Lambda) < d_0 \ for \ j \in \mathbb{Z}, \ n \geq 0, \qquad (3.36)$$

then there exists a point $x \in \mathbb{R}^{2sk}$ (independent of n) with the property

$$|(z_{j-s}(n), \dots, z_{j+s-1}(n)) - \phi^j(x)| \leq Ld \ for \ j \in \mathbb{Z}, \ n \geq 0.$$

Proof. Apply Theorem 1.2.3 to find numbers $L', d' > 0$ such that ϕ has the LpSP on the set $U = N_{d'}(\Lambda)$ with constants L', d'. It is shown after the remark to Theorem 1.2.3 that ϕ is expansive on a neighborhood of Λ, we assume that ϕ is expansive on U with an expansivity constant b. Let y be coordinate in $\mathbb{R}^{2sk}$ and let w be coordinate in $\mathbb{R}^k$.

There exists $N > 0$ such that $\Lambda \subset \{|y| \leq N\}$. Let

$$K = \{|y| \leq N + d'\} \times \{|w| \leq N + d'\} \subset \mathbb{R}^{2sk} \times \mathbb{R}^k.$$

Apply Theorem 3.3.2 to find a number $c_0 = c_0(K)$ for the mapping $f = F$ (with $\Phi = G$) and the compact set K. Set

$$L = 2c_0 L', \ d_0 = \min\left(\frac{d'}{L+1}, \frac{d'}{2c_0}, \frac{b}{2L+2s}\right). \qquad (3.37)$$

Fix $d < d_0$ and assume that $\xi = \{z_j(n)\}$ is a d-static d-pseudotrajectory for $(\mathcal{T}^n, \mathcal{B})$. Obviuosly, we have

$$z_j(n) = F(z_{j-s}(n), \dots, z_{j+s}(n)) + (z_j(n) - z_j(n+1)) +$$

$$+ (z_j(n+1) - F(z_{j-s}(n), \dots, z_{j+s}(n)).$$

It follows from (3.30) and (3.31) that

$$F(z_{j-s}(n), \dots, z_{j+s}(n)) = z_j(n) + F_1, \qquad (3.38)$$

where

$$|F_1| < 2d \tag{3.39}$$

for all j, n. Let $\zeta = G(z_{j-s}(n), \ldots, z_{j-s+1}(n))$.

It follows from (3.36) and from the choice of N that

$$|z_j(n)| \leq N + d', \quad |z_j(n) + F_1| \leq N + d',$$

hence

$$(z_{j-s}(n), \ldots, z_{j+s-1}(n), z_j(n)), (z_{j-s}(n), \ldots, z_{j+s-1}(n), z_j(n) + F_1) \in K.$$

Compare (3.38) with the equality

$$F(z_{j-s}(n), \ldots, z_{j+s-1}(n), \zeta) = z_j(n)$$

and apply the second statement of Theorem 3.3.2 and estimate (3.39) to show that

$$z_{j+s}(n) = G(z_{j-s}(n), \ldots, z_{j+s-1}(n)) + F_2, \quad |F_2| < 2dc_0. \tag{3.40}$$

Now we fix n and introduce

$$Z_j = (Z_j^{(1)}, \ldots, Z_j^{(2s)}) \in \mathbb{R}^{2sk}$$

setting

$$Z_j^{(1)} = z_{j-s}(n), \ldots, Z_j^{(2s)} = z_{j+s-1}(n).$$

We deduce from (3.35) that

$$\phi(Z_j) = (Z_j^{(2)}, \ldots, G(Z_j^{(1)}, \ldots, Z_j^{(2s)})) =$$

$$= (z_{j+1-s}(n), \ldots, G(z_{j-s}(n), \ldots, z_{j+s-1}(n))).$$

Note that

$$Z_{j+1} = (Z_{j+1}^{(1)}, \ldots, Z_{j+1}^{(2s)}) = (z_{j+1-s}(n), \ldots, z_{j+s}(n)),$$

so that

$$|Z_{j+1} - \phi(Z_j)| = |G(z_{j-s}(n), \ldots, z_{j+s-1}(n)) - z_{j+s}(n)|,$$

hence (3.40) implies that $\{Z_j\}$ is a $2dc_0$-pseudotrajectory for ϕ. By condition (3.36), $Z_j \in N_{d_0}(U)$, and it follows from (3.37) that there exists a point $x \in \mathbb{R}^{2sk}$ such that

$$|Z_j - \phi^j(x)| =$$

$$= |(z_{j-s}(n), \ldots, z_{j+s-1}(n)) - \phi^j(x)| \leq 2c_0 L' d = Ld \text{ for } j \in \mathbb{Z}. \tag{3.41}$$

There is a logical possibility that the point x for which (3.41) holds depends on n. Denote it $x(n)$. Apply the same procedure as above to find $x(n+1)$ with

$$|(z_{j-s}(n+1), \ldots, z_{j+s-1}(n+1)) - \phi^j(x(n+1))| < Ld, \quad j \in \mathbb{Z}.$$

Let us estimate

$$|\phi^j(x(n+1)) - \phi^j(x(n))| \le |(z_{j-s}(n+1), \dots, z_{j+s-1}(n+1)) - \phi^j(x(n+1))|+$$

$$+|(z_{j-s}(n+1), \dots, z_{j+s-1}(n+1)) - (z_{j-s}(n), \dots, z_{j+s-1}(n))|+$$

$$+|(z_{j-s}(n), \dots, z_{j+s-1}(n)) - \phi^j(x(n))| \le (2L+2s)d < b, \ j \in \mathbb{Z}$$

(we apply (3.31) to estimate the second term). Since

$$\text{dist}(Z_j, \Lambda) < d_0, \ |Z_j - \phi^j(x(n))| < Ld_0$$

(and a similar inequality holds for $x(n+1)$), the inequality $(L+1)d_0 \le d'$ (see (3.37)) implies the inclusions

$$\phi^j(x(n)), \phi^j(x(n+1)) \in U,$$

hence it follows from the choice of b that $x(n) = x(n+1)$. This completes the proof.   □

Remark. Of course, a statement similar to Theorem 3.3.3 holds if ϕ is a diffeomorphism on a neigborhood of its hyperbolic set.

Example 3.1 Consider the following one-dimensional lattice dynamical system:

$$u_j(n+1) = u_j(n) + \kappa(u_{j+1}(n) - 2u_j(n) + u_{j-1}(n)) + \alpha f(u_j(n)) \qquad (3.42)$$

(it is called a system with diffusion coupling). Here κ is the diffusion coefficient, α is a parameter, and $f(u) = u(u-a)(1-u)$ with $0 < a < 1$. System (3.42) has form (3.25) with

$$F(z_1, z_2, z_3) = \kappa z_1 + (1 - 2\kappa)z_2 + \alpha f(z_2) + \kappa z_3,$$

and it is easy to see that if $\kappa \ne 0$, then $F \in \text{GHC}(2,1)$.

It is shown in [Af1, Cho6] that, for certain values of κ, α, a, the corresponding Henon-type diffeomorphism

$$x_{j+1}^{(1)} = x_j^{(2)}, \ x_{j+1}^{(2)} = 2x_j^{(2)} - \frac{\alpha}{\kappa}f(x_j^{(2)}) - x_j^{(1)}$$

has a hyperbolic set (it acts like a "Smale horseshoe diffeomorphism" on some rectangle), hence Theorem 3.3.3 applies to system (3.42).

Shadowing of d-travelling pseudotrajectories

Fix $l, m \in \mathbb{Z}$. We define a "travelling wave" coordinate $q = lj - ls + mn$, then (3.28) can be rewritten in the form

$$v(q + ls + m) = F(v(q), v(q+l), \dots, v(q + 2ls)). \qquad (3.43)$$

Let us begin with the case

$$l > 0, \ m > ls. \tag{3.44}$$

If we introduce, for $q \in \mathbb{Z}$,

$$x_q^{(1)} = v(q), \dots, x_q^{(1+l)} = v(q+l), \dots,$$

$$x_q^{(1+2ls)} = v(q+2ls), \dots, x_q^{(ls+m)} = v(q+ls+m-1),$$

then Eq. (3.43) may be represented as a dynamical system (with time q) on the set of (l, m)-travelling waves,

$$x_{q+1}^{(1)} = x_q^{(2)}, \ \dots, \ x_{q+1}^{(ls+m-1)} = x_q^{(ls+m)},$$

$$x_{q+1}^{(ls+m)} = F(x_q^{(1)}, \dots, x_q^{(1+2ls)}). \tag{3.45}$$

We assume that F satisfies the generalized Hadamard conditions, so that (3.45) defines a diffeomorphism

$$\phi : \mathbb{R}^{(ls+m)k} \to \mathbb{R}^{(ls+m)k}.$$

Now we consider the following equation:

$$l\gamma + m\nu = 1, \ \gamma, \nu \in \mathbb{Z}. \tag{3.46}$$

We assume that (3.46) has a solution (γ, ν) with $\nu \geq 0$ (for example, if l, m are relatively prime, Eq. (3.46) has an infinite set of solutions with $\nu \geq 0$). Let for such a solution

$$\kappa = \max(|\gamma|, \nu).$$

Consider a d-(m, l)-travelling wave pseudotrajectory $\{z_j(n)\}$, fix a solution (γ, ν) of Eq. (3.46) with $\nu \geq 0$, fix $j \in \mathbb{Z}, n \geq 0$, and for integer $t \geq 0$ define vectors

$$Z_t = (Z_t^{(1)}, \dots, Z_t^{(ls+m)}) \in \mathbb{R}^{(ls+m)k}$$

by setting

$$Z_t^{(i)} = z_{\alpha(i,t)}(\beta(i,t)), \tag{3.47}$$

where

$$\alpha(i,t) = j + (i+t-1)\gamma, \ \beta(i,t) = n + (i+t-1)\nu \tag{3.48}$$

for $i, t \geq 0$.

Theorem 3.3.4. *Assume that the diffeomorphism ϕ defined by (3.45) has a hyperbolic set Λ and take $L, d_0 > 0$ such that ϕ has the LpSP with these constants on $U = N_{d_0}(\Lambda)$ (see Theorem 1.2.3). There exist numbers $L_1, L_2 > 0$ having the following property. If*

$$d(L_1 + L_2\kappa) < d_0 \tag{3.49}$$

and if for a d-(l, m)-travelling wave pseudotrajectory $\{z_j(n)\}$ and for a solution (γ, ν) of (3.46) with $\nu \geq 0$ there exist j, n such that for the vectors Z_t we have

$$\text{dist}(Z_t, \Lambda) < d_0, \ t \geq 0, \tag{3.50}$$

then there exists $x \in \mathbb{R}^{(ls+m)k}$ such that

$$|Z_t - \phi^t(x)| \leq d(L_1 + L_2\kappa)L, \ t \geq 0.$$

Proof. The hyperbolic set Λ is compact, hence there exists $N > 0$ such that the $s(s+1)d_0$-neighborhood of Λ is a subset of the set

$$\{(x^{(1)}, \ldots, x^{(ls+m)}) \in \mathbb{R}^{(ls+m)k} : |x^{(i)}| \leq N\}.$$

Since $F \in C^1(\mathbb{R}^{(2s+1)k}, \mathbb{R}^k)$, there is $M > 0$ such that

$$\max_{1 \leq j \leq 2s+1} \left\| \frac{\partial F}{\partial z_j} \right\| \leq M$$

for all $(z_1, \ldots, z_{2s+1})$ with $|z_i| \leq N$.

Let

$$L_1 = Ms(2s+1) + 1, \ L_2 = 1 + s.$$

Consider a d-(l, m)-travelling wave pseudotrajectory $\{z_j(n)\}$ and j, n such that (3.49) and (3.50) hold. If we show that $\{Z_t, t \geq 0\}$ is a $d(L_1 + L_2\kappa)$-pseudotrajectory of ϕ, this will reduce our theorem to Theorem 1.2.3.

Since

$$\phi(Z_t) = (Z_t^{(2)}, \ldots, Z_t^{(ls+m)}, F(Z_t^{(1)}, \ldots, Z_t^{(1+2ls)})),$$

we see that

$$|Z_{t+1} - \phi(Z_t)| = |z_{\alpha(ls+m+1,t)}(\beta(ls+m+1,t)) -$$

$$- F(z_{\alpha(1,t)}(\beta(1,t)), \ldots, z_{\alpha(1+2ls,t)}(\beta(1+2ls,t)))|. \tag{3.51}$$

It follows from (3.32) that

$$|z_{j-pm}(n+pl) - z_j(n)| < |p|d$$

for all $p \in \mathbb{Z}$ such that $n, n + p \geq 0$. Hence, for $i = 2, \ldots, 2s$ we have

$$|z_{\alpha(1+li,t)}(\beta(1+li,t)) - z_{\alpha(1+li,t)+i\nu m}(\beta(1+li,t) - i\nu l)| < i\nu d.$$

Since

$$\alpha(1+li,t) + i\nu m = j + t\gamma + l\gamma i + i\nu m = \alpha(1,t) + i$$

and

$$\beta(1+li,t) - i\nu l = n + t\nu + i\nu m - i\nu m = \beta(1,t),$$

we see that

$$|z_{\alpha(1+li,t)}(\beta(1+li,t)) - z_{\alpha(1,t)+i}(\beta(1,t))| < i\nu d.$$

Now it follows from the choice of M that

$$|F(z_{\alpha(1,t)}(\beta(1,t)), \ldots, z_{\alpha(1+2ls,t)}(\beta(1+2ls,t)))-$$

$$-F(z_\alpha(\beta), z_{\alpha+1}(\beta), \ldots, z_{\alpha+2s}(\beta))| <$$

$$< Md(1 + s + \ldots + 2s) = Mds(s+1) \qquad (3.52)$$

(in the estimates above, $\alpha = \alpha(1,t), \beta = \beta(1,t)$).

Similar arguments show that if

$$\alpha' = \alpha(ls + m + 1, t), \quad \beta' = \beta(ls + m + 1, t),$$

then

$$|z_{\alpha'}(\beta') - z_{\alpha'+m(\nu s-\gamma)}(\beta' - l(\nu s - \gamma))| < d|\nu s - \gamma| \le \kappa d(s+1), \qquad (3.53)$$

and

$$\alpha' + m(\nu s - \gamma) = \alpha + s, \quad \beta' - l(\nu s - \gamma) = \beta + 1.$$

It follows from (3.30),(3.51)-(3.53) that

$$|Z_{t+1} - \phi(Z_t)| < |z_{\alpha+s}(\beta + 1) - F(z_\alpha(\beta), \ldots, z_{\alpha+2s}(\beta))|+$$

$$+Mds(s+1) + \kappa d(s+1) = d(L_1 + L_2\kappa).$$

This completes the proof.   □

The cases of m, l which do not satisfy (3.44) (but for which Eq. (3.46) has solutions (γ, ν) with $\nu \ge 0$) can be treated similarly. We discuss here in detail only the case $m = l = 1$. In this case, the "travelling coordinate" is $q = j+s-m$, and Eq. (3.43) has the form

$$v(q + s + 1) = F(v(q), v(q + 1), \ldots, v(q + 2s)). \qquad (3.54)$$

For brevity, we consider here the case $k = 1$. If $s = 1$, we assume that

$$\left|\frac{\partial F}{\partial z_3}(z) - 1\right| \ge c_0 > 0 \qquad (3.55)$$

for all $z \in \mathbb{R}^3$, and if $s > 1$, we assume that

$$\left|\frac{\partial F}{\partial z_{2s+1}}(z)\right| \ge c_0 > 0.$$

Under these conditions, there exists a mapping G such that the equality

$$v_{s+2} = F(y_1, \ldots, y_{2s+1})$$

is equivalent to

$$y_{2s+1} = G(y_1, \ldots, y_{2s}).$$

Hence, Eq. (3.54) is equivalent to

$$v(q + 2s) = G(v(q), v(q + 1), \ldots, v(q + 2s - 1)). \tag{3.56}$$

Let

$$x_q^{(1)} = v(q), \ldots, x_q^{(2s)} = v(q + 2s - 1), \quad q \in \mathbb{Z},$$

then Eq. (3.56) may be represented as a dynamical system with time q,

$$x_{q+1}^{(1)} = x_q^{(2)}, \quad \ldots, \quad x_q^{(2s)} = G(x_q^{(1)}, \ldots, x_q^{(2s)}). \tag{3.57}$$

We assume, in addition, that

$$\left| \frac{\partial G}{\partial x_q^{(1)}}(z) \right| \geq K > 0,$$

then it follows from Theorem 3.3.1 that (3.57) defines a diffeomorphism ϕ : $\mathbb{R}^{2s} \to \mathbb{R}^{2s}$.

For $l = m = 1$, Eq. (3.46) has the form

$$\gamma + \nu = 1.$$

Obviously, its solutions with $\nu \geq 0$ form the set

$$\{(1 - \nu, \nu) : \nu \geq 0\}.$$

Fix $j \in \mathbb{Z}, n, \nu \in \mathbb{Z}_+$, define $\beta(i, t)$ for $i, t \geq 0$ similarly to (3.48), and set

$$\alpha(i, t) = j + (i + t - 1)(1 - \nu).$$

Define $Z_t^{(i)}$ by (3.47), and consider

$$Z_t = (Z_t^{(1)}, \ldots, Z_t^{(2s)}) \in \mathbb{R}^{2s}.$$

Theorem 3.3.5. *Assume that the diffeomorphism ϕ defined by (3.57) has a hyperbolic set Λ and take $L, d_0 > 0$ such that ϕ has the LpSP with these constants on $U = N_{d_0}(\Lambda)$ (see Theorem 1.2.3). There exist numbers $L_1, L_2 > 0$ having the following property. If for a d-$(1,1)$-travelling wave pseudotrajectory $\{z_j(n)\}$ and for some $\nu \geq 0$*
  (a) $d(L_1 + L_2\kappa) < d_0$;
  (b) there exist j, n such that for the vectors Z_t inequalities (3.50) hold,
then there exists $x \in \mathbb{R}^{2s}$ such that

$$|Z_t - \phi^t(x)| \leq d(L_1 + L_2\nu)L, \quad t \geq 0.$$

Proof. The proof is based on the same ideas as the proofs of two previous theorems, so we give only a sketch for the case $s = 1$.

Due to the compactness of the hyperbolic set Λ, we can find $M > 0$ such that in all our considerations below we have

$$\left|\frac{\partial F}{\partial z_j}\right| \le M.$$

Let

$$L_1 = \frac{M+1}{c_0} + 1, \; L_2 = \frac{M}{c_0} + 2.$$

By (3.50), it is sufficient to show that $\{Z_t\}$ is a $d(L_1 + L_2\nu)$-pseudotrajectory of ϕ. We have

$$|Z_{t+1} - \phi(Z_t)| = |Z_{t+1}^{(2)} - G(Z_t^{(1)}, Z_t^{(2)})|.$$

Denote $\zeta = G(Z_t^{(1)}, Z_t^{(2)})$, then

$$\zeta = F(Z_t^{(1)}, Z_t^{(2)}, \zeta).$$

Set $\alpha = \alpha(1,t), \beta = \beta(1,t)$. It follows from the equalities

$$\alpha(2,t) = \alpha + 1 - \nu, \; \beta(2,t) = \beta + \nu, \; Z_t^{(2)} = z_{\alpha(2,t)}(\beta(2,t)),$$

and from (3.32) that

$$|Z_t^{(2)} - z_{\alpha+1}(\beta)| \le \nu d.$$

Hence,

$$\zeta = F(z_\alpha(\beta), z_{\alpha+1}(\beta), \zeta) + F_1, \tag{3.58}$$

where $|F_1| \le M\nu d$. By the definition of a d-pseudotrajectory, we can write

$$z_{\alpha+1}(\beta+1) = F(z_\alpha(\beta), z_{\alpha+1}(\beta), z_{\alpha+2}(\beta)) + F_2, \tag{3.59}$$

where $|F_2| < d$. Inequality (3.32) implies that

$$|z_{\alpha+2}(\beta) - z_{\alpha+1}(\beta+1)| < d,$$

and we deduce from (3.59) that

$$z_{\alpha+1}(\beta+1) = F(z_\alpha(\beta), z_{\alpha+1}(\beta), z_{\alpha+1}(\beta+1)) + F_3, \tag{3.60}$$

where $|F_3| < (M+1)d$. Comparing (3.58), (3.60), and taking (3.55) into account, we can show that

$$|\zeta - z_{\alpha+1}(\beta+1)| < \frac{d}{c_0}(1 + M + M\nu).$$

Since

$$Z_{t+1}^{(2)} = z_{\alpha(2,t+1)}(\beta(2,t+1)),$$

we see that

$$|z_{\alpha+1}(\beta+1) - Z_{t+1}^{(2)}| < |\nu - 1|d \le (2\nu + 1)d,$$

and finally we have

$$|Z_{t+1} - \phi(Z_t)| = |Z_t^{(2)} - \zeta| < d(L_1 + L_2\nu).$$

This completes the proof.   □

Example 3.2 Consider again system (3.42). Here $s = 1$. For $\kappa \neq 1$ the conditions of the last theorem are satisfied. The corresponding system for $(1, 1)$-travelling waves is generated by the following Henon-type diffeomorphism ϕ:

$$x_{k+1}^{(1)} = x_k^{(2)}, \ x_{k+1}^{(2)} = \frac{\kappa}{\kappa - 1} x_k^{(1)} + \frac{(1 - 2\kappa)x_k^{(2)} + \alpha f(x_k^{(2)})}{1 - \kappa}.$$

It is shown in [Af2, Cho6] that in the space of parameters (κ, α, a) there is a nonempty region for which ϕ has a hyperbolic set. Hence, Theorem 3.3.5 applies to system (3.42).

Shadowing of d-homogeneous pseudotrajectories

Let us write Eq. (3.29) in the form

$$v(n + 1) = G(v(n)), \tag{3.61}$$

where $G(u) = F(u, \ldots, u)$. We assume that (3.61) satisfies conditions similar to the generalized Hadamard conditions, so that (3.61) defines a diffeomorphism $\phi : \mathbb{R}^k \to \mathbb{R}^k$.

Theorem 3.3.6. *Assume that the diffeomorphism ϕ defined by (3.61) has a hyperbolic set Λ. Then there exist constants L, d_0 such that if*
  (a) $\{z_j(n)\}$ is a d-homogeneous d-pseudotrajectory of ϕ with $d < d_0$;
  (b) there exists $j \in \mathbb{Z}$ such that for all $n \geq 0$ we have

$$\mathrm{dist}(z_j(n), \Lambda) < d_0,$$

then there exists $x \in \mathbb{R}^k$ such that

$$|z_j(n) - \phi^n(x)| \leq Ld \text{ for } n \geq 0.$$

The proof of this theorem mostly repeats the proof of Theorem 3.3.3, and we do not give it here.

3.4 Global Attractors for Evolution Systems

Theory of global attractors for infinite-dimensional dynamical systems was intensively developed in the last two decades (see [Bab, Hal, Lad1]). The main objects of application of this theory are partial differential equations. We discuss in this book applications of the shadowing theory to evolution systems generated by semilinear parabolic equations, the mostly studied class of evolution systems (see [He1]).

Note that the shadowing approach was applied to study qualitative behavior of discretizations for parabolic equations near hyperbolic fixed points by Larsson and Sanz-Serna ([Lar2, Lar3]). Their works developed methods of Beyn [Bey] who studied multi-step approximations of autonomous systems of ordinary differential equations near a hyperbolic rest point, and of Alouges-Debussche [Alo] who investigated pure time discretizations for parabolic equations.

In this section, we apply analogs of shadowing results for structurally stable diffeomorphisms (Subsect. 2.2.2) to establish a kind of Lipschitz shadowing in a neighborhood of the global attractor for an evolution system generated by a parabolic equation [Lar4]. We show that a wide class of semilinear parabolic equations satisfy our conditions.

Consider a semilinear parabolic equation

$$u_t = u_{xx} + f(u), \ u \in \mathbb{R}, \ x \in [0, 1], \tag{3.62}$$

with the Dirichlet boundary conditions

$$u(0, t) = u(1, t) = 0. \tag{3.63}$$

It is assumed that $f \in C^2$.

Let C_0^∞ be the set of functions $v(x)$ of class C^∞ on $(0, 1)$ with compact support in $(0, 1)$. We consider two standard Hilbert spaces, L^2, the closure of C_0^∞ in the norm

$$\|v\|_{L^2} = \left(\int_0^1 |v|^2 dx \right)^{1/2},$$

and H_0^1, the closure of C_0^∞ in the norm

$$\|v\|_{H_0^1} = \|v\|_{L^2} + \left(\int_0^1 |v_x|^2 \, dx \right)^{1/2}.$$

Below we denote H_0^1 by $\mathcal{H}$, we write $|v|$ instead of $\|v\|_{H_0^1}$, and dist(Y, X) is the corresponding distance between $Y, X \subset \mathcal{H}$.

We assume that problem (3.62)-(3.63) generates a semigroup of operators $S(t)u_0, \ t > 0, u_0 \in \mathcal{H}$, defining solutions of (3.62). It is well known (see [Hel]) that it is enough, for example, to assume that the following condition is satisfied: there is a constant C such that

$$uf(u) \le C. \tag{3.64}$$

It is also known that under our conditions $S(t)u_0$ depends smoothly on u_0 and that for a fixed $t > 0$ and a bounded set $B \subset \mathcal{H}$ there exists $C(t, B)$ such that

$$|S(t)v - S(t)v'| \le C(t, B)|v - v'| \tag{3.65}$$

for $v, v' \in B$.

Now we describe the following two properties (a) and (b) of $S(t)$.

Property (a): $S(t)$ has a *global attractor* $\mathcal{A}$ in $\mathcal{H}$, i.e., a compact subset of $\mathcal{H}$ such that (a1) $\mathcal{A}$ is invariant, i.e., $S(t)\mathcal{A} = \mathcal{A}$ for $t \in \mathbb{R}$;

(a2) $\mathcal{A}$ is uniformly globally attractive, i.e., for any $\epsilon > 0$ and for any bounded set B in $\mathcal{H}$ there exists $T > 0$ such that

$$\text{dist}(S(t)B, \mathcal{A}) < \epsilon \text{ for } t \geq T;$$

(a3) $\mathcal{A}$ is Lyapunov stable, i.e., for any neighborhood W of $\mathcal{A}$ there is a neighborhood U of $\mathcal{A}$ such that for $u \in U$ we have $S(t)u \in W, t \geq 0$.

Note that (a3) is a consequence of (a2).

Condition (3.64) implies [He1] that $S(t)$ has a global attractor. The existence of a global attractor implies the existence of a bounded open set $\mathcal{H}_0$ such that for any bounded set $B \subset \mathcal{H}$ there is a time $T(B)$ such that $S(t) \subset \mathcal{H}_0$ for $t \geq T(B)$ (a set with this property is called *absorbing*).

Property (b): $S(t)$ has *Morse-Smale structure* on $\mathcal{A}$, i.e., conditions (a4)–(a7) below are satisfied. First we give some definitions.

Note that a fixed point $p = p(x)$ of $S(t)$ is a solution of the following boundary-value problem:

$$p_{xx} + f(p) = 0, \; p(0) = p(1) = 0. \tag{3.66}$$

As usual, we say that a fixed point p is a hyperbolic fixed point of $S(t)$ if the spectrum of the derivative $DS(t)(p)$ does not intersect the unit circle [He1]. One can reformulate this condition as follows. A fixed point $p = p(x)$ of $S(t)$ is hyperbolic if and only if 0 is not an eigenvalue of the linear variational operator,

$$v \mapsto v_{xx} + f'(p(x))v,$$

with the Dirichlet boundary conditions. In this case, we call the solution $p(x)$ hyperbolic.

If p is hyperbolic, then the stable manifold of p with respect to $S(t)$ defined by

$$W^s_S(p) = \{u \in \mathcal{H} : S(t)u \to p \text{ as } t \to \infty\}$$

and the unstable manifold of p defined by

$$W^u_S(p) = \{u \in \mathcal{H} : S(t)u \text{ exists for } t \leq 0 \text{ and } S(t)u \to p \text{ as } t \to -\infty\}$$

are smooth immersed submanifolds of $\mathcal{H}$ [He1]. If p, q are hyperbolic fixed points of $S(t)$, we say that $W^u_S(p)$ and $W^s_S(q)$ are transverse if, for any $v \in W^u_S(p) \cap W^s_S(q)$, the sum of $T_v W^u_S(p)$ and $T_v W^s_S(q)$ equals $\mathcal{H}$ [He2].

Now we formulate the definition of the Morse-Smale structure on $\mathcal{A}$.

(a4) $\mathcal{A}$ contains a finite number of fixed points $\pi_1, \ldots, \pi_N$ of $S(t)$, and these points are hyperbolic;

(a5)

$$\mathcal{H} = \bigcup_{1 \leq i \leq N} W^s_S(\pi_i) \tag{3.67}$$

(every trajectory of $S(t)$ tends to a fixed point);

(a6)

$$\mathcal{A} = \bigcup_{1 \le i \le N} W_S^u(\pi_i) \tag{3.68}$$

(the attractor $\mathcal{A}$ is the union of unstable manifolds of the fixed points of $S(t)$);
  (a7) $W_S^s(\pi_i)$ and $W_S^u(\pi_j)$ are transverse, $i, j \in \{1, \ldots, N\}$.

It is known that, for a generic nonlinearity f, all fixed points of $S(t)$ are hyperbolic. Let us explain the meaning of this statement.

Fix an integer $q \ge 0$. We introduce the C^q strong Whitney topology on the set of functions $f : \mathbb{R} \to \mathbb{R}$ of class C^q as follows. For two functions f, g and for a compact set $K \subset \mathbb{R}$ define the number

$$\rho_K^q(f, g) = \sum_{r=0}^q \sup_{v \in K} \left| \frac{\partial^r f}{\partial v^r} - \frac{\partial^r g}{\partial v^r} \right|. \tag{3.69}$$

The base of neighborhoods of a function f in the C^q strong Whitney topology consists of the sets

$$\{g : \rho_{K_n}^q(f, g) < \epsilon_n\},$$

where $\{K_n\}$ is a countable family of compact subsets of $\mathbb{R}$ such that any point of $\mathbb{R}$ has a neighborhood that intersects a finite number of the sets K_n, the equality

$$\bigcup_n K_n = \mathbb{R}$$

holds, and $\{\epsilon_n\}$ is a sequence of positive numbers.

It is known [Hirs1] that every residual subset of the space of functions $f : \mathbb{R} \to \mathbb{R}$ of class C^q with the C^q strong Whitney topology is dense in this space.

Brunovsky and Chow [Bru] showed that there exists a residual subset $\mathcal{G}$ of the set of functions of class C^2 on $\mathbb{R}$ with the C^2 strong Whitney topology having the following two properties.
  (1) If $f \in \mathcal{G}$, then any solution $p(x)$ of (3.66) is hyperbolic.
  (2) If $f \in \mathcal{G}$, then the set of $\mu > 0$ for which the problem

$$p_{xx} + \mu f(p) = 0, \ p(0) = p(1) = 0, \tag{3.70}$$

has nonhyperbolic solutions is countable.

For a special class of functions f, it is possible to give an explicit description of the set $\{\mu\}$ that correspond to nonhyperbolic solutions of problem (3.70). Chaffee and Infante [Cha] considered Eq. (3.62) with a nonlinearity $\mu f \in C^2$ and f satisfying the conditions
  (1) $f(0) = 0, f'(0) = 1$;
  (2) $u f''(u) < 0$ for $u \ne 0$;
  (3)

$$\overline{\lim}_{u \to \pm\infty} \frac{f(u)}{u} \le 0.$$

It is shown in [Hel] that problem (3.70) has nonhyperbolic solutions only if $\mu = n^2 \pi^2, n = 1, 2, \ldots$.

Thus, for a C^2-generic nonlinearity all fixed points of $S(t)$ are hyperbolic. It follows from results of Henry [He2] and Angenent [Ang] that for hyperbolic fixed points their stable and unstable manifolds are always transverse. Hence, for a generic nonlinearity f in Eq. (3.62) satisfying condition (3.64), the global attractor $\mathcal{A}$ has properties (a4) and (a7). It is easy to establish also properties (a5) and (a6).

Denote $\sigma(u) = S(1)u, u \in \mathcal{H}$. Our goal is to study conditions under which the semi-dynamical system σ has a variant of the LpSP$_+$ on a neighborhood of the global attractor of $S(t)$. Fix $d > 0$.

Our proof of the shadowing property is based on the existence of a finite-dimensional smooth *inertial manifold* constructed as follows. There exists an orthogonal projection P with finite-dimensional $P\mathcal{H}$, and a C^1-mapping Φ : $P\mathcal{H} \to Q\mathcal{H}$ (where $Q = I - P$) such that $\mathcal{M}$, the graph of Φ, has the following two properties:

(m1) $S(t)\mathcal{M} \subset \mathcal{M}, t \geq 0$;

(m2) $\mathcal{M}$ is exponentially attractive, i.e., for any bounded set $B \subset \mathcal{H}$ there exist positive constants C, a (depending on B) such that

$$\text{dist}(S(t)u, \mathcal{M}) \leq C \exp(-at)\text{dist}(u, \mathcal{M}) \tag{3.71}$$

for $u \in B$ and $t \geq 0$.

Obviously, it follows from our assumptions that $\mathcal{A} \subset \mathcal{M}$.

Let us fix an absorbing set $\mathcal{H}_0$ for $\mathcal{A}$. It is known that there is a bounded set $\mathcal{H}'$ such that

$$S(t)\mathcal{H}_0 \subset \mathcal{H}' \text{ for } t \geq 0.$$

We denote by $\mathcal{H}^*$ a ball in $\mathcal{H}$ centered at the origin and containing the 1-neighborhood of $\mathcal{H}'$.

Write problem (3.62)-(3.63) as an evolution equation,

$$\dot{u} = Au + R(u), \tag{3.72}$$

on $\mathcal{H}$. It is known [Fo, Cho5] that it is possible to modify the nonlinearity R in Eq. (3.72) outside the ball $\mathcal{H}^*$ so that the modified nonlinearity vanishes on a neighborhood of infinity, and the system has an inertial manifold. It follows that the trajectories of the modified system beginning at points of $\mathcal{H}_0$ coincide with the corresponding trajectories of the original system.

Below we work with the modified system preserving notation (3.72) for it. Let inequality (3.71) be satisfied for $u \in \mathcal{H}^*$.

Take an integer $T > 0$ such that

$$\nu = C \exp(-aT) < \frac{1}{2} \tag{3.73}$$

(below we impose one more restriction on T; this restriction is "absolute", i.e., it depends only on the attractor $\mathcal{A}$).

Theorem 3.4.1. *Assume that $S(t)$ has properties (a) and (b). Then there exist constants $d_0, L_0 > 0$ and a neighborhood W of A in $\mathcal{H}$ such that if for a sequence $\{u_k : k \geq 0\} \subset W$ we have*

$$|\sigma(u_k) - u_{k+1}| \leq d, \; \text{dist}(u_0, \mathcal{M}) \leq 2d, \qquad (3.74)$$

and $d \leq d_0$, then there is a point $u \in \mathcal{M}$ such that

$$|\sigma^k(u) - u_k| \leq L_0 d, \; k \geq 0.$$

Remark. As was mentioned, it is enough to assume that $S(t)$ has property (a), and every fixed point of $S(t)$ is hyperbolic.

To prove this theorem, first we reduce our shadowing problem to an analogous problem on $\mathcal{M}$, and then we establish the desired shadowing property on $\mathcal{M}$.

Reduction

For $u_0 \in \mathcal{H}$ let $p_0 = Pu_0$ and denote by $p(t, p_0)$ the solution of the finite-dimensional system on $P\mathcal{H}$,

$$\dot{p} = Ap + PR(p + \Phi(p)) \qquad (3.75)$$

such that $p(0, p_0) = p_0$. Then (see [Fo]) for $u_0 \in \mathcal{M}$ we have

$$u(t) = S(t)u_0 = p(t, p_0) + \Phi(p(t, p_0)). \qquad (3.76)$$

We introduce the following notation. For $p_0 = Pu_0, u_0 \in \mathcal{H}$, we set

$$\sigma_1^*(p_0) = p(1, p_0), \;\; \sigma_T^*(p_0) = p(T, p_0)$$

(the number T was fixed by (3.73)), and for $m_0 \in \mathcal{M}$ we set

$$\phi_1(m_0) = \sigma_0^*(p_0) + \Phi(\sigma_0^*(p_0)), \;\; \phi(m_0) = \sigma_T^*(p_0) + \Phi(\sigma_T^*(p_0)),$$

where $p_0 = Pm_0$. It follows from the inclusions $f, \Phi \in C^1$ that σ_1^*, σ_T^* (and ϕ_1, ϕ) are C^1-mappings of $P\mathcal{H}$ (of $\mathcal{M}$, respectively) to itself. In addition, by (3.76) for $u_0 \in \mathcal{M}$ we have

$$\phi_1(u_0) = S(1)u_0 = \sigma(u_0),$$

i.e., ϕ_1 is the restriction of σ on $\mathcal{M}$.

Obviously, A is the global attractor for σ in $\mathcal{H}$ and for ϕ_1 in $\mathcal{M}$ (the definitions are parallel to the one for σ).

Set $C_1 = C(1, \mathcal{H}_0), C_T = C(T, \mathcal{H}_0)$ (see (3.65)).

Fix a neighborhood $\mathcal{M}_0 \subset \mathcal{H}_0$ of A in $\mathcal{M}$. Since A is Lyapunov stable, there exists a neighborhood $\mathcal{M}_1 \subset \mathcal{M}_0$ of A such that $S(t)u \in \mathcal{M}_0$ for $u \in \mathcal{M}_1$ and $t \geq 0$. It follows that for $p_0 \in \Pi_1 = P\mathcal{M}_1$ we have $p(t, p_0) \in \Pi_0 = P\mathcal{M}_0$ for

$t \geq 0$ and that ϕ is a diffeomorphism of class C^1 of $\mathcal{M}_1$ onto its image. Let K_1 be a Lipschitz constant of $\sigma_T^* = p(T, .)$ on Π_1.

Due to the same reason there exists $d^0 > 0$ such that the inequality $\mathrm{dist}(u, \mathcal{A}) \leq d^0$ implies the inclusions $\sigma^n(u) \in \mathcal{H}_0$ for $n \geq 0$. Let W_1 be the d^0-neighborhood of $\mathcal{A}$. Assume, in addition, that the d^0-neighborhood of W_1 is a subset of $\mathcal{H}_0$ and that P projects this neighborhood of W_1 into Π_1.

Take a sequence $\{u_k\} \subset W_1$ such that (3.74) holds with $d \leq d^0$. Denote $z_k = u_{Tk}$.

Lemma 3.4.1. *Let*

$$C_2 = 1 + C_1 + \ldots + C_1^{T-1}, \; C_3 = \frac{2C_2}{1 - \nu}.$$

Then
(1) $|\sigma^T(z_k) - z_{k+1}| \leq C_2 d;$
(2) if $\mathrm{dist}(z_0, \mathcal{M}) \leq 2d$, *then* $\mathrm{dist}(z_k, \mathcal{M}) \leq C_3 d, \; k \geq 0.$

Proof. First note that the choice of d^0 and W_1 implies that $\sigma^n(u_k) \in \mathcal{H}_0$ for $n \geq 0$. Since C_1 is a Lipschitz constant of σ on $\mathcal{H}_0$, statement (1) follows from Lemma 1.1.3.

To prove statement (2), denote $b_k = \mathrm{dist}(z_k, \mathcal{M})$. By assumption, we have $b_0 \leq 2d$. Now we estimate

$$b_{k+1} \leq \mathrm{dist}(\sigma^T(z_k), z_{k+1}) + \mathrm{dist}(\sigma^T(z_k), \mathcal{M}) \leq$$

$$\leq C_2 d + \nu \mathrm{dist}(z_k, \mathcal{M}) = C_2 d + \nu b_k$$

(we refer to (3.71) and (3.73) here). It follows that

$$b_0 \leq 2d \leq 2C_2 d, \; b_1 \leq C_2 d + \nu b_0 \leq 2C_2 d(1 + \nu), \; \ldots,$$

$$b_n \leq 2C_2 d(1 + \nu + \ldots + \nu^n) < \frac{2C_2 d}{1 - \nu} = C_3 d.$$

This completes the proof. $\square$

Lemma 3.4.2. *Assume that for some u we have*

$$\sigma^n(u) \in \mathcal{H}_0, \; n \geq 0, \; \text{and} \; |\sigma^{kT}(u) - z_k| \leq L'd, \; k \geq 0.$$

Then

$$|\sigma^k(u) - u_k| \leq C_5 d, \; k \geq 0,$$

where

$$C_5 = 1 + C_4 + \ldots + C_4^T, \; C_4 = C_1 L'.$$

This lemma is proved similarly to statement (2) of Lemma 1.3.1. Lemmas 3.4.1 and 3.4.2 show that our problem of Lipschitz shadowing for σ is reduced to the same problem for σ^T.

We need below also the following statement.

Lemma 3.4.3. *Let the ball $B = \{|u| \leq R\}$ be a subset of $\mathcal{H}_0$. For any $d > 0$ there exists a number $N = N(R, d)$ such that if $\{u_k\} \subset \mathcal{H}_0$, $u_0 \in B$, and*

$$|\sigma^T(u_k) - u_{k+1}| \leq d,$$

then dist$(u_n, \mathcal{M}) \leq 2d$ *for some* $n \in [0, N]$.

Proof. Denote $b_k =$ dist$(u_k, \mathcal{M})$ and set $\nu_1 = \nu + 1/2$. It follows from (3.73) that $\nu_1 < 1$. Find $N > 0$ such that

$$\nu_1^N R < 2d.$$

To obtain a contradiction, assume that $b_k > 2d$ for $0 \leq k \leq N$. For $0 \leq k \leq N - 1$ we have

$$b_{k+1} \leq \text{dist}(\sigma^T(u_k), \mathcal{M}) + |\sigma^T(u_k) - u_{k+1}| \leq \nu b_k + d.$$

By our assumption, $b_k > 2d$, hence $d < b_k/2$, and

$$b_{k+1} \leq (\nu + \frac{1}{2})b_k = \nu_1 b_k \text{ for } 0 \leq k \leq N - 1.$$

It follows that

$$b_N \leq \nu_1^N b_0 \leq \nu_1^N R < 2d.$$

The obtained contradiction proves our lemma.   □

We proceed with a sequence $\{z_k\}$ that satisfies the assumptions stated before Lemma 3.4.1. Take $d^1 = d^0/C_3$. We consider below $d \leq d^1$.

Denote $z_k' = Pz_k$ and set $v_k = z_k' + \Phi(z_k') \in \mathcal{M}$. We shall prove that $\{v_k\}$ satisfies

$$|\phi(v_k) - v_{k+1}| \leq C'd$$

with C' independent of d.

Since dist$(z_k, \mathcal{M}) \leq C_3 d$ (Lemma 3.4.1), we can find $w_k \in \mathcal{M}$ such that

$$|w_k - z_k| \leq C_3 d. \tag{3.77}$$

We deduce from the choice of d^1 that $w_k \in \mathcal{H}_0$. Now it follows from (3.77) that

$$|\sigma^T(w_k) - \sigma^T(z_k)| \leq C_6 d, \tag{3.78}$$

where $C_6 = C_T C_3$. From Lemma 3.4.1 and from (3.78) we conclude that

$$|\sigma^T(w_k) - z_{k+1}| \leq |\sigma^T(z_k) - z_{k+1}| + |\sigma^T(w_k) - \sigma^T(z_k)| \leq C_7 d, \tag{3.79}$$

where $C_7 = C_2 + C_6$. Denote $w_k' = Pw_k$. By our choice, $z_k', w_k' \in \Pi_1$. It follows from the definition of ϕ that

$$P\sigma^T(w_k) = P\phi(w_k) = \sigma^*_T(Pw_k) = \sigma^*_T(w'_k). \qquad (3.80)$$

We want to estimate the value

$$|\sigma^*_T(z'_k) - z'_{k+1}| \le |\sigma^*_T(z'_k) - \sigma^*_T(w'_k)| + |\sigma^*_T(w'_k) - Pz_{k+1}|. \qquad (3.81)$$

Since

$$|z'_k - w'_k| = |Pz_k - Pw_k| \le C_3 d$$

(see (3.77)), and K_1 is a Lipschitz constant of σ^*_T on Π_1, the first term in (3.81) does not exceed $K_1 C_3 d$.

By (3.80),

$$|\sigma^*_T(w'_k) - Pz_{k+1}| = |P\sigma^T(w_k) - Pz_{k+1}|.$$

Hence, the second term in (3.81) does not exceed

$$|\sigma^T(w_k) - z_{k+1}| \le C_7 d$$

(see (3.79)). Thus, we obtain the estimate

$$|\sigma^*_T(z'_k) - z'_{k+1}| \le C_8 d, \qquad (3.82)$$

where $C_8 = K_1 C_3 + C_7$.

Let K_2 be a Lipschitz constant of Φ on Π_1. We obtain from (3.82) the estimate

$$|v_{k+1} - \phi(v_k)| \le |z'_{k+1} - \sigma^*_T(z'_k)| + |\Phi(z'_{k+1}) - \Phi(\sigma^*_T(z'_k))| \le C_9 d, \qquad (3.83)$$

where $C_9 = (1 + K_2)C_8$.

Denote by r the metric on $\mathcal{M}$ generated by the distance in $\mathcal{H}$. Obviously, there is a constant $K_3 > 0$ such that, for $v, v' \in \mathcal{M}_1$, the inequalities

$$\frac{r(v, v')}{K_3} \le |v - v'| \le r(v, v')$$

hold.

It follows from (3.83) that $r(v_{k+1}, \phi(v_k)) \le K_3 C_9 d$. Let us show that Theorem 3.4.1 is a corollary of the following statement.

Theorem 3.4.2. *There exists a neighborhood M of $\mathcal{A}$ in $\mathcal{M}$ and numbers $d', L > 0$ such that if $\{v_k\} \subset M$ and $r(\phi(v_k), v_{k+1}) \le d \le d'$, then there is a point $v \in M$ with the property*

$$r(\phi^k(v), v_k) \le Ld, \quad k \ge 0.$$

Indeed, assume that there exists $v \in \mathcal{M}$ such that

$$r(\phi^k(v), v_k) \le Ld, \quad k \ge 0.$$

Then

$$|\phi^k(v) - v_k| \leq Ld, \ k \geq 0. \tag{3.84}$$

Let us estimate

$$|z_k - v_k| \leq |z_k - w_k| + |w_k - v_k|.$$

By (3.77), the first term (and also the value $|z_k' - w_k'|$) does not exceed $C_3 d$. Since $w_k = w_k' + \Phi(w_k'), v_k = z_k' + \Phi(z_k')$, we estimate the second term by $C_3 d(1 + K_2)$. Now we obtain from (3.84) the inequality

$$|z_k - \phi^k(v)| = |z_k - \sigma^{Tk}(v)| \leq (L + (C_3(2 + K_2)))d.$$

This proves our reduction statement.

Shadowing on the inertial manifold

Properties (a4)-(a7) of the attractor $\mathcal{A}$ of σ imply the following properties of ϕ_1 on $\mathcal{A}$.

(a4)$'$ $\mathcal{A}$ contains fixed points $\pi_1, \ldots, \pi_N$ of ϕ_1, and they are hyperbolic (we denote by $W^s(\pi_i), W^u(\pi_i)$ their stable and unstable manifolds);

(a5)$'$
$$\mathcal{M} = \bigcup_{1 \leq i \leq N} W^s(\pi_i); \tag{3.85}$$

(a6)$'$
$$\mathcal{A} = \bigcup_{1 \leq i \leq N} W^u(\pi_i); \tag{3.86}$$

(a7)$'$ $W^s(\pi_i)$ and $W^u(\pi_j)$ are transverse, $i, j \in \{1, \ldots, N\}$.

Let us prove only (a7)$'$. Take a point

$$x \in W^s(\pi_i) \cap W^u(\pi_j).$$

As previously, we denote by $T_p N$ the tangent space of N at p. Let

$$T_1^s = T_x W^s(\pi_i), \ T_2^s = T_x W_S^s(\pi_i),$$

$$T_1^u = T_x W^u(\pi_j), \ T_2^u = T_x W_S^u(\pi_j).$$

By (a7), $W_S^s(\pi_i)$ and $W_S^u(\pi_j)$ are transverse at x, i.e.,

$$T_2^s + T_2^u = \mathcal{H}. \tag{3.87}$$

We claim that

$$T_1^z = T_x \mathcal{M} \cap T_2^z, \ z = s, u. \tag{3.88}$$

Note that the norm in $T_p \mathcal{M}, p \in \mathcal{M}$, generated by the metric r on $\mathcal{M}$ coincides with the norm in $T_p \mathcal{M}$ induced from $\mathcal{H}$.

The equality

$$T_1^s = \{w \in T_p \mathcal{M} : |D\phi_1^k(x)w| \to 0, k \to \infty\}$$

is proved in [Pi1], Chap. 13. Similarly one proves that

$$T_2^s = \{w \in \mathcal{H} : |D\sigma^k(x)w| \to 0, k \to \infty\}.$$

Obviously, these equalities imply (3.88) for $z = s$. For $z = u$ the proof is similar. Now it follows from (3.87) that

$$T_1^s + T_1^u = T_x\mathcal{M},$$

i.e., $W^s(\pi_i)$ and $W^u(\pi_j)$ are transverse at x.

Since any fixed point π_i of ϕ_1 is hyperbolic, there exist "stable" and "unstable" subspaces E_i^s and E_i^u of $T_{\pi_i}\mathcal{M}$ with the standard properties,

- $E_i^z, z = s, u$, are $D\phi_1$-invariant;
- $E_i^s \oplus E_i^u = T_{\pi_i}\mathcal{M}$;
- there are $K_i > 0, \lambda_i \in (0,1)$ such that

$$|D\phi_1^m(\pi_i)v| \leq K_i\lambda_i^m|v|, \; v \in E_i^s, \; m \geq 0, \tag{3.89}$$

$$|D\phi_1^{-m}(\pi_i)v| \leq K_i\lambda_i^m|v|, \; v \in E_i^u, \; m \geq 0. \tag{3.90}$$

Take $\lambda_0 \in (0,1)$ and a natural number T_0 such that

$$K_i\lambda_i^{T_0} < \lambda_0, \; 1 \leq i \leq N.$$

We fix T satisfying (3.73) so that $T \geq T_0$ (note that T_0 is an "absolute" constant).

Since $\phi = \phi_1^T$, it follows from (3.89) and (3.90) that

$$\|D\phi|_{E_i^s}\| \leq \lambda_0, \; \|D\phi^{-1}|_{E_i^u}\| \leq \lambda_0. \tag{3.91}$$

As previously, for a set $X \subset \mathcal{M}$ we denote

$$O^+(X) = \bigcup_{k \geq 0} \phi^k(X), \; O^-(X) = \bigcup_{k \leq 0} \phi^k(X), \; O(X) = O^+(X) \cup O^-(X).$$

The following statement is proved similarly to Lemma 2.2.9. We denote by $T\mathcal{M}$ the tangent bundle of $\mathcal{M}$.

Lemma 3.4.4. *There exists $\lambda_1 \in (\lambda_0, 1)$, neighborhoods Z_i of π_i in $\mathcal{M}$, and continuous subbundles S_i, U_i of $T\mathcal{M}$ on $\overline{Z}_i \cup O(Z_i)$ such that*
(1) $S_i(x) \oplus U_i(x) = T_x\mathcal{M}$ for $x \in Z_i$;
(2) S_i, U_i are $D\phi$-invariant on $O(Z_i)$ (consequently, the equalities from (1) hold for $x \in O(Z_i)$);
(3)

$$\|D\phi|_{S_i(x)}\| \leq \lambda_1, \; \|D\phi^{-1}|_{U_i(x)}\| \leq \lambda_1$$

for $x \in Z_i$;
(4)

$$S_i(x) \subset S_j(x), \; U_j(x) \subset U_i(x)$$

for $x \in O^+(Z_i) \cap O^-(Z_j)$.

It is shown in [Pi3] that there exists a neighborhood $\mathcal{M}_2 \subset \mathcal{M}_1$ of $\mathcal{A}$ such that $\phi(\overline{\mathcal{M}}_2) \subset \mathcal{M}_2$ (the neighborhood $\mathcal{M}_1$ was introduced during the reduction of Theorem 3.4.1 to Theorem 3.4.2). We denote

$$Z = \bigcup_{1 \leq i \leq N} Z_i.$$

Let τ be a Birkhoff constant for ϕ on $\overline{\mathcal{M}}_2$, i.e., a positive number with the property

$$\text{card}\{k \in \mathbb{Z} : \phi^k(x) \in \overline{\mathcal{M}}_2 \setminus Z\} \leq \tau$$

for $x \in \overline{\mathcal{M}}_2$ (here card is the cardinality). Let $\mathcal{M}_3 = \phi(\mathcal{M}_2)$.
    Denote

$$N_0 = \max_{x \in \overline{\mathcal{M}}_2} (\|D\phi(x)\|, \|D\phi^{-1}(x)\|).$$

Take $p \in \overline{\mathcal{M}}_2$, there exists $m_0 \in [0, \tau]$ such that $q - \phi^{m_0}(p) \subset Z_i$ and $\phi^m(p) \notin Z_l$ with $i \neq l$ for $0 \leq m < m_0$. Define linear subspaces $S(p), U(p)$ of $T_p \mathcal{M}$ by

$$W(p) = D\phi^{-m_0}(q)W_i(q), \quad W = S, U.$$

Now we define a mapping $e_p : T_p \mathcal{M} \to \mathcal{M}$ for $p \in \mathcal{M}$ by

$$e_p(v) = P(p + v) + \Phi(P(p + v)), \quad v \in T_p \mathcal{M}.$$

Since Φ is of class C^1, e_p is also of class C^1.
    Straightforward calculation shows that

$$De_p(0) = I. \tag{3.92}$$

The following statement is proved similarly to Lemma 2.2.10.

Lemma 3.4.5. *For $p \in \overline{\mathcal{M}}_2$, the spaces $S(p), U(p)$ have the properties*
(1)

$$S(p) \oplus U(p) = T_p \mathcal{M};$$

(2) there exists $N_1 > 0$ such that, for $Q^s(p), Q^u(p)$, the projectors onto $S(p), U(p)$ parallel to $U(p), S(p)$, respectively, the inequalities

$$\|Q^s(p)\|, \|Q^u(p)\| \leq N_1$$

hold;
(3)

$$D\phi(p)S(p) \subset S(\phi(p)), \quad D\phi^{-1}(p)U(p) \subset U(\phi^{-1}(p))$$

(the second inclusion holds if, in addition, $p \in \overline{\mathcal{M}}_3$);
    (4) given $\mu, a_1 > 0$ there exists $a_2 > 0$ such that if $p, z \in \phi^\mu(\overline{\mathcal{M}}_2)$ and $r(z, \phi^\mu(p)) < a_2$, then there is a linear isomorphism $\Pi(p, z) : T_z \mathcal{M} \to T_z \mathcal{M}$ with the properties

$$\|\Pi(p, z) - I\| < a_1, \quad \Pi(p, z)[De_z^{-1}(q)D\phi^\mu(p)S(p)] \subset S(z),$$

and a linear isomorphism $\Theta(p, z) : T_pM \to T_pM$ with the properties

$$\|\Theta(p, z) - I\| < a_1, \ \Theta(p, z)[De_p^{-1}(t)D\phi^{-\mu}(z)U(z)] \subset U(p),$$

where $q = \phi^\mu(p), t = \phi^{-\mu}(z)$.

It is easy to see that for $v \in S_i(x)$ we have

$$|D\phi^m(x)v| \leq K\lambda_1^m|v|, \ m \geq 0,$$

and for $v \in U_i(x)$ we have

$$|D\phi^m(x)v| \leq K\lambda_1^{-m}|v|$$

if $m \leq 0$ and $\phi^k(x) \in \overline{M}_2$ for $k \in [m, 0]$, where $K = (N_0/\lambda_1)^\tau$ (and τ is the Birkhoff constant chosen above). Take $\mu \geq \tau$ such that $K\lambda_1^\mu < \lambda_1$. Denote $\xi = \phi^\mu$.

Lemma 1.1.3 shows that ϕ has the LpSP in a neighborhood of $\mathcal{A}$ if and only if ξ has. Thus, we work below with ξ.

Obviously, Lemma 3.4.5 remains true for ξ instead of ϕ, but in statement (3) $\mathcal{M}_3$ is to be replaced by $\mathcal{M}_4$, where

$$\mathcal{M}_4 = \xi(\mathcal{M}_2).$$

In addition, by the choice of μ, the following inequalities hold:

$$|D\xi(p)v| \leq \lambda_1|v|, \ v \in S(p), \ p \in \mathcal{M}_2, \tag{3.93}$$

$$|D\xi^{-1}(p)v| \leq \lambda_1|v|, \ v \in U(p), \ p \in \mathcal{M}_4. \tag{3.94}$$

For $a > 0$ we denote

$$\mathcal{E}_a(p) = \{v \in T_pM : |v| < a\}, \ \mathcal{D}_a(p) = e_p(\mathcal{E}_a(p)).$$

Obviously, there exists a neighborhood $\mathcal{M}_5 \subset \mathcal{M}_4$ of $\mathcal{A}$ and a number $c > 0$ such that for $p \in \overline{\mathcal{M}}_5$ we have

$$\mathcal{D}_c(p), \xi(\mathcal{D}_c(p)), \xi^{-1}(\mathcal{D}_c(p)) \subset \mathcal{M}_4,$$

and e_p is a diffeomorphism of $\mathcal{E}_c(p)$ onto $\mathcal{D}_c(p)$ with uniform estimates of $\|De_p\|, \|De_p^{-1}\|$.

Denote $M = \mathcal{M}_5$. Take a sequence $\{v_k : k \geq 0\} \subset M$ such that

$$|\xi(v_k) - v_{k+1}| \leq d.$$

Below we denote by d' positive constants that depend on properties of ξ on M and do not depend on $\{v_k\}$. At each step of the proof, we consider d that does not exceed the minimal d' previously chosen. Since we choose d' finitely many times, no generality is lost.

Fix $k \geq 0$ and denote $p = v_k, z = v_{k+1}, H_k = T_{v_k}\mathcal{M}$. Take $d' < c/2$ and such that the inequality

$$r(z, \xi(p)) \leq d' \tag{3.95}$$

implies the inclusion

$$\xi^{-1}(z) \in \mathcal{D}_c(p). \tag{3.96}$$

Find $0 < b < c$ such that

$$\xi(\mathcal{D}_b(x)) \subset \mathcal{D}_{d'}(\xi(x)), \ x \in M$$

(obviously, b depends only on ξ). It follows from (3.95) that

$$\xi(\mathcal{D}_b(p)) \subset \mathcal{D}_c(z).$$

Thus, the mapping $\psi_k : \mathcal{E}_b(p) \to H_{k+1}$ given by

$$\psi_k(w) = e_z^{-1} \circ \xi \circ e_p(w)$$

is properly defined. Let us introduce the following notation:

$$q = \xi(p), \ q' = e_z^{-1}(q), \ t = \xi^{-1}(z), \ t' = e_p^{-1}(t),$$

$$D = D\psi_k(0), \ D' = D\xi(t)De_p(t'), \ G = De_p^{-1}(t)D\xi^{-1}(z).$$

Note that t', D', G are well-defined since (3.96) holds.

Fix $N_2 > 0$ such that

$$\|D\xi(x)\|, \|D\xi^{-1}(x)\| \leq N_2, \ x \in M.$$

Let $N = \max(N_1, N_2)$ (N_1 is defined in Lemma 3.4.5).

Find $\nu_0 \in (0, 1)$ such that

$$\lambda = (1 + \nu_0)^2 \lambda_1 < 1$$

and let

$$N_1 = N\frac{1 + \lambda}{1 - \lambda}.$$

Take $\kappa > 0$ such that $\kappa N_1 < 1$ and find $\nu < \nu_0$ with the property

$$N(4N + 1)\nu < \frac{\kappa}{2}. \tag{3.97}$$

It follows from (3.92) that

$$D = De_z^{-1}(q)D\xi(p). \tag{3.98}$$

Now we find d' such that inequality (3.95) implies the inequalities

$$|De_p(t')w| \leq (1 + \nu)|w|, \ w \in T_p\mathcal{M}, \tag{3.99}$$

$$|De_z^{-1}(q)w| \leq (1 + \nu)|w|, \ w \in T_q\mathcal{M}, \tag{3.100}$$

$$|De_p^{-1}(t)w| \leq (1 + \nu)|w|, \ w \in T_t\mathcal{M} \tag{3.101}$$

(we apply (3.92) and the uniform continuity of De_x, De_x^{-1}),

$$||D - D'|| < \nu \tag{3.102}$$

(compare (3.98), the definition of D', and take into account the previous argument), and

$$||\Pi(p, z) - I||, ||\Theta(p, z) - I||, ||\Theta^{-1}(p, z) - I|| < \nu \tag{3.103}$$

(see Lemma 3.4.5).

Define $A_k : H_k \to H_{k+1}$ by

$$A_k = \Pi(p, z)DQ^s(p) + D'\Theta^{-1}(p, z)Q^u(p).$$

For $w \in S(p)$ we have $w^s = A_kw = \Pi(p, z)Dw \in S(z)$ (see Lemma 3.4.5), hence

$$A_kS(p) \subset S(z). \tag{3.104}$$

Since $|D\xi(p)w| \leq \lambda_1|w|$ by (3.93), it follows from (3.100) and (3.103) that $|w^s| \leq (1 + \nu)^2\lambda_1|w| = \lambda|w|$, hence

$$||A_k|_{S(p)}|| \leq \lambda. \tag{3.105}$$

Now we consider a mapping $B_k : U(z) \to H_k$ defined as follows: $B_kw = \Theta(p, z)Gw$. Apply Lemma 3.4.5 to show that $w^u = B_kw \in U(p)$, hence

$$B_kU(z) \subset U(p), \tag{3.106}$$

and $|w^u| \leq \lambda|w|$, hence

$$||B_k|_{U(z)}|| \leq \lambda. \tag{3.107}$$

It follows from

$$A_kw^u = D'\Theta^{-1}(p, z)\Theta(p, z)Gw = D'Gw = w$$

that

$$A_kB_k|_{U(z)} = I. \tag{3.108}$$

Represent

$$\psi_k(w) = Dw + \rho(w).$$

Obviously there exists d' such that

$$|\rho(w) - \rho(w')| \leq \frac{\kappa}{2}|w - w'| \tag{3.109}$$

for $|w|, |w'| \leq d'$.

Now we represent

$$\psi_k(w) = A_kw + \chi(w), \tag{3.110}$$

where $\chi(w) = (D - A_k)w + \rho(w), \chi(0) = \psi_k(0)$.

Let us estimate $||D - A_k||$,

$$||D - A_k|| = ||D(Q^s(p) + Q^u(p)) - A_k|| \le ||DQ^s(p) - \Pi(p, z)DQ^s(p)|| +$$

$$+||DQ^u(p) - D'Q^u(p)|| + ||D'Q^u(p) - D'\Theta^{-1}(p, z)Q^u(p)||. \qquad (3.111)$$

Since $||Q^s(p)|| \le N, ||D|| \le (1+\nu)N \le 2N$, and $||\Pi(p, z)-I|| < \nu$ (see (3.103)), we see that the first term on the right in (3.111) does not exceed $2N^2\nu$.

The same reasons (and inequality (3.102)) show that the second term is estimated by $N\nu$. Since $||D'|| < 2N$, the third term is estimated by $2N^2\nu$. This gives the inequality

$$||D - A_k|| \le N(4N + 1)\nu \le \frac{\kappa}{2}$$

(see (3.97)). Combined with (3.109), the last inequality shows that

$$|\chi(w) - \chi(w')| \le \kappa|w - w'| \text{ for } |w|, |w'| \le d'. \qquad (3.112)$$

Since the derivatives $De_x(y), De_x^{-1}(y')$ are uniformly bounded for

$$x \in M, \ y \in \mathcal{E}_c(x), \ y' \in \mathcal{D}_c(x), \qquad (3.113)$$

there exists $N^* > 0$ such that for x, y, y' that satisfy (3.113) we have

$$r(x, e_x(y)) \le N^*|x - y|, \ |x - e_x^{-1}(y')| \le N^*r(x, y').$$

Since $\psi_k(0) = e_z^{-1}(\xi(p))$, it follows from $r(\xi(p), z) < d$ that

$$|\psi_k(0)| \le N^*r(z, \xi(p)) \le N^*d. \qquad (3.114)$$

Now it follows from Lemma 3.4.5 and from (3.104)-(3.108), (3.112), and (3.114) that ψ_k satisfy all the conditions of Theorem 1.3.1 (with obvious change of notation).

Hence, there exist $d', L' > 0$ (depending only on ξ and M) and a sequence $w_k \in H_k$ such that $\psi_k(w_k) = w_{k+1}$ and $|w_k| \le L'd$ (if $d \le d'$).

Set $w'_k = e_{v_k}(w_k)$, then it follows from the definition of ψ_k that $w'_{k+1} = \xi(w'_k)$, hence the sequence w'_k is a ξ-trajectory of $v = w'_0$, i.e., $w'_k = \xi^k(v)$.

Now we deduce from the definition of N^* that there exists d' such that if $r(\xi(v_k), v_{k+1}) \le d \le d'$, then the inequalities

$$r(v_k, \xi^k(v)) = r(v_k, w'_k) \le N^*|w_k| = N^*L'd = Ld$$

hold for $k \ge 0$. This completes the proof of Theorem 3.4.2.

4. Numerical Applications of Shadowing

As was mentioned, one of the goals of this book is to describe some "numerically oriented" applications of the shadowing theory. There is a lot of expository papers devoted to this topic. Let us mention, for example, the papers [Po, San] devoted to applications of shadowing in the study of chaotic dynamical systems and the review paper [Coo5], where the use of shadowing results in numerical computations is demonstrated.

4.1 Finite Shadowing

Various methods were developed to establish the existence of a real trajectory near a computed one (see [Coo1-Coo4, Ham, Sau1-Sau2, V], and others). Here we prove two theorems of Chow and Palmer [Cho2, Cho3]. These results allow us to estimate how far a numerically computed finite trajectory is from a true one. Note that Hadeler [Had] applied the Newton-Kantorovich method in a similar situation.

The first theorem we prove deals with one-dimensional semi-dynamical systems generated by mappings of a segment into itself. Let $f : [0,1] \to [0,1]$ be a function of class C^2. Consider a finite sequence $X = \{x_0, \ldots, x_{N+1}\} \subset [0,1]$. We assume that $Df(x_n) \neq 0$ for $0 \leq n \leq N$, and introduce the values

$$\Pi_n^m = Df^{-1}(x_n)Df^{-1}(x_{n+1}) \ldots Df^{-1}(x_m) \text{ for } n \leq m \leq N$$

(note that $\Pi_n^n = Df^{-1}(x_n)$ according to this notation).

Theorem 4.1.1 [Cho2]. *Let*

$$M = \max_{x \in [0,1]} |D^2 f(x)|.$$

Assume that, for the values

$$\sigma = \max_{0 \leq n \leq N} \sum_{m=n}^{N} |\Pi_n^m|$$

and

$$\tau = \max_{0 \leq n \leq N} |\sum_{m=n}^{N} \Pi_n^m(x_{m+1} - f(x_m))|,$$

the inequality

$$2M\sigma\tau \leq 1 \qquad (4.1)$$

holds.

    If $X \subset [\epsilon, 1 - \epsilon]$, where

$$\epsilon = \frac{2\tau}{1 + \sqrt{1 - 2M\sigma\tau}},$$

then there exists a point $x \in [0, 1]$ such that

$$\tau\left(1 + \frac{1}{2(1 + \sqrt{1 - 2M\sigma\tau})}\right)^{-1} \leq \max_{0 \leq n \leq N} |f^n(x) - x_n| \leq \epsilon. \qquad (4.2)$$

Proof. Take a point $x \in [0, 1]$ and set $z_n = f^n(x) - x_n, 0 \leq n \leq N$. Then z_n satisfy the relations

$$z_{n+1} = f(x_n) - x_{n+1} + Df(x_n)z_n + g_n(z_n) \text{ for } 0 \leq n \leq N - 1, \qquad (4.3)$$

where

$$g_n(z) = f(x_n + z) - f(x_n) - Df(x_n)z.$$

We want to find a sequence $\{z_n\}$ satisfying (4.3) and such that $|z_n| \leq \epsilon$. Let $\mathcal{S}$ be the set of sequences $\bar{z} = \{z_n : |z_n| \leq \epsilon, 0 \leq n \leq N\}$. $\mathcal{S}$ can be identified with a compact convex susbet of $\mathbb{R}^{N+1}$.

    Define a mapping $\mathcal{T} : \mathcal{S} \to \mathbb{R}^{N+1}$ as follows: for $\bar{z} = \{z_n\} \in \mathcal{S}$ set

$$(\mathcal{T}\bar{z})_n = -\sum_{m=n}^{N} \Pi_n^m(f(x_m) - x_{m+1} + g_m(z_m)), \ n = 0, \ldots, N.$$

Since $x_n \in [\epsilon, 1 - \epsilon]$ and $|z_n| \leq \epsilon$, we have $x_n + z_n \in [0, 1]$, and it follows from the definition of g_n that $\mathcal{T}$ is a continuous mapping and that

$$|g_n(z_n)| \leq \frac{M}{2}|z_n|^2. \qquad (4.4)$$

Let us show that $\mathcal{T}$ maps $\mathcal{S}$ into itself. Applying (4.4), we have, for $\bar{z} \in \mathcal{S}$,

$$|(\mathcal{T}(z))_n| \leq \left|\sum_{m=n}^{N} \Pi_n^m(f(x_m) - x_{m+1})\right| + \sum_{m=n}^{N} |\Pi_n^m|\frac{M}{2}|g_m(z_m)|^2 \leq$$

$$\leq \tau + \frac{M\sigma\epsilon^2}{2} = \epsilon, \ 0 \leq n \leq N$$

(we take into account that ϵ satisfies the equation $M\sigma\epsilon^2 - 2\epsilon + 2\tau = 0$). Hence, $\mathcal{T}$ maps $\mathcal{S}$ into itself, and Brower's fixed point theorem implies the existence of a fixed point $\bar{z}^*$ of $\mathcal{T}$ in $\mathcal{S}$.

    It follows from the definition of $\mathcal{T}$ that

$$z_n^* = (\mathcal{T}z^*)_n = -Df^{-1}(x_n) \sum_{m=n+1}^{N} \Pi_n^m (f(x_m) - x_{m+1} + g_m(z_m^*)) -$$

$$-Df^{-1}(x_n)(f(x_n) - x_{n+1} + g_n(z_n^*)) =$$

$$= Df^{-1}(x_n)z_{n+1}^* - Df^{-1}(x_n)(f(x_n) - x_{n+1} + g_n(z_n^*)),$$

hence the sequence $\bar{z}^*$ is a solution of (4.3).

To prove our theorem, it remains to establish the estimate from below in (4.2). Set

$$m = \max_{0 \le n \le N} |z_n^*|.$$

It follows from

$$\sum_{m=n}^{N} \Pi_n^m (f(x_m) - x_{m+1}) = z_n^* - \sum_{m=n}^{N} \Pi_n^m g_m(z_m^*)$$

that

$$\tau \le m + \frac{M\sigma m^2}{2} \le m + \frac{M\sigma \epsilon m}{2} \le$$

(we take into account that $m \le \epsilon$)

$$\le \left(1 + \frac{\epsilon}{4\tau}\right) m =$$

(we apply (4.1))

$$= \left(1 + \frac{1}{2(1 + \sqrt{1 - 2M\sigma\tau})}\right) m.$$

This gives the desired estimate from below.   □

In [Cho2], problems of estimation of the values σ and τ are discussed, and the result is applied to the quadratic mapping $f(x) = ax(1 - x)$ with $a = 3.8$ and $x_0 = .6$.

The finite shadowing result obtained in [Cho3] for the multidimensional case is based on similar ideas. Let $f : \mathbb{R}^k \to \mathbb{R}^k$ be a mapping of class C^2. We consider a finite sequence $X = \{x_0, \ldots, x_N\}$, where $x_n \in \mathbb{R}^k$ and the values $|f(x_n) - x_{n+1}|$ are small.

For any sequence $\{h_n \in \mathbb{R}^k : 0 \le n \le N - 1\}$, the difference equation

$$u_{n+1} = Df(x_n)u_n + h_n, \ 0 \le n \le N - 1, \tag{4.5}$$

is obviously solvable. Thus, the linear operator $\mathcal{L} : \mathbb{R}^{k(N+1)} \to \mathbb{R}^{kN}$ defined for $\bar{u} = \{u_n \in \mathbb{R}^k : 0 \le n \le N\}$ by

$$(\mathcal{L}\bar{u})_n = u_{n+1} - Df(x_n)u_n, \ 0 \le n \le N - 1,$$

is onto. Hence, it has right inverses. We choose any such right inverse of $\mathcal{L}$ and denote it by $\mathcal{L}^{-1}$ in the following theorem.

Remark. If we assume that the matrices $Df(x_n)$ are invertible, then we can solve Eq. (4.5) setting $u_N = 0$ and consequently finding

$$u_{N-1} = -Df^{-1}(x_{N-1})h_{N-1}$$

and so on. Obviously, this process gives a right inverse of $\mathcal{L}$ corresponding to the operator applied to define $\mathcal{T}$ in the proof of Theorem 4.1.1 above.

Theorem 4.1.2 [Cho3]. *Let*

$$M = \sup_{x \in \mathbb{R}^k} ||D^2 f(x)||.$$

Assume that

$$|x_{n+1} - f(x_n)| \le d \ for \ 0 \le n \le N - 1$$

and the inequality

$$2M||\mathcal{L}^{-1}||^2 d \le 1$$

holds.

Then there exists a point $x \in \mathbb{R}^k$ *such that*

$$|f^n(x) - x_n| \le 2||\mathcal{L}^{-1}||\frac{d}{1 + \sqrt{1 - 2M||\mathcal{L}^{-1}||^2 d}} \ for \ 0 \le n \le N.$$

Proof. Take a point $x \in \mathbb{R}^k$ and set $z_n = f^n(x) - x_n, 0 \le n \le N$. Then z_n satisfy the relations

$$z_{n+1} = Df(x_n)z_n + g_n(z_n) \ for \ 0 \le n \le N - 1, \tag{4.6}$$

where

$$g_n(z) = f(x_n) - x_{n+1} + f(x_n + z) - f(x_n) - Df(x_n)z.$$

We want to find a sequence $\{z_n\}$ satisfying (4.6) and such that

$$|z_n| \le \epsilon = 2||\mathcal{L}^{-1}||\frac{d}{1 + \sqrt{1 - 2M||\mathcal{L}^{-1}||^2 d}} \ for \ 0 \le n \le N.$$

Let $\mathcal{S}$ be the set of sequences $\bar{z} = \{z_n \in \mathbb{R}^k : |z_n| \le \epsilon, 0 \le n \le N\}$. $\mathcal{S}$ can be identified with a compact convex susbet of $\mathbb{R}^{k(N+1)}$.

We define $g(\bar{z}) \in \mathbb{R}^{kN}$ setting

$$(g(\bar{z}))_n = g_n(z_n), \ 0 \le n \le N - 1,$$

write Eq. (4.6) in the form

$$\mathcal{L}\bar{z} = g(\bar{z}),$$

and introduce the operator $\mathcal{T} : \mathcal{S} \to \mathbb{R}^{k(N+1)}$,

$$T(\bar{z}) = \mathcal{L}^{-1}g(\bar{z}).$$

Obviously, the operator T is continuous. Let us show that T maps $\mathcal{S}$ into itself. We first note that

$$|g_n(z_n)| \le |g_n(0)| + |f(x_n + z_n) - f(x_n) - Df(x_n)z_n| \le d + \frac{M|z_n|^2}{2}.$$

It follows from the definition of ϵ that if $\bar{z} \in \mathcal{S}$, then

$$|(T(\bar{z}))_n| \le ||\mathcal{L}^{-1}|| \left(d + \frac{M\epsilon^2}{2}\right) = \epsilon.$$

Hence, T maps $\mathcal{S}$ into itself, and Brower's theorem implies the existence of a fixed point $\bar{z}^*$ of T in $\mathcal{S}$. Apply $\mathcal{L}$ from the left to the equality

$$\bar{z}^* = \mathcal{L}^{-1}g(\bar{z}^*)$$

to show that $\bar{z}^*$ gives a solution of Eq. (4.6). It follows that the point $x = x_0 + z_0$ has the desired properties. □

The main attention of the authors in [Cho3] is paid to the choice of $\mathcal{L}^{-1}$ with not very large $||\mathcal{L}^{-1}||$ and to the influence of round-off errors in the process of computation.

4.2 Periodic Shadowing for Flows

In computer investigation of systems of ordinary differential equations arising in practice, it is important to be sure that a numerical solution reflects the real behavior of the system under investigation. It is shown in Sect. 1.5 that in a neighborhood of a hyperbolic set any approximate trajectory is shadowed by a real one. But from the practical viewpoint, it is a difficult problem to check the existence of a hyperbolic set. Thus, one needs other tools to shadow numerically computed orbits.

One of the basic problems in qualitative theory of differential equations is the search of closed trajectories, i.e., the trajectories of nonconstant periodic solutions. A lot of computer investigation of specific systems was connected with this problem, let us mention the papers of De Gregorio [DG], Franke and Selgrade [Fr2], Schwartz [Sc], and Sinai and Vul [Si].

We devote this section to a method of periodic shadowing developed by Coomes, Koçak, and Palmer [Coo2]. This method provides a possibility to establish the existence of a real closed trajectory near an approximate one and gives error bounds for the distance between the true and the approximate closed trajectories in terms of computable quantities associated with the variational flow along the approximate trajectory.

The method of periodic shadowing we describe is a part of the theory of "practical" shadowing for ordinary differential equations created by Coomes, Koçak, and Palmer (see also [Coo1–Coo6]).

Let us consider system (1.178) assuming that the vector field X is of class C^2 in $\mathbb{R}^n$. We again denote by $\Xi(t, x)$ the trajectory of system (1.178) such that $\Xi(0, x) = x$, and by $D\Xi(t, x)$ the corresponding variational trajectory.

Fix $d > 0$.

Definition 4.1 *A finite set*

$$(Y, h) = (\{y_k \in \mathbb{R}^n : 0 \le k \le N\}, \{h_k > 0 : 0 \le k \le N\})$$

is called a "periodic discrete (d, h)-pseudotrajectory" (pd(d, h)-pseudotrajectory below) for system (1.178) if the inequalities

$$|y_{k+1} - \Xi(h_k, y_k)| \le d, \ 0 \le k \le N, \tag{4.7}$$

and

$$X(y_k) \neq 0, \ 0 \le k \le N, \tag{4.8}$$

hold.

Here and below, to make our formulas shorter, for any variable a_k with index $0 \le k \le N$, we introduce $a_{N+1} = a_0$, so that, for $k = N$, the inequality in (4.7) has the form

$$|y_0 - \Xi(h_N, y_N)| \le d$$

(and this is the reason to call (Y, h) a periodic pseudotrajectory).

A pd(d, h)-pseudotrajectory is a natural model for the computer output in the process of computer realisation of a one-step method for numerical approximation of solutions for an autonomous system of ODE near a real closed trajectory.

Below we describe a simplified version of the main result in [Coo2] (some estimates in our Theorem 4.2.1 are more "rough" than in [Coo2]; this enables us to make the proof more "readable").

Let us begin with some additional notation. We will simultaneously explain the meaning of the defined objects.

Take a pd(d, h)-pseudotrajectory (Y, h). Assume that we are given a finite sequence $\{Y_k : 0 \le k \le N\}$ of $n \times n$ matrices such that

$$\|Y_k - D\Xi(h_k, y_k)\| \le d \text{ for } 0 \le k \le N. \tag{4.9}$$

The matrices Y_k are approximations for the matrices $D\Xi(h_k, y_k)$. Since the variational flow $\Theta(t) = D\Xi(t, y_k)$ satisfies the variational initial-value problem

$$\frac{d}{dt}\Theta = DX(\Xi(t, y_k))\Theta, \ \Theta(0) = I, \tag{4.10}$$

a natural way to find Y_k is to apply a one-step method to approximate simultaneously the problem

$$\dot{x} = X(x), \ x(0) = y_k,$$

and (4.10) for $t \in [0, h_k]$.

Let us denote by Σ_k the subspace of $\mathbb{R}^n$ orthogonal to the vector $X(y_k)$.

For $0 \leq k \leq N$, let S_k be an $n \times (n-1)$ matrix chosen so that its columns form an "almost orthonormal" basis in Σ_k, i.e., the inequalities

$$|S_k^* X(y_k)|, \|S_k^* S_k - I\| \leq d_1 \tag{4.11}$$

hold with some $d_1 > 0$, and $*$ denotes transpose.

Next we choose $(n-1) \times (n-1)$ matrices $A_k, 0 \leq k \leq N$, so that

$$\|A_k - S_{k+1}^* Y_k S_k\| \leq d_1, \ 0 \leq k \leq N. \tag{4.12}$$

The quantity d_1 in inequalities (4.11) and (4.12) is introduced to account for possible round-off errors in the necessary matrix computations.

Now we define some constants involved in the proof of Theorem 4.2.1. Set

$$h_{\min} = \min_{0 \leq k \leq N} h_k, \ h_{\max} = \max_{0 \leq k \leq N} h_k.$$

Let U be a convex bounded open set containing the set

$$\bigcup_{k=0}^{N} \Xi([0, h_{\max}], y_k).$$

For this set U, we define

$$M_0 = \sup_{x \in U} |X(x)|, \ M_1 = \sup_{x \in U} \|DX(x)\|, \ M_2 = \sup_{x \in U} \|D^2 X(x)\|$$

(here $D^2 X$ is the second derivative of X).

Find a number $\epsilon_0 \in (0, h_{\min})$ with the following property: if $|x - y_k| \leq \epsilon_0$, then the solution $\Xi(t, x)$ is defined for $[0, \chi]$, and

$$\Xi(t, x) \in U \text{ for } t \in [0, \chi],$$

where

$$\chi = h_{\max} + \epsilon_0.$$

Finally, we define

$$\Delta = \min_{0 \leq k \leq N} |X(y_k)|, \ \theta = \max_{0 \leq k \leq N-1} \|Y_k\|.$$

Theorem 4.2.1. *Assume that (Y, h) is a pd(d, h)-pseudotrajectory for system (1.178) such that the matrix*

$$L = I - A_N \ldots A_0$$

is invertible. Let

$$C_1 = \max_{0 \le k \le N} \left(||A_{k-1} \dots A_0|| \, ||L^{-1}|| \left(1 + \sum_{m=1}^{N} ||A_N \dots A_m|| \right) + \sum_{m=1}^{k} ||A_{k-1} \dots A_m|| \right),$$

$$C = \max \left(\frac{\theta C_1 (1 + d_1) + 1}{\Delta}, C_1 \sqrt{1 + d_1} \right),$$

$$d_K = C \left(d(M_1 + \sqrt{1 + d_1}) + 3d_1 \frac{\sqrt{1 + d_1} + \Delta^{-1}}{1 - d_1(1 + \Delta^{-2})} \right),$$

and

$$\overline{M} = M_0 M_1 + 2M_1 \exp(M_1 \chi) \sqrt{1 + d_1} + M_2 \exp(2M_1 \chi)(1 + d_1).$$

Assume that for the introduced quantities the inequalities
(i) $(1 + \Delta^{-2})d_1 < 1$;
(ii) $d_K < 1$;
(iii) $2Cd(1 - d_K)^{-1}\sqrt{1 + d_1} < \epsilon_0$;
(iv) $2\overline{M}C^2 d(1 - d_K)^{-2} < 1$
hold.
Then there exists a point x_0 *and numbers* $t_k, 0 \le k \le N$, *such that the trajectory of* x_0 *is closed, and*

$$|t_k - h_k|, |x_k - y_k| \le L_0 d \text{ for } 0 \le k \le N, \tag{4.13}$$

where $x_{k+1} = \Xi(t_k, x_k)$,

$$L_0 = \frac{2C}{1 - d_K} \sqrt{1 + d_1}.$$

To prove Theorem 4.2.1, we first establish a technical statement.

Lemma 4.2.1. *Let* $\mathcal{X}$ *and* $\mathcal{Y}$ *be finite-dimensional vector spaces of the same dimension, let* B *be an open subset of* $\mathcal{X}$, *and let* $\mathcal{G} : B \to \mathcal{Y}$ *be a mapping of class* C^2 *having the following properties.*
(1) The derivative $D\mathcal{G}(v_0), v_0 \in B$, *has an inverse* $\mathcal{K}$.
(2) The set B *contains the closed ball* $\mathcal{X}_0$ *centered at* v_0 *with radius*

$$\epsilon = 2||\mathcal{K}|| \, |\mathcal{G}(v_0)|.$$

(3) The inequality

$$2M||\mathcal{K}||^2 |\mathcal{G}(v_0)| < 1$$

holds, where

$$M = \max_{v \in \mathcal{X}_0} ||D^2 \mathcal{G}(v)||.$$

Then the equation

$$\mathcal{G}(v) = 0 \tag{4.14}$$

has a unique solution in the ball $\mathcal{X}_0$.

Proof. Consider the mapping

$$T(v) = v_0 - \mathcal{K}(\mathcal{G}(v) - D\mathcal{G}(v_0)(v - v_0)).$$

Obviously, a point $v \in \mathcal{X}_0$ is a fixed point of T if and only if v is a solution of (4.14). Hence, to prove our lemma it is enough to show that T maps the ball $\mathcal{X}_0$ into itself and contracts on it.

Take $v \in \mathcal{X}_0$ and estimate

$$|T(v) - v_0| \leq ||\mathcal{K}||(|\mathcal{G}(v_0)| + |\mathcal{G}(v) - \mathcal{G}(v_0) - D\mathcal{G}(v_0)(v - v_0)|) \leq$$

$$\leq ||\mathcal{K}|| \left(|\mathcal{G}(v_0)| + \frac{M}{2}|v - v_0|^2 \right) \leq \frac{1}{2}(\epsilon + M||\mathcal{K}||\epsilon^2) <$$

$$< \frac{\epsilon}{2} + M||\mathcal{K}||^2|\mathcal{G}(v_0)|\epsilon < \epsilon.$$

This shows that $T(\mathcal{X}_0) \subset \mathcal{X}_0$.

For $v, v' \in \mathcal{X}_0$ we have

$$|T(v) - T(v')| \leq ||\mathcal{K}|| \, |\mathcal{G}(v) - \mathcal{G}(v') - D\mathcal{G}(v_0)(v - v')| \leq$$

$$\leq \frac{||\mathcal{K}||M}{2}|v - v'|^2 \leq ||\mathcal{K}||M\epsilon|v - v'| =$$

$$= 2||\mathcal{K}||^2 M|\mathcal{G}(v_0)| \, |v - v'|,$$

and it follows from condition (3) that T contracts on $\mathcal{X}_0$. This completes the proof. $\square$

Now let us prove Theorem 4.2.1. Consider the hyperplanes H_k in $\mathbb{R}^n$ which are the images of $\mathbb{R}^{n-1}$ under the mappings $z \mapsto y_k + S_k z$. These hyperplanes are approximately orthogonal to the vectors $X(y_k)$.

We want to find the sequence of points x_k described in our theorem so that $x_k \in H_k$. Thus, we are looking for a sequence of times $\{t_k : 0 \leq k \leq N\}$ and a sequence of points $\{z_k \in \mathbb{R}^{n-1} : 0 \leq k \leq N\}$ such that

$$y_{k+1} + S_{k+1}z_{k+1} = \Xi(t_k, y_k + S_k z_k), \quad 0 \leq k \leq N \qquad (4.15)$$

(let us recall that, by our convention, $y_{N+1} = y_0$ etc).

We introduce the space

$$\mathcal{X} = \mathbb{R}^{N+1} \times (\mathbb{R}^{n-1})^{N+1}$$

with the norm

$$|\{s_k \in \mathbb{R} : 0 \leq k \leq N\}, \{w_k \in \mathbb{R}^{n-1} : 0 \leq k \leq N\}| =$$

$$= \max \left(\max_{0 \leq k \leq N} |s_k|, \max_{0 \leq k \leq N} |w_k| \right),$$

and the space

$$\mathcal{Y} = \mathbb{R}^{n(N+1)}$$

with the norm

$$|\{g_k \in \mathbb{R}^n : 0 \leq k \leq N\}| = \max_{0 \leq k \leq N} |g_k|.$$

Let B be the open set in $\mathcal{X}$ defined as follows: $v = (\{s_k\}, \{w_k\})$ is in B if and only if

$$|s_k - h_k| < \epsilon_0 \text{ and } |w_k| < \frac{\epsilon_0}{1 + d_1}, \ 0 \leq k \leq N.$$

Now we define a mapping $\mathcal{G} : B \to \mathcal{Y}$ by

$$(\mathcal{G}(v))_k = y_{k+1} + S_{k+1} w_{k+1} - \Xi(s_k, y_k + S_k w_k), \ 0 \leq k \leq N. \quad (4.16)$$

It follows from (4.11) that

$$\|S_k\| \leq \sqrt{1 + d_1},$$

hence the choice of ϵ_0 implies that the mapping $\mathcal{G}$ is well-defined and belongs to the class C^2. Since for $x_k = y_k + S_k w_k$ we have

$$|x_k - y_k| \leq \sqrt{1 + d_1}|w_k|,$$

to prove Theorem 4.2.1 it is enough to find a solution $v = (\{t_k\}, \{z_k\})$ of Eq. (4.14) in the closed ball of radius $(L_0 d)/\sqrt{1 + d_1}$ centered at

$$v_0 = (\{h_k\}, 0).$$

Thus, it remains to show that $\mathcal{G}$ satisfies the conditions of Lemma 4.2.1.

Condition (1). The main technical problem is to construct the operator K inverse to $D\mathcal{G}(v_0)$. Note that, for $u = (\{\tau_k\}, \{\xi_k\}) \in \mathcal{X}$, the value of $D\mathcal{G}(v_0)u$ is given by

$$(D\mathcal{G}(v_0)u)_k = -\tau_k X(\Xi(h_k, y_k)) + S_{k+1}\xi_{k+1} - D\Xi(h_k, y_k)S_k\xi_k, \ 0 \leq k \leq N.$$

We will approximate $D\mathcal{G}(v_0)$ by an operator T. To define T, we first prove the invertibility of the operator

$$J_k = \begin{bmatrix} X(y_k)^* \\ S_k^* \end{bmatrix}.$$

We consider this operator acting from $\mathbb{R}^n$ equipped with the Euclidean norm to $\mathbb{R} \times \mathbb{R}^{n-1}$ equipped with the norm $|(s, w)| = \max(|s|, |w|)$.

Note that

$$J_k \left[\frac{X(y_k)}{|X(y_k)|^2}, S_k \right] = I - H, \quad (4.17)$$

where

$$H = \begin{bmatrix} 0 & -X(y_k)^* S_k \\ -S_k^* \frac{X(y_k)}{|X(y_k)|^2} & I - S_k^* S_k \end{bmatrix}.$$

We consider H as an operator from $\mathbb{R} \times \mathbb{R}^{n-1}$ into itself with the norm defined above. Condition (i) of our theorem implies that

$$\|H\| \leq d_1(1 + \Delta^{-2}) < 1,$$

hence Lemma 1.2.3 shows that the operator $I - H$ is invertible, and

$$\|(I - H)^{-1}\| \leq \frac{1}{1 - d_1(1 + \Delta^{-2})}.$$

Now it follows from equality (4.17) that the operator J_k is invertible, and

$$J_k^{-1} = \left[\frac{X(y_k)}{|X(y_k)|^2}, S_k \right] (I - H)^{-1}.$$

This leads to the following estimate:

$$\|J_k^{-1}\| \leq \left(\|S_k\| + \frac{1}{|X(y_k)|} \right) \frac{1}{1 - d_1(1 + \Delta^{-2})} \leq$$

$$\leq \frac{\sqrt{1 + d_1} + \Delta^{-1}}{1 - d_1(1 + \Delta^{-2})}. \tag{4.18}$$

We define the operator T mentioned above for $u \in \mathcal{X}$ by

$$(Tu)_k = J_{k+1}^{-1} \left(\begin{array}{c} -\tau_k |X(y_{k+1})|^2 - X(y_{k+1})^* Y_k S_k \xi_k \\ \xi_{k+1} - A_k \xi_k \end{array} \right), \ 0 \leq k \leq N.$$

For the operator T one easily obtains the explicit expression of the inverse operator $\mathcal{T}$ as follows. Take $g \in \mathcal{Y}$ and define $v = (\{\tau_k\}, \{\xi_k\}) = \mathcal{T}g$ by the formulas

$$\tau_k = -\frac{1}{|X(y_{k+1})|^2} X(y_{k+1})^* (Y_k S_k \xi_k + g_k), \ 0 \leq k \leq N,$$

$$\xi_0 = L^{-1}(S_0^* g_N + \sum_{m=1}^{N} A_N \ldots A_m S_m^* g_{m-1}),$$

and

$$\xi_k = A_{k-1} \ldots A_0 \xi_0 + \sum_{m=1}^{k} A_{k-1} \ldots A_m S_m^* g_{m-1}, \ 0 < k \leq N.$$

Let us show that for any g we have $T\mathcal{T}g = g$. Obviously, it is enough to check the equalities $(T\mathcal{T}g)_k = g_k$ which are equivalent to

$$\left(\begin{array}{c} -\tau_k |X(y_{k+1})|^2 - X(y_{k+1})^* Y_k S_k \xi_k \\ \xi_{k+1} - A_k \xi_k \end{array} \right) = J_{k+1} g_k, \ 0 \leq k \leq N. \tag{4.19}$$

We consider only the case $k \neq 0$ (it is useful for the reader to consider the remaining case). The "first component" on the left in (4.19) is equal to

$$X(y_{k+1})^* \{ (Y_k S_k \xi_k + g_k) - Y_k S_k \xi_k \} = X(y_{k+1})^* g_k.$$

The "second component" is equal to

$$A_k \dots A_0 \xi_0 + \sum_{m=1}^{k+1} A_k \dots A_m S_m^* g_{m-1} -$$

$$-A_k \dots A_0 \xi_0 - A_k \sum_{m=1}^{k} A_{k-1} \dots A_m S_m^* g_{m-1} = S_{k+1}^* g_k.$$

These equalities prove (4.19).

It follows from the formulas defining $\mathcal{T}$ that we can estimate

$$\|\mathcal{T}\| \le C, \tag{4.20}$$

where the constant C is introduced in the statement of Theorem 4.2.1.

Now we use $\mathcal{T}$ to construct the inverse $\mathcal{K}$ of $D\mathcal{G}(v_0)$. We claim that

$$\|\mathcal{T}(D\mathcal{G}(v_0) - \mathcal{T})\| \le d_K. \tag{4.21}$$

If inequality (4.21) holds, then we deduce from condition (ii) of our theorem and from Lemma 1.2.3 that the operator

$$K = (I + \mathcal{T}(D\mathcal{G}(v_0) - \mathcal{T}))^{-1}$$

exists. In this case, it follows from the relations

$$(K\mathcal{T})D\mathcal{G}(v_0) = (I + \mathcal{T}(D\mathcal{G}(v_0) - \mathcal{T}))^{-1} \mathcal{T} D\mathcal{G}(v_0) =$$

$$= (I + \mathcal{T} D\mathcal{G}(v_0) - I)^{-1} \mathcal{T} D\mathcal{G}(v_0) = I$$

that $\mathcal{K} = K\mathcal{T}$ is the inverse of $D\mathcal{G}(v_0)$. Now Lemma 1.2.3 and inequalities (4.20), (4.21) imply the estimate

$$\|\mathcal{K}\| \le \frac{C}{1 - d_K}. \tag{4.22}$$

To establish (4.21), we introduce an auxiliary operator $T_0 : \mathcal{X} \to \mathcal{Y}$ by

$$(T_0 u)_k = -\tau_k X(y_{k+1}) + S_{k+1} \xi_{k+1} - Y_k S_k \xi_k, \ 0 \le k \le N.$$

Now we can estimate the left-hand side of (4.21) as follows:

$$\|\mathcal{T}(D\mathcal{G}(v_0) - \mathcal{T})\| \le \|\mathcal{T}\|(\|D\mathcal{G}(v_0) - T_0)\| + \|T_0 - \mathcal{T}\|). \tag{4.23}$$

For $u \in \mathcal{X}$ and $0 \le k \le N$ we have

$$|((D\mathcal{G}(v_0) - T_0)u)_k| \le$$

$$\le (|X(\Xi(h_k, y_k)) - X(y_{k+1})| + \|(D\Xi(h_k, y_k) - Y_k)S_k\|) \, |u| \le$$

$$\leq \left(M_1 |\Xi(h_k, y_k) - y_{k+1}| + \sqrt{1 + d_1} \|D\Xi(h_k, y_k) - Y_k\| \right) |u| \leq$$

$$\leq \left(M_1 + \sqrt{1 + d_1} \right) d|u|.$$

This shows that

$$\|D\mathcal{G}(v_0) - T_0)\| \leq \left(M_1 + \sqrt{1 + d_1} \right) d. \tag{4.24}$$

We can write

$$((T_0 - T)u)_k = J_{k+1}^{-1} J_{k+1} \left(-\tau_k X(y_{k+1}) + S_{k+1}\xi_{k+1} - Y_k S_k \xi_k \right) - (Tu)_k =$$

$$= J_{k+1}^{-1} \left(\begin{array}{c} X(y_{k+1})^* S_{k+1}\xi_{k+1} \\ (A_k - S_{k+1}^* Y_k S_k)\xi_k - S_{k+1}^* X(y_{k+1})\tau_k - (I - S_{k+1}^* S_{k+1})\xi_{k+1} \end{array} \right),$$

hence it follows from (4.11), (4.12), and (4.18) that

$$\|T_0 - T\| \leq 3d_1 \frac{\sqrt{1 + d_1} + \Delta^{-1}}{1 - d_1(1 + \Delta^{-2})}. \tag{4.25}$$

Combining estimates (4.20), (4.23), (4.24), and (4.25), we obtain the desired inequality (4.21). This shows that the operator $\mathcal{K}$ is the inverse of $D\mathcal{G}(v_0)$, and inequality (4.22) holds. Thus, condition (1) of Lemma 4.2.1 is satisfied.

Condition (2). Take

$$\epsilon = 2\|\mathcal{K}\| |\mathcal{G}(v_0)|.$$

It follows from (4.7) and (4.16) that $|\mathcal{G}(v_0)| \leq d$. Now we deduce from (4.22) and from condition (iii) of our theorem that

$$\epsilon \leq \frac{L_0 d}{\sqrt{1 + d_1}} = \frac{2Cd}{1 - d_K} < \frac{\epsilon_0}{\sqrt{1 + d_1}}, \tag{4.26}$$

hence the closed ball $\mathcal{X}_0$ of radius ϵ centered at v_0 is contained in the open set B. Thus, to prove our theorem, it remains to check condition (3) of Lemma 4.2.1.

Condition (3). Take

$$v = (\{s_k\}, \{w_k\}), \quad u = (\{\tau_k\}, \{\xi_k\}), \quad u' = (\{\tau_k'\}, \{\xi_k'\}) \in \mathcal{X}.$$

Direct calculation shows that

$$(D^2\mathcal{G}(v)uu')_k =$$

$$-\tau_k\tau_k' DX(\Xi(s_k, y_k + S_k w_k))X(\Xi(s_k, y_k + S_k w_k)) -$$

$$-\tau_k DX(\Xi(s_k, y_k + S_k w_k))D\Xi(s_k, y_k + S_k w_k)S_k\xi_k' -$$

$$-\tau_k' DX(\Xi(s_k, y_k + S_k w_k))D\Xi(s_k, y_k + S_k w_k)S_k\xi_k -$$

$$-D^2\Xi(s_k, y_k + S_k w_k)(S_k\xi_k)(S_k\xi_k').$$

It follows from our assumptions that, for $v \in \mathcal{X}_0$, the inclusion

$$\Xi(t, y_k + S_k w_k) \in U$$

holds for $t \in [0, s_k]$, hence all the arguments in the expression for $D^2 \mathcal{G}(v)$ above are in U.

Since $\Theta(t) = D\Xi(t, y_k + S_k w_k)$ satisfies a system analogous to (4.10), we immediately obtain from the Gronwall lemma that

$$\|D\Xi(t, y_k + S_k w_k)\| \leq \exp(M_1 t) \tag{4.27}$$

for $t \in [0, s_k]$.

Differentiating (4.10), we see that $\Phi(t) = D^2 \Xi(t, y_k + S_k w_k)$ is a solution of the following initial-value problem:

$$\frac{d}{dt} \Phi = DX(\Xi(t, y_k + S_k w_k))\Phi + D^2 X(\Xi(t, y_k + S_k w_k))\Theta(t)\Theta(t), \quad \Phi(0) = 0,$$

hence

$$\Phi(t) = \int_0^t \Theta(t - s) D^2 X(\Xi(s, y_k + S_k w_k))\Theta(s)\Theta(s) ds$$

for $t \in [0, s_k]$.

Taking (4.27) into account, we obtain the estimate

$$\|D^2 \Xi(t, y_k + S_k w_k)\| \leq$$

$$\leq \int_0^t \exp(M_1(t - s)) M_2 \exp(2M_1 s) ds = M_2 \exp(M_1 t) \frac{\exp(M_1 t) - 1}{M_1} \leq$$

$$\leq t M_2 \exp(2M_1 t)$$

(we apply the elementary inequality

$$\exp(s) - 1 \leq s \exp(s) \text{ for } s \geq 0$$

in the last estimate).

These estimates and the expression for $D^2 \mathcal{G}(v)$ above show that

$$\|D^2 \mathcal{G}(v)\| \leq \overline{M} \text{ for } v \in \mathcal{X}_0.$$

Thus, to complete the proof of Theorem 4.2.1 it remains to apply Lemma 4.2.1 having in mind estimates (4.22), $|\mathcal{G}(v_0)| < d$, and condition (iv) of the theorem.
$\square$

In [Coo2], the described method is applied to rigorously establish the existence of one asymptotically stable and one hyperbolic "saddle" periodic orbit of the Lorenz system (1.177) for certain values of the parameters σ, r, and β.

4.3 Approximations of Spectral Characteristics

In the previous sections, the shadowing approach was applied to study "finite-time" problems. In the study of infinite-time behavior of a dynamical system, its "spectral" characteristics (for example, Lyapunov exponents of trajectories) are of great interest. We investigate two types of spectral characteristics in this section. In Subsect. 4.3.1, we show that application of numerical methods to evaluate upper Lyapunov exponents near a hyperbolic set leads to resulting errors proportional to the errors of the method and to round-off errors [Cor1]. Note that the problem of evaluation of Lyapunov exponents is also discussed in [Coo5].

In Subsect. 4.3.2, we introduce symbolic images of a dynamical system generated by partitions of its phase space. It is shown that this method allows us to approximate the Morse spectrum of the investigated dynamical system [Os3].

4.3.1 Evaluation of Upper Lyapunov Exponents

Let ϕ be a diffeomorphism of $\mathbb{R}^n$ of class C^1. Fix a point $x \in \mathbb{R}^n$. We define the *upper Lyapunov exponent* of the positive semi-trajectory $O^+(x)$ by the usual formula

$$\mu(O^+(x)) = \max_{v \in \mathbb{R}^n, |v|=1} \overline{\lim}_{m \to \infty} \frac{1}{m} \log |D\phi^m(x)v| \qquad (4.28)$$

(here $\overline{\lim}$ denotes lim sup). Below we often write $\mu(x)$ instead of $\mu(O^+(x))$.

Note that the value $|D\phi^m(x)v|$ can be written as

$$|D\phi(\phi^{m-1}(x))D\phi(\phi^{m-2}(x))\ldots D\phi(x)v|. \qquad (4.29)$$

Even if we know the semi-trajectory $O^+(x)$ exactly, due to round-off errors we really compute not the value (4.29) but an approximate value

$$|F(\phi^{m-1}(x))F(\phi^{m-2}(x))\ldots F(x)v|, \qquad (4.30)$$

where $F(x)$ is an approximation of $D\phi(x)$. It is known that Lyapunov exponents are not always stable with respect to small perurbations of the linear operators $D\phi(x)$ [By], so that if we substitute (4.30) in (4.28) instead of (4.29), this may lead to a significant change of the value $\mu(x)$ (even if the values $||F(x) - D\phi(x)||$ are uniformly small).

Below we describe conditions [Cor1] under which approximate values of upper Lyapunov exponents are close to the exact ones.

Fix $d > 0$. Consider a mapping $F : \mathbb{R}^n \to \mathrm{GL}(n, R)$ such that

$$||D\phi(x) - F(x)|| \le d \qquad (4.31)$$

for all $x \in \mathbb{R}^n$. We treat $F(x)$ as an approximation of the value $D\phi(x)$ (for example, given by a numerical method).

Let $\xi = \{x_k : k \ge 0\}$ be a d-pseudotrajectory of ϕ. We introduce the number

$$\mu(\xi, F) = \max_{v \in \mathbb{R}^n, |v|=1} \overline{\lim}_{m \to \infty} \frac{1}{m} \log |F(x_{m-1})F(x_{m-2})\dots F(x_0)v|. \qquad (4.32)$$

Theorem 4.3.1. *Let $\phi : \mathbb{R}^n \to \mathbb{R}^n$ be a diffeomorphism of class C^2. Assume that Λ is a locally maximal hyperbolic set of ϕ such that $\dim U(p) = 1$ for $p \in \Lambda$. There exists a neighborhood W of Λ and numbers $L_0, d_0 > 0$ with the following property. If $\xi \subset W$ is a d-pseudotrajectory of ϕ such that $d \leq d_0$,*

$$\text{dist}(x_0, \Lambda) < d,$$

and inequality (4.31) holds for $x \in W$, then there is a point $p \in \Lambda$ such that

$$|\mu(p) - \mu(\xi, F)| \leq L_0 d. \qquad (4.33)$$

Before we prove Theorem 4.3.1, we state an auxiliary result. We consider a sequence of linear mappings $f_i : \mathbb{R}^n \to \mathbb{R}^n, i \geq 0$, having the form

$$f_i(x) = A_i x + g_i x, \qquad (4.34)$$

where $A_i \in GL(n, \mathbb{R})$, and g_i are small $n \times n$ matrices.

It is assumed that there exist constants $M, C_0, l > 0$, and $\lambda_0 \in (0, 1)$ such that

(a1) $\|A_i\|, \|A_i^{-1}\| \leq M, i \geq 0$;

(a2) there exist linear subspaces S_i, U_i of $\mathbb{R}^n$ such that

(a2.1) $S_i \oplus U_i = \mathbb{R}^n, i \geq 0$;

(a2.2) $A_i T_i = T_{i+1}$ for $i \geq 0, T = S, U$;

(a2.3) if $v \in S_i$, then

$$|A_{i+m-1} \dots A_i v| \leq C_0 \lambda_0^m |v| \text{ for } m \geq 0;$$

(a2.4) if $v \in U_i$, then

$$|A_{i-m}^{-1} \dots A_i^{-1} v| \leq C_0 \lambda_0^m |v| \text{ for } m, i - m \geq 0;$$

(a3)

$$\|g_i\| \leq l, \; i \geq 0. \qquad (4.35)$$

Remark. It follows from the remark after Definition 1.15 that under the above conditions there exists a positive number a (depending on M, C_0, and λ) such that $\angle(S_i, U_i) \geq a, i \geq 0$.

The statement below is a special variant of the stable manifold theorem (Theorem 1.2.1) for small linear perturbations of hyperolic mappings, and it can be proved using Perron's method (see the proof of Theorem 12.5 in [Pi1]).

For fixed $i \geq 0$, we can introduce coordinates $x = (y, z)$ with respect to the representation of $\mathbb{R}^n$ in (a2.1) so that y are coordinates in S_i and z are coordinates in U_i.

Theorem 4.3.2. *Assume that mappings (4.34) satisfy conditions (a1)-(a3). Given $C > C_0, \lambda \in (0, \lambda_0), b > 0$, there exist numbers $l_0 = l_0(M, C, \lambda)$ and $K_0 = K_0(M, C, \lambda)$ such that if g_i satisfy (4.35) with $l \leq l_0$, then there exist linear subspaces $\mathcal{S}_i, \mathcal{U}_i$ of $\mathbb{R}^n$ such that*
(1) $\mathcal{S}_i, \mathcal{U}_i$ *are given by*

$$z = \xi_i y \text{ and } y = \eta_i z,$$

respectively, and

$$\|\xi_i\|, \|\eta_i\| \leq K_0 l;$$

(2) *if $x_0 \in \mathcal{S}_i$, then*

$$|f_{i+m-1} \circ \ldots \circ f_m(x_0)| \leq C\lambda^m |x_0| \text{ for } m \geq 0;$$

(3) *if $x_0 \in \mathcal{U}_i$, then*

$$|f_{i+m-1} \circ \ldots \circ f_m(x_0)| \geq \frac{\lambda^{-m}}{C} |x_0| \text{ for } m \geq 0;$$

(4) $f_i(\mathcal{S}_i) = \mathcal{S}_{i+1}$ *and* $f_i(\mathcal{U}_i) = \mathcal{U}_{i+1}$.

Now we prove Theorem 4.3.1.

Proof. Denote by C_0, λ_0 the hyperbolicity constants of Λ, let $S(p), U(p), p \in \Lambda$, be the hyperbolic structure on Λ. Find numbers M, a such that

$$\|D\phi(p)\|, \|D\phi^{-1}(p)\| \leq M$$

and

$$\angle(S(p), U(p)) \geq a \text{ for } p \in \Lambda.$$

Apply Theorem 1.2.3 to find a neighborhood U of Λ such that ϕ has the LpSP on U with constants L, d_0'. Since the set Λ is locally maximal, we may assume that the inclusion $O(p) \subset U$ implies $p \in \Lambda$. Let $\mathcal{L} \geq 1$ and $\mathcal{L}_1$ be Lipschitz constants of ϕ and $D\phi$ on U. Find a neighborhood W of Λ and a number $\Delta > 0$ such that $N_\Delta(W) \subset U$.

Set $L_1 = L\mathcal{L}$ and find $d_1 > 0$ such that

$$\mathcal{L}d_1 \leq d_0' \text{ and } L_1 d_1 < \Delta.$$

Take a d-pseudotrajectory $\xi = \{x_k : k \geq 0\} \subset W$ such that $d \leq d_1$ and $\text{dist}(x_0, \Lambda) < d$. Find a point $y \in \Lambda$ with $|x_0 - y| < d$. Set $\xi' = \{y_k : k \in \mathbb{Z}\}$, where $y_k = \phi^k(y)$ for $k \leq 0$ and $y_k = x_k$ for $k \geq 1$. Since $x_0, y_0 \in W$, the set ξ' is an $\mathcal{L}d$-pseudotrajectory of ϕ. It follows from the choice of W and d_1 that there is a point p such that

$$|\phi^k(p) - y_k| \leq L\mathcal{L}d, \ k \in \mathbb{Z}.$$

The inclusion

$$O(p) \subset \overline{N_{L_1 d}(W)}$$

implies that $p \in \Lambda$ (and hence $O(p) \subset \Lambda$). Set $p_i = \phi^i(p)$.

We can represent $F(x_i)$ in the form $F(x_i) = D\phi(p_i) + H_i$. The inequalities

$$||D\phi(x_i) - D\phi(p_i)|| \leq L_1 \mathcal{L}_1 d$$

and $||F(x_i) - D\phi(x_i)|| < d$ imply that

$$||H_i|| < L_2 d, \tag{4.36}$$

where $L_2 = L_1 \mathcal{L}_1 + 1$.

Set $S_i = S(p_i), U_i = U(p_i)$. Obviously, the mappings

$$f_i(x) = A_i x + G_i x,$$

where $A_i = D\phi(p_i)$ and $G_i = H_i$, satisfy the conditions of Theorem 4.3.2. Fix $C > C_0, \lambda \in (\lambda_0, 1)$, and apply this theorem to find the corresponding $l_0 = l_0(M, C, \lambda), K_0 = K_0(M, C, \lambda)$. Take $d_2 \in (0, d_1)$ such that $L_2 d_2 \leq l_0$, then it follows from (4.36) that for $d \leq d_2$ there exist the corresponding subspaces $\mathcal{S}_i$ and $\mathcal{U}_i$ such that for the matrices ξ_i, η_i we have the estimates

$$||\xi_i||, ||\eta_i|| \leq K_0 L_2 d, \ i \geq 0.$$

It is geometrically obvious that there exists a constant $L_3 > 0$ such that

$$\angle(\mathcal{S}_i, S_i), \angle(\mathcal{U}_i, U_i) \leq L_3 d. \tag{4.37}$$

Take $n - 1$ linearly independent unit vectors $v_1^s, \dots, v_{n-1}^s \in S_0$ and a unit vector $v^u \in U_0$. By the definition of a hyperbolic set,

$$|D\phi^m(p)v_k^s| \leq C_0 \lambda_0^m |v_k^s| \text{ for } k = 1, \dots, n-1,$$

hence

$$\overline{\lim}_{m \to \infty} \frac{1}{m} \log |D\phi^m(p)v_k^s| \leq \log \lambda_0 < 0 \text{ for } k = 1, \dots, n-1.$$

Since

$$|D\phi^m(p)v^u| \geq (C_0)^{-1} \lambda_0^{-m} |v^u|,$$

we obtain the inequality

$$\overline{\lim}_{m \to \infty} \frac{1}{m} \log |D\phi^m(p)v^u| \geq -\log \lambda_0 > 0,$$

and it follows that

$$\mu(p) = \overline{\lim}_{m \to \infty} \frac{1}{m} \log |D\phi^m(p)v^u|.$$

Similar reasons based on statements (2) and (3) of Theorem 4.3.2 show that if w^u is a unit vector in $\mathcal{U}_0$, then

$$\mu(\xi, F) = \overline{\lim}_{m \to \infty} \frac{1}{m} \log |F(x_{m-1}) \dots F(x_0) w^u|.$$

Fix $m \geq 0$, let $v(m)$ be a unit vector in U_m, and let $w(m)$ be a unit vector in $\mathcal{U}_m$. It follows from (4.37) that if

$$\langle v(m), w(m) \rangle > 0, \tag{4.38}$$

(here $\langle, \rangle$ is the scalar product), then $|v(m) - w(m)| \leq L_3 d$.

Denote

$$a_m = |D\phi(p_m)v(m)|, \quad a'_m = |F(x_m)w(m)|.$$

Let

$$v_m = D\phi^m(p)v^u, \quad v_{m+1} = D\phi^{m+1}(p)v^u.$$

Since the space U_m is one-dimensional, there exists a nonzero number b such that $v_m = bv(m)$. Then

$$v_{m+1} = D\phi(p_m)v_m = bD\phi(p_m)v(m),$$

hence

$$\frac{|v_{m+1}|}{|v_m|} = \frac{|v_{m+1}|}{|b|} = |D\phi(p_m)v(m)| = a_m.$$

It follows that

$$|v_{m+1}| = a_m|v_m| = a_m \dots a_0|v^u|,$$

so that

$$\mu(p) = \overline{\lim}_{m \to \infty} \frac{1}{m} \log(a_{m-1} \dots a_0).$$

Similarly,

$$\mu(\xi, F) = \overline{\lim}_{m \to \infty} \frac{1}{m} \log(a'_{m-1} \dots a'_0).$$

Obviously, we can take $v(m), w(m)$ in the definitions of a_m, a'_m so that inequalities (4.38) hold.

Set $\delta = w(m) - v(m)$. Take $d_2 = \min(d_1, 1)$, then (4.36) implies for $d \leq d_2$ the inequalities $\|H_m\| \leq L_2$ and $\|F(x_m)\| \leq M + L_2$. Since

$$F(x_m)w(m) = (D\phi(p_m) + H_m)(v(m) + \delta),$$

we can estimate

$$|a_m - a'_m| = |\|F(x_m)w(m)\| - |D\phi(p_m)v(m)\|| \leq |F(x_m)w(m) - D\phi(p_m)v(m)| \leq$$

$$\leq |F(x_m)\delta| + |(F(x_m) - D\phi(p_m))v(m)| \leq L_4 d, \tag{4.39}$$

where $L_4 = (M + L_2)L_3 + L_2$ (we apply (4.36) to estimate the second term).

Take an arbitrary point $q \in \Lambda$, fix a unit vector $v \in U(q)$, and set $a(q) = |D\phi(q)v|$. Since $U(q)$ contains two unit vectors, v and $-v$, the function $a(q)$ is properly defined. The unstable spaces $U(q)$ of the hyperbolic structure are continuous with respect to $q \in \Lambda$ (see Subsect. 1.2.1), hence the function a is

continuous. Since the set Λ is compact, there exist numbers a_- and a_+ such that
$$a_- \leq a(q) \leq a_+ \text{ for } q \in \Lambda.$$
Take $d_0 \in (0, d_2)$ such that $2L_4 d_0 < a_-$. Since $2/a_-$ is a Lipschitz constant of the function $\log t$ on $[a_-/2, 2a_+]$, it follows from (4.39) that if $d \leq d_0$, then for $m \geq 0$ we have
$$|\log a_m - \log a'_m| \leq L_0 d,$$
where $L_0 = 2a_- L_4$. Hence, we obtain the inequalities
$$\frac{1}{m}|\log(a_{m-1} \ldots a_0) - \log(a'_{m-1} \ldots a'_0)| \leq L_0 d,$$

and the desired inequality (4.33) follows. $\square$

Remark. We applied the assumption $\phi \in C^2$ to obtain estimate (4.36). If ϕ is of class C^1, then for given $\epsilon > 0$ one can find d_0 such that for $d \leq d_0$ the estimate $\|H_i\| \leq \epsilon$ holds. Hence, for a diffeomorphism ϕ of class C^1 one can prove an analog of Theorem 4.3.1 of the following form: given $\epsilon > 0$ there exists $d_0 > 0$ such that if $\xi \subset W$ is a d-pseudotrajectory of ϕ with $d \leq d_0$, dist$(x_0, \Lambda) < d$, and inequality (4.31) holds for $x \in W$, then there is a point $p \in \Lambda$ such that
$$|\mu(p) - \mu(\xi, F)| \leq \epsilon.$$

4.3.2 Approximation of the Morse Spectrum

Morse spectrum.

We begin with the definiton of a *chain recurrent set* [Con]. Let ϕ be a homeomorphism of a metric space (X, r). Fix $d > 0$ and denote by $P(d)$ the set of periodic d-pseudotrajectories of ϕ, i.e., of d-pseudotrajectories $\xi = \{x_k : k \in \mathbb{Z}\}$ such that there exists a number p with the property $x_k = x_{k+p}$ for any $k \in \mathbb{Z}$ (this number p is called a period of ξ). The chain recurrent set $CR(\phi)$ is defined by the equality
$$CR(\phi) = \bigcap_{d>0} P(d).$$
It is easy to show that this set is closed and ϕ-invariant.

Now let Λ be a compact invariant set of a diffeomorphism $\phi : \mathbb{R}^n \to \mathbb{R}^n$. We construct a special dynamical system connected with the pair (Λ, ϕ) as follows.

Recall that the $(n-1)$-dimensional real projective space $P_{n-1}\mathbb{R}$ is defined by identification of one-dimensional subspaces of $\mathbb{R}^n$. For $y \in \mathbb{R}^n, y \neq 0$, we denote by $[y] \in P_{n-1}\mathbb{R}$ the class of equivalence of the line $\{ky : k \in \mathbb{R}\}$. Since for nonzero $y_1, y_2 \in \mathbb{R}^n$ with $[y_1] = [y_2]$ (i.e., for $y_1 = ky_2$ with $k \neq 0$) and for any x we have $D\phi(x)y_1 = kD\phi(x)y_2$, we can define a mapping $F(x) : P_{n-1}\mathbb{R} \to P_{n-1}\mathbb{R}$ by $F(x)[y] = [D\phi(x)y]$. Below we denote points of $P_{n-1}\mathbb{R}$ by v.

Denote $T\Lambda = \Lambda \times P_{n-1}\mathbb{R}$. We fix a metric ρ on $P_{n-1}\mathbb{R}$ and introduce the corresponding metric

$$r((x,v),(x',v')) = |x - x'| + \rho(v,v')$$

on $T\Lambda$.

Now we define a mapping $\Phi(x,v) = (\phi(x), F(x)v)$. Obviously, Φ is a homeomorphism of $T\Lambda$. Fix a point $(x,v) \in T\Lambda$, take a vector $w \in \mathbb{R}^n$ such that $|w| = 1$ and $[w] = v$, and define the number

$$a(x,v) = |D\phi(x)w|.$$

The number $a(x,v)$ is properly defined, since there are two vectors with the described properties, w and $-w$. Consider a sequence $(x_k, v_k) \in T\Lambda$ such that $(x_k, v_k) \to (x,v)$. Obviously, we can choose vectors $w_k, w \in \mathbb{R}^n$ so that $|w_k| = |w| = 1$, $[w_k] = v_k$, $[w] = v$, and $w_k \to w$. In this case, we have $-w_k \to -w$, and it follows that the mapping

$$a : T\Lambda \to \mathbb{R}_+$$

defined above is continuous.

Let $\xi = \{(x_k, v_k) : k \geq 0\} \subset T\Lambda$ be a d-pseudotrajectory of Φ, as usual, this means that the inequalities

$$r((x_{k+1}, v_{k+1}), \Phi(x_k, v_k)) < d, \ k \geq 0,$$

are fulfilled. Fix a natural m and set

$$\lambda(m,\xi) = \frac{1}{m} \sum_{k=1}^{m-1} \log a(x_k, v_k).$$

The *Morse spectrum* of the dynamical system Φ on the chain recurrent set $CR(\Phi)$ is defined as follows:

$$\Sigma(\Phi) = \{\lambda = \lim_{k \to \infty} \lambda(m_k, \xi_k)\},$$

where $m_k \to \infty$ as $k \to \infty$, and $\xi_k \subset CR(\Phi)$ are d_k-pseudotrajectories of Φ with $d_k \to 0$ as $k \to \infty$. Note that we give here the definition of the Morse spectrum in the simplest possible case of a diffeomorphism of $\mathbb{R}^n$, for more general definitions see [Os3].

If $\xi = \{(x_k, v_k)\} \subset T\Lambda$ is a periodic d-pseudotrajectory of Φ of period p, we define the number

$$\lambda(\xi) = \frac{1}{p} \sum_{k=1}^{p} \log a(x_k, v_k). \tag{4.40}$$

Now we introduce the *periodic Morse spectrum*

$$\Sigma_P(\Phi) = \{\lambda = \lim_{k \to \infty} \lambda(\xi_k)\},$$

where $\xi_k \subset T\Lambda$ are periodic d_k-pseudotrajectories of Φ with $d_k \to 0$ as $k \to \infty$. Colonius and Kliemann [Col] showed that $\Sigma(\Phi) = \Sigma_P(\Phi)$. Hence, to investigate the Morse spectrum it is enough to study periodic d_k-pseudotrajectories of Φ with $d_k \to 0$.

Symbolic image.

Let us describe the concept of *symbolic image* of a dynamical system [Os1]. Let again ϕ be a homeomorphism of a compact metric space (X, r). Consider a finite covering

$$\mathcal{D} = \{\mathcal{D}(1), \ldots, \mathcal{D}(s)\} \tag{4.41}$$

of X by closed sets. For $i = 1, \ldots, s$, we introduce the sets

$$c(i) = \{j \in \{1, \ldots, s\} : \phi(\mathcal{D}(i)) \cap \mathcal{D}(j) \neq \emptyset\}.$$

The symbolic image of ϕ corresponding to the covering $\mathcal{D}$ is a graph $\mathcal{G}$ with directed edges and with s vertices (the vertices are denoted by numbers $1, \ldots, s$). The graph $\mathcal{G}$ contains an edge $i \to j$ if and only if $j \in c(i)$.

We characterize the covering $\mathcal{D}$ by two numbers,

$$\delta(\mathcal{D}) = \max_{1 \leq i \leq s} \text{diam} \mathcal{D}(i)$$

and

$$\rho(\mathcal{D}) = \min \{\text{dist}(\phi(\mathcal{D}(i)), \mathcal{D}(j)) : 1 \leq i \leq s, j \notin c(i)\}.$$

A sequence $z = \{z_k : k \in \mathbb{Z}\}$ of vertices of the graph $\mathcal{G}$ is called a *path* if $z_{k+1} \in c(z_k)$ for all k. We are mostly interested in periodic paths. If $z = \{z_k\}$ is a periodic path of period p, we write it as $z = \{z_1, \ldots, z_p\}$. A periodic path $\{z_1, \ldots, z_p\}$ is called *simple* if $z_i \neq z_j$ for $1 \leq i < j \leq p$.

A vertex of $\mathcal{G}$ is called *recurrent* if it belongs to a periodic path. Two recurrent vertices, i and j, are called *equivalent* if there is a periodic path containing both i and j. It is easy to see that the set of recurrent vertices decomposes into classes $\{\mathcal{H}_k\}$ of equivalent vertices, and each periodic path z determines a unique class $\mathcal{H}_k = \mathcal{H}(z)$ it belongs to.

Remark. An approach based on the concept of symbolic image was applied by Osipenko [Os2] to approximate the set of periodic points of a dynamical system. A close method for investigation of dynamical systems was developed by Diamond, Kloeden, and Pokrovskii [D].

Approximation process.

Now we fix a covering (4.41) of the set $T\Lambda$. For any pair of sets $\mathcal{D}(i)$ and $\mathcal{D}(j)$ with $j \in c(i)$ (what is the same, for any edge $i \to j$ of the corresponding graph $\mathcal{G}$), we fix a point $\chi(i, j) = (x(i, j), v(i, j)) \in \mathcal{D}(i)$ such that $\Phi(\chi(i, j)) \in \mathcal{D}(j)$, and introduce the number $A(i, j) = a(\chi(i, j))$.

For a periodic path $z = \{z_1, \ldots, z_p\}$ of period p we introduce the number

$$\mu(z) = \frac{1}{p} \sum_{k=1}^{p} \log A(z_k, z_{k+1}). \tag{4.42}$$

It is easy to see that if a sequence $z = \{z_k\}$ is a periodic path with two different periods p' and p'', then the corresponding numbers $\mu(z)$ obtained by formula (4.42) with $p = p'$ and $p = p''$ coincide.

Let $\mathcal{H}$ be a class of equivalent recurrent vertices of $\mathcal{G}$ and let $z^1, \ldots, z^q$ be all simple periodic paths in $\mathcal{H}$. Set

$$\mu_{\min}(\mathcal{H}) = \min\{\mu(z^i) : 1 \le i \le q\},$$

$$\mu_{\max}(\mathcal{H}) = \max\{\mu(z^i) : 1 \le i \le q\}.$$

Lemma 4.3.1. *For any periodic path $z \in \mathcal{H}$,*

$$\mu(z) \in [\mu_{\min}(\mathcal{H}), \mu_{\max}(\mathcal{H})].$$

Proof. Consider a periodic path $z = \{z_1, \ldots, z_p\}$ of period p. If the path z is not simple, there exist different indices i and j, $1 \le i, j \le p$, such that $z_i = z_j$. Since z is a periodic path, we may assume that $i = 1$. Introduce two sets of vertices of $\mathcal{G}$,

$$z' = \{z_1, \ldots, z_{j-1}\} \text{ and } z'' = \{z_j, \ldots, z_p\}.$$

It follows from

$$z_{j-1} \to z_j = z_1 \text{ and } z_p \to z_1 = z_j$$

that z' and z'' are periodic paths of periods $p' = j - 1$ and $p'' = p - j + 1$, respectively. In this case, we write $z = z' + z''$. Repeating this process, we finally represent the path z in the form

$$z = \kappa_1 z^1 + \ldots + \kappa_q z^q,$$

where κ_j are natural numbers, $z^1, \ldots, z^q$ are simple periodic paths of periods $p_1, \ldots, p_q$, and $\kappa_1 p_1 + \ldots + \kappa_q p_q = p$.

Let $z^j = \{z_1^j, \ldots, z_{p_j}^j\}$. Set $\pi_j = \kappa_j p_j / p$. We can write

$$\mu(z) = \frac{1}{p} \sum_{k=1}^{p} \log A(z_k, z_{k+1}) = \frac{1}{p} \sum_{j=1}^{q} \kappa_j \sum_{k=1}^{p_j} \log A(z_k^j, z_{k+1}^j) =$$

$$= \frac{1}{p} \sum_{j=1}^{q} \kappa_j p_j \mu(z^j) = \sum_{j=1}^{q} \pi_j \mu(z^j).$$

It follows that, for any periodic path z, the number $\mu(z)$ is a linear combination of numbers $\mu(z^j)$, where z^j are simple periodic paths. Since $\pi_1 + \ldots + \pi_q = 1$, our lemma is proved. $\qquad \square$

Now we consider the set

$$\Sigma(\mathcal{D}) = \bigcup_{\mathcal{H}} [\mu_{\min}(\mathcal{H}), \mu_{\max}(\mathcal{H})],$$

where the union is taken over all classes $\mathcal{H}$ of equivalent recurrent vertices of the graph $\mathcal{G}$.

For $b > 0$ we define the value

$$\eta(b) = \sup\{|a(x,v) - a(x',v')| : (x,v), (x',v') \in T\Lambda, r((x,v),(x',v')) \leq b\}.$$

Since the mapping a is continuous and the set $T\Lambda$ is compact, we have

$$\eta(b) \to 0 \text{ as } b \to 0.$$

The values of a are positive, hence we can find a number $\nu > 0$ such that

$$\frac{1}{\nu} = \min_{(x,v) \in T\Lambda} a(x,v).$$

Denote $\delta = \delta(\mathcal{D})$. Consider a finite sequence $\xi = \{h_k : 1 \leq k \leq p\} \subset T\Lambda$, let $h_k = (x_k, v_k)$. Find for this sequence the value $\lambda(\xi)$ by formula (4.40). Let us assume that there exist vertices $z_1, \ldots, z_p$ of $\mathcal{G}$ such that $h_k \in \mathcal{D}(z_k)$ for $1 \leq k \leq p$, and

$$z_1 \to z_2 \to \ldots \to z_p \to z_1, \tag{4.43}$$

i.e., $z = \{z_1, \ldots, z_p\}$ is a periodic path of period p.

Lemma 4.3.2. *The inequality*

$$|\lambda(\xi) - \mu(z)| \leq \nu\eta(\delta) \tag{4.44}$$

holds.

Proof. Since the points h_k and $H_k = \chi(z_k, z_{k+1})$ (chosen above to define the values $A(z_k, z_{k+1})$) belong to the same set $\mathcal{D}(z_k)$, we have

$$r(h_k, H_k) \leq \operatorname{diam}\mathcal{D}(z_k) \leq \delta(\mathcal{D}) = \delta.$$

Since $a(x,v) \geq 1/\nu$ for $(x,v) \in T\Lambda$, the inequalities

$$|\log a(x,v) - \log a(x',v')| \leq \nu|a(x,v) - a(x',v')|$$

hold for $(x,v), (x',v') \in T\Lambda$. Hence,

$$|\log A(z_k, z_{k+1}) - \log a(x_k, v_k)| \leq \nu|a(H_k) - a(h_k)| \leq \nu\eta(\delta).$$

Obviously, this inequality proves (4.44). □

Theorem 4.3.3. *For any covering (4.41) we have*

$$\Sigma(\varPhi) \subset N_{2\nu\eta(\delta)}(\Sigma(\mathcal{D})), \tag{4.45}$$

where $\delta = \delta(\mathcal{D})$.

Proof. Fix $\lambda \in \Sigma(\varPhi) = \Sigma_P(\varPhi)$. By definition, there exists a sequence ξ^m of periodic d_m-pseudotrajectories of $\varPhi$ such that $\lambda(\xi^m) \to \lambda$ and $d_m \to 0$ as $m \to \infty$. Find m_0 such that $d_m < \rho(\mathcal{D})$ for $m \geq m_0$, below we work with $m \geq m_0$.

Let $\xi^m = \{h_1^m, \ldots, h_p^m\}$. Take vertices $z_k, 1 \leq k \leq p$, of $\mathcal{G}$ such that $h_k^m \in \mathcal{D}(z_k)$. Since

$$r(\varPhi(h_k^m), h_{k+1}^m) < d_m < \rho(\mathcal{D}),$$

we have

$$\mathrm{dist}(\varPhi(\mathcal{D}(z_k)), \mathcal{D}(z_{k+1})) < \rho(\mathcal{D}),$$

and it follows that the graph $\mathcal{G}$ contains an edge $z_k \to z_{k+1}$. This means that, for $z = \{z_1, \ldots, z_p\}$, (4.43) is satisfied, i.e., z is a periodic path of period p, and we can apply Lemma 4.3.2. By inequality (4.44),

$$|\lambda(\xi^m) - \mu(z)| \leq \nu\eta(\delta).$$

By Lemma 4.3.1, $\mu(z) \in \Sigma(\mathcal{D})$, and it follows that

$$\mathrm{dist}(\lambda(\xi^m), \Sigma(\mathcal{D})) \leq \nu\eta(\delta).$$

Passing to the limit as $m \to \infty$ in this inequality, we see that

$$\mathrm{dist}(\lambda, \Sigma(\mathcal{D})) \leq \nu\eta(\delta) < 2\nu\eta(\delta),$$

and our theorem is proved. $\qquad\qquad\qquad\qquad\qquad\qquad\qquad\qquad\qquad\square$

Let us recall the definition of the *Hausdorff distance* for a metric space (X, r). Take two sets $X_1, X_2 \subset X$ and set

$$D(X_1, X_2) = \inf_{x \in X_1} \mathrm{dist}(x, X_2),$$

$$R(X_1, X_2) = \max(D(X_1, X_2), D(X_1, X_2)).$$

The number $R(X_1, X_2)$ is called the Hausdorff distance between the sets X_1 and X_2.

Theorem 4.3.4. *Let $\mathcal{D}^m$ be a sequence of coverings (4.41) such that $\delta_m = \delta(\mathcal{D}^m) \to 0$ as $m \to \infty$. Then*

$$R(\Sigma(\mathcal{D}^m), \Sigma(\varPhi)) \to 0 \text{ as } m \to \infty.$$

Proof. It follows from Theorem 4.3.3 that

$$D(\Sigma(\varPhi), \Sigma(\mathcal{D}^m)) \to 0 \text{ as } m \to \infty,$$

and it remains to show that

$$D(\Sigma(\mathcal{D}^m), \Sigma(\Phi)) \to 0 \text{ as } m \to \infty.$$

To obtain a contradiction, assume that there exists $c > 0$ and points $\mu_m \in \Sigma(\mathcal{D}^m)$ such that

$$\text{dist}(\mu_m, \Sigma(\Phi)) \geq c, \ m \geq 0.$$

Since $\Sigma(\mathcal{D}^m)$ is a finite union of closed segments, we may assume that the points μ_m are endpoints of the corresponding segments. Hence, there exist periodic paths z^m in the graphs $\mathcal{G}^m$ (constructed by $\mathcal{D}^m$) such that

$$\text{dist}(\mu(z^m), \Sigma(\Phi)) \geq c, \ m \geq 0. \tag{4.46}$$

Let $z^m = \{z_1^m, \ldots, z_p^m\}$. Take $H_k^m = \chi(z_k^m, z_{k+1}^m) \in \mathcal{D}(z_k^m)$ chosen to define $A(z_k^m, z_{k+1}^m)$. Since $\Phi(H_k^m) \in \mathcal{D}(z_{k+1}^m)$ and $\text{diam}\mathcal{D}(z_{k+1}^m) \leq \delta_m$, a sequence $\xi^m = \{h_k^m\}$, where $h_{k+lp}^m = H_k^m$ for $1 \leq k \leq p$ and $l \in \mathbb{Z}$, is a periodic $2\delta_m$-pseudotrajectory of Φ of period p.

The equalities $a(h_k^m) = A(z_k^m, z_{k+1}^m)$ imply that $\lambda(\xi^m) = \mu(z^m)$. Obviously, the sequence $\{\lambda(\xi^m)\}$ is bounded, let λ be its limit point. By the definition of $\Sigma(\Phi)$, $\lambda \in \Sigma(\Phi)$. We see that the relations $\mu(z_m) \to \lambda$ and (4.46) are contradictory. This proves our theorem. $\square$

Theorems 4.3.3 and 4.3.4 form a base of a numerical method for approximation of the Morse spectrum $\Sigma(\Phi)$ [Os3].

4.4 Discretizations of PDEs

In this section, we study shadowing properties of discretizations of a parabolic equation.

Consider a parabolic equation (3.62) with the Dirichlet boundary conditions (3.63). The following semi-implicit discretization of (3.62) is studied. Fix a natural N, let $D = 1/(N+1)$, and let $h > 0$ be the time step. We approximate the values $u(mD, nh)$ of a solution of (3.62) by $v_m^n, n \geq 0, m = 0, \ldots, N+1$, given by the system of equations

$$\Delta v^{n+1} = Av^{n+1} + \underline{f}(v^n), \ n \geq 0 \tag{4.47}$$

where

$$v^n = (v_1^n, \ldots, v_N^n) \in \mathbb{R}^N, \ \underline{f}(v) = (f(v_1), \ldots, f(v_N)),$$

and

$$\Delta v^{n+1} = \frac{v^{n+1} - v^n}{h}, \ (Av)_m = \frac{v_{m+1} - 2v_m + v_{m-1}}{D^2}$$

with $v_0 = v_{N+1} = 0$. System (4.47) defines a mapping $\phi = \phi_{h,N} : \mathbb{R}^N \to \mathbb{R}^N$ such that $v^{n+1} = \phi(v^n)$, by

$$\phi(v) = J^{-1}(v + h\underline{f}(v)), \qquad (4.48)$$

where $J = I - hA$, and I is the unit matrix.

In Subsect. 4.4.1, we study the finite-dimensional diffeomorphism ϕ and show that, for a generic nonlinearity f, it has Morse-Smale structure on its global attractor [Ei1]. This allows us to estimate the influence of round-off errors.

In Subsect. 4.4.2, shadowing results of Sect. 3.4 are applied to estimate differences between approximate and exact solutions on infinite time intervals in terms of h and D [Lar4]

4.4.1 Shadowing in Discretizations

The dynamical system ϕ defined by (4.48) was investigated by Oliva, Kuhn, and Magalhães in [Ol]. It was shown there that if

$$f \in C^1, \ |f'(u)| \le M, \text{ and } hM < 1, \qquad (4.49)$$

then ϕ is a diffeomorphism of class C^1. We assume everywhere in this subsection that conditions (4.49) are satisfied.

Due to round-off errors, we get from a computer not the sequence $\{v^n\}$ generated by (4.47) but a sequence $\{w^n \in \mathbb{R}^N : n \ge 0\}$ such that the inequalities

$$|\Delta w^{n+1} - Aw^{n+1} - \underline{f}(w^n)| < d \qquad (4.50)$$

hold with some $d > 0$ (we refine the choice of a norm in $\mathbb{R}^N$ below).

Lemma 4.4.1. *There exists $l = l(h, N)$ such that any sequence $\{w^n\}$ which satisfies (4.50) is an ld-pseudotrajectory of $\phi = \phi_{h,N}$.*

Proof. Write (4.50) as

$$\Delta w^{n+1} = Aw^{n+1} + \underline{f}(w^n) + z^n,$$

where $|z^n| < d$. Since

$$w^{n+1} = J^{-1}(w^n + h\underline{f}(w^n)) + hJ^{-1}z^n = \phi(w^n) + hJ^{-1}z^n,$$

our lemma is proved with $l = h\|J^{-1}\|$. $\qquad \square$

We study the problem of shadowing of sequences $\{w^n\}$ satisfying (4.50) near global attractors $\mathcal{A} = \mathcal{A}_{h,N}$ of diffeomorphisms ϕ. It follows from Theorem 4.4.5 below that, for a generic nonlinearity f, ϕ has the LpSP$_+$ on a neighborhood of $\mathcal{A}$. Hence, for such f and for sequences $\{w^n\}$ satisfying (4.50) with small d, there exist trajectories $\{v^n\}$ of our discretization such that

$$|v^n - w^n| \le L(h, N)d \text{ for } n \ge 0.$$

First we establish some properties of $\phi_{h,N}$ we need. In [Ol], the following statement concerning $\phi = \phi_{h,N}$ was proved.

Theorem 4.4.1.
   (1) There exists a continuous function $\mathcal{V}(v)$ on $\mathbb{R}^N$ such that $\mathcal{V}(\phi(v)) \leq \mathcal{V}(v)$, and the equality holds if and only if v is a fixed point of ϕ.
   (2) If p, q are hyperbolic fixed points of ϕ, then the stable manifold of p is transverse to the unstable manifold of q.

Let us begin with some definitions. For vectors $v, w \in \mathbb{R}^N$, let

$$\langle v, w \rangle = D \sum_{m=1}^{N} u_m v_m, \ |v|^2 = \langle v, v \rangle.$$

We say that a diffeomorphism $\phi : \mathbb{R}^N \to \mathbb{R}^N$ is *dissipative* (in the sense of Levinson) if there exists a bounded set $B \subset \mathbb{R}^N$ such that for any $v^0 \in \mathbb{R}^N$ there exists n_0 with the following property:

$$\phi^n(v^0) \in B, \ n \geq n^0.$$

It is well known (see [Ha], for example) that if ϕ is dissipative, then ϕ has a global attractor $\mathcal{A}$, and $\mathcal{A} \subset \overline{B}$ (see the definition of a global attractor in Sect. 3.4).
   Let $\mathcal{B}_{a_0,a_1}$ be the set of functions $f \in C^1(\mathbb{R})$ such that for any $u \in \mathbb{R}$ we have

$$u f(u) \leq a_0 + a_1 u^2. \tag{4.51}$$

Obviously, for $f \in \mathcal{B}_{a_0,a_1}$, the inequality

$$\langle f(v), v \rangle \leq a_0 + a_1 |v|^2 \tag{4.52}$$

holds for any $N \geq 1$ and for any $v \in \mathbb{R}^N$.

Theorem 4.4.2 *Assume that $f \in \mathcal{B}_{a_0,a_1}$ with $a_1 < \pi^2$. Then there exist numbers $h_0, N_0, \rho > 0$ depending on a_1 and M (in (4.49)) and such that*

$$\overline{\lim}_{n \to \infty} |\phi^n(v)| \leq \rho$$

for all $v \in \mathbb{R}^N, h \in (0, h_0], N \geq N_0$.

Easy calculation shows that the following statement holds.

Lemma 4.4.2. *The eigenvalues of the matrix A are*

$$-\left(\frac{4}{D^2}\right) \sin^2\left(\frac{m\pi D}{2}\right), \ m = 1, \ldots, N,$$

and the corresponding eigenvectors are

$$(\sin m\pi D, \sin 2m\pi D, \ldots, \sin N\pi D).$$

Now let us prove Theorem 4.4.2.

Proof.

Substitute

$$v^{n+1} = \frac{v^{n+1} - v^n}{2} + \frac{v^{n+1} + v^n}{2}$$

into the left-hand side of

$$\langle \Delta v^{n+1}, v^{n+1} \rangle = \langle A v^{n+1} + \underline{f}(v^n), v^{n+1} \rangle.$$

We obtain

$$\frac{|v^{n+1}|^2 - |v^n|^2}{2h} + \frac{|v^{n+1} - v^n|^2}{2h} = \langle A v^{n+1}, v^{n+1} \rangle + \langle \underline{f}(v^n), v^{n+1} \rangle. \qquad (4.53)$$

Since $|f(u_1) - f(u_2)| \le M|u_1 - u_2|$ by (4.49), it follows from the Cauchy inequality that

$$\langle \underline{f}(v^n), v^{n+1} \rangle = \langle \underline{f}(v^{n+1}), v^{n+1} \rangle + \langle \underline{f}(v^n) - \underline{f}(v^{n+1}), v^{n+1} \rangle \le$$

$$\le \langle \underline{f}(v^{n+1}), v^{n+1} \rangle + M|v^{n+1} - v^n| \cdot |v^{n+1}|. \qquad (4.54)$$

Obviously, for any $a > 0$ we have

$$|v^{n+1} - v^n| \cdot |v^{n+1}| \le \frac{a}{2}|v^{n+1}|^2 + \frac{1}{2a}|v^{n+1} - v^n|^2. \qquad (4.55)$$

Fix $b > 0$ such that $a_1 + 2b < \pi^2$. Since the matrix A is symmetric, it follows from Lemma 4.4.2 that for all $v \in \mathbb{R}^N$ we have

$$\langle A v, v \rangle \le - \left(\frac{4}{D^2} \right) \sin^2 \left(\frac{\pi D}{2} \right) |v|^2 \le -(a_1 + 2b)|v|^2,$$

where the last inequality holds for D small enough (we assume that it holds for $N \ge N_0$).

Hence, for $N \ge N_0$ and for all $v \in \mathbb{R}^N$ we have

$$\langle A v, v \rangle + \langle \underline{f}(v), v \rangle \le a_0 - 2b|v|^2. \qquad (4.56)$$

Take $a > 0$ such that $Ma < 2b$, then we deduce from (4.54)-(4.56) that the right-hand side of (4.53) does not exceed

$$a_0 - b|v^{n+1}|^2 + \frac{M}{2a}|v^{n+1} - v^n|^2. \qquad (4.57)$$

Now if we take $h_0 = a/M$, then (4.53) and (4.57) imply the inequality

$$\frac{1}{2h}[|v^{n+1}|^2 - |v^n|^2] \le a_0 - b|v^{n+1}|^2 \qquad (4.58)$$

for $n \geq 0, N \geq N_0, h \in (0, h_0]$.

It follows from (4.58) that

$$|v^{n+1}|^2(1 + 2bh) \leq |v^n|^2 + 2a_0 h,$$

therefore we have

$$|v^n|^2 \leq (1 + 2bh)^{-n}|v^0|^2 + 2a_0 h \sum_{m=1}^{n} (1 + 2bh)^{-m},$$

and this gives

$$\overline{\lim_{n \to \infty}} |v^n|^2 \leq 2a_0 h \sum_{m=1}^{\infty} (1 + 2bh_0)^{-m} = \frac{a_0}{b}.$$

We take $\rho^2 = a_0/b$, this completes the proof.   □

Thus, condition (4.51) guarantees the existence of the global attractor $\mathcal{A}(h, N)$ of $\phi_{h,N}$.

We will consider the set of functions $f : \mathbb{R} \to \mathbb{R}$ of class $C^q, q \geq 1$, with two topologies. One of them is the C^q strong Whitney topology introduced in Sect. 3.4, we denote the corresponding functional space by $\mathcal{F}_s^q$. Another considered topology is the standard *topology of uniform C^q-convergence on compact subsets* of $\mathbb{R}$. The base of neighborhoods of a function f in this topology consists of sets

$$\{g : \rho_K^q(f, g) < \epsilon\}$$

for compact sets $K \subset \mathbb{R}$ and positive numbers ϵ (the numbers $\rho_K^q(f, g)$ are defined by (3.69)). We denote the corresponding functional space by $\mathcal{F}_w^q$.

Fix a compact set $K \subset \mathbb{R}^N$ and a diffeomorphism $\phi = \phi_{h,N}$. Let $\mathcal{F}^q$ be one of the spaces $\mathcal{F}_s^q$ or $\mathcal{F}_w^q$.

Theorem 4.4.3. *For $q \geq 1$, the set*

$$\mathcal{H}^q(K) = \{f \in \mathcal{F}^q : \text{fixed points of } \phi \text{ in } K \text{ are hyperbolic}\}$$

is residual in $\mathcal{F}^q$.

Proof. Fix $q \geq 1$ and $L > 0$ such that

$$K \subset \{v : -L < v_m < L, m = 1, \ldots, N\}.$$

As usual, we say that a fixed point v of ϕ is *simple* if

$$\det(D\phi(v) - I) \neq 0. \tag{4.59}$$

Define

$$\mathcal{H} = \{f \in \mathcal{F}^q : \text{all fixed points of } \phi \text{ in } K \text{ are simple}\}.$$

We claim that $\mathcal{H}$ is a residual subset of $\mathcal{F}^q$. A point v is a fixed point of ϕ if and only if

$$\Phi(v) := Av + \underline{f}(v) = 0.$$

For a mapping $\Psi : \mathbb{R}^N \to \mathbb{R}^N$ of class C^1 we say that a point x is *critical* if

$$\operatorname{rank} D\Psi(x) < N.$$

We denote by $S(\Psi)$ the set of critical points of Ψ.

Since

$$\Phi(v) = \frac{1}{h}J(\phi(v) - v), \quad D\Phi(v) = \frac{1}{h}J(D\phi(v) - I),$$

and $\det J \neq 0$, for a fixed point v of ϕ, (4.59) is equivalent to $v \notin S(\Phi)$.

First consider the set

$$\mathbb{R}_0^N = \{v \in \mathbb{R}^N : v_i \neq v_j \text{ for } i \neq j\},$$

and define $K_0 = \mathbb{R}_0^N \cup K$. The case of $\mathbb{R}_0^N$ will be a particular case of $\mathbb{R}_A^N$ (see below), but we treat it separately to clarify the main idea.

Let $\pi = (\pi_1, \ldots, \pi_N)$ be a permutation of $\{1, \ldots, N\}$. Take $l > 0, l_1, \ldots, l_N \in \mathbb{R}$ such that

$$-L < l_1 - 2l < l_1 + 2l < l_2 - 2l < \ldots, l_N + 2l < L, \tag{4.60}$$

and denote $\mu = (l, l_1, \ldots, l_N)$. Set

$$R_{0,\pi,\mu} = \{v \in \mathbb{R}_0^N : l_i - l \leq v_{\pi(i)} \leq l_i + l, i = 1, \ldots, N\},$$

and

$$\mathcal{H}_{0,\pi,\mu} = \{f \in \mathcal{F}^q : \text{ fixed points of } \phi \text{ in } R_{0,\pi,\mu} \text{ are simple}\}.$$

Simple fixed points are isolated, hence there is only a finite number of them in a compact set. This implies that the set $\mathcal{H}_{0,\pi,\mu}$ is open.

Let us prove that $\mathcal{H}_{0,\pi,\mu}$ is dense. By Sard's theorem (see [Hirs2]),

$$\operatorname{mes} \Phi(S(\Phi)) = 0$$

(here mes is Lebesgue measure in $\mathbb{R}^N$).

Therefore, given $\epsilon > 0$ there exists $a \in \mathbb{R}^N$, $|a| < \epsilon$, such that

$$a \notin \Phi(S(\Phi)). \tag{4.61}$$

Consider $f_1 \in \mathcal{F}^q$ such that

$$f_1(u) = f(u) - a_{\pi(m)}$$

for

$$l_m - l \leq u \leq l_m + l, \, m = 1, \ldots, N$$

(here a_{π_m} is the π_m-component of a). Since (4.60) holds, l is fixed, and ϵ is arbitrary, f_1 can be found in an arbitrary $\mathcal{F}^q$-neighborhood of f. Set

$$\Phi_1(v) = Av + \underline{f}_1(v). \tag{4.62}$$

Obviously, for $v \in R_{0,\pi,\mu}$,

$$\Phi_1(v) = \Phi(v) - a, \ D\Phi_1(v) = D\Phi(v). \tag{4.63}$$

Hence, if $v \in R_{0,\pi,\mu}$ and $\Phi_1(v) = 0$, then $\Phi(v) = a$. It follows from (4.61) and (4.63) that v is a simple fixed point of

$$\phi_1(v) = J^{-1}(v + h\underline{f}_1(v)).$$

We see that the set $\mathcal{H}_{0,\pi,\mu}$ is dense. Since there exists a countable family of sets $\mathcal{H}_{0,\pi,\mu}$ such that their union contains K_0, the set

$$\mathcal{H}_0 = \cap \mathcal{H}_{0,\pi,\mu}$$

(we take the intersection corresponding to the family) is residual in $\mathcal{F}^q$, and, for $f \in \mathcal{H}_0$, all fixed points of ϕ in K_0 are simple.

Now consider a decomposition of $\{1, \ldots, N\}$ into disjoint subsets $\Lambda_1, \ldots, \Lambda_m$ numbered according to the lexicographical order, that is

$$m_1 < m_2 \text{ iff } \min\{i \in \Lambda_{m_1}\} < \min\{i \in \Lambda_{m_2}\}.$$

Let

$$\Lambda = \{\Lambda_1, \ldots, \Lambda_m\}.$$

Define the set

$$\mathbb{R}_\Lambda^N = \{v \in \mathbb{R}^N : v_i = v_j \text{ iff there exists } k \text{ with } i, j \in \Lambda_k\}.$$

The set $\mathbb{R}_0^N$ defined earlier corresponds to

$$\Lambda = \{\{1\}, \ldots, \{N\}\}.$$

Let $\pi = (\pi_1, \ldots, \pi_m)$ be a permutation of $\{1, \ldots, m\}$. Take $l > 0, l_1, \ldots, l_m \in \mathbb{R}$ such that

$$-L < l_1 - 2l < l_1 + 2l < l_2 - 2l < \ldots, l_m + 2l < L,$$

denote $\mu = (l, l_1, \ldots, l_m)$,

$$R_{\Lambda,\pi,\mu}^N = \{v \in \mathbb{R}_\Lambda^N : l_i - l \le v_j \le l_i + l \text{ if } j \in \Lambda_{\pi(i)}, i = 1, \ldots, m\},$$

and

$$\mathcal{H}_{\Lambda,\pi,\mu} = \{f \in \mathcal{F}^q : \text{ fixed points of } \phi \text{ in } R_{\Lambda,\pi,\mu} \text{ are simple}\}.$$

Obviously, any $\mathcal{H}_{\Lambda,\pi,\mu}$ is open in $\mathcal{F}^q$. Let us show that any $\mathcal{H}_{\Lambda,\pi,\mu}$ is dense. Denote by p_Λ a projection of $\mathbb{R}^N$ onto the linear subspace

$$\mathcal{R}_\Lambda = \overline{\mathbb{R}_\Lambda^N} = \{v \in \mathbb{R}^N : v_i = v_j \text{ if there exists } k \text{ with } i, j \in \Lambda_k\}.$$

Note that for $v \in \mathcal{R}_\Lambda$ we have $\underline{f}(v) \in \mathcal{R}_\Lambda$.

Consider the mapping

$$\Phi_\Lambda = p_\Lambda \circ \Phi|_{\mathcal{R}_\Lambda}$$

which is obviously of class C^1.

It follows from Sard's theorem (we take into account that Φ_Λ acts from $\mathcal{R}_\Lambda$ into $\mathcal{R}_\Lambda$) that

$$\mathrm{mes}_\Lambda \Phi_\Lambda(S(\Phi_\Lambda)) = 0$$

(here mes_Λ is Lebesgue measure in $\mathcal{R}_\Lambda$). Given $\epsilon > 0$ there exists $a \in \mathcal{R}_\Lambda$ such that $|a| < \epsilon$ and

$$a \notin \Phi_\Lambda(S(\Phi_\Lambda)).$$

Consider $f_1 \in \mathcal{F}^q$ such that

$$f_1(u) = f(u) - a_j$$

for $u \in [l_i - l, l_i + l]$ if $j \in \Lambda_{\pi(i)}$, $i = 1, \ldots, m$. We see that, for $v \in \mathcal{R}_\Lambda$,

$$\underline{f}_1(v) = \underline{f}(v) - a.$$

The same reasons as in the case of $R_{0,\pi,\mu}$ show that for $v \in \mathcal{R}_{\Lambda,\pi,\mu}$ with $\Phi_{1,\Lambda}(v) = 0$ we have

$$\mathrm{rank} D\Phi_{1,\Lambda}(v) = \dim \mathcal{R}_\Lambda, \tag{4.64}$$

where

$$\Phi_{1,\Lambda} = p_\Lambda(Av + \underline{f}_1(v)).$$

Therefore, the corresponding ϕ_1 has a finite number of fixed points in $R_{\Lambda,\pi,\mu}$, and for these fixed points (4.64) holds. Obviously, f_1 can be found in an arbitrary $\mathcal{F}^q$ neighborhood of f.

Now we show that these fixed points can be done simple. For simplicity of notation we assume that ϕ itself has a finite number of fixed points, $w^1, \ldots, w^s$, in $R_{\Lambda,\pi,\mu}$, and that (4.64) holds for Φ instead of $\Phi_{1,\Lambda}$. It follows that given a small neighborhood V_i of w^i (in $\mathcal{R}_\Lambda$) we can find a neighborhood W_i of f in $\mathcal{F}^q$ such that if $f_1 \in W_i$ and

$$f_1(w^i_m) = f(w^i_m), \quad m = 1, \ldots, N, \tag{4.65}$$

then w^i is the unique fixed point of ϕ_1 in V_i.

Take f_1 such that (4.65) holds for $w^1, \ldots, w^s$, and

$$f_1'(w^k_m) = f'(w^k_m) + \mu, \quad k = 1, \ldots, s, \ m = 1, \ldots, N, \tag{4.66}$$

where $\mu \in \mathbb{R}$. Then, for any w^k,

$$p_N^k(\mu) := \det(D\phi_1(w^k) - I) =$$

$$= \det[J^{-1}(I + h \ \mathrm{diag}(f'(w^i_1) + \mu, \ldots, f'(w^i_N) + \mu)) - I],$$

and easy computation shows that

$$p_N^k(\mu) = \frac{h^N \mu^N}{\det J} + \text{lower-degree terms.}$$

Take small disjoint neighborhoods V_k of points w^k, find the corresponding neighborhoods W_k of f in $\mathcal{F}^q$. There exists a neighborhood W_0 of f in $\mathcal{F}^q$ such that, for $f_1 \in W_0$, the corresponding diffeomorphism ϕ_1 has no fixed points in

$$R_{\Lambda,\pi,\mu} \setminus (V_1 \cup \ldots \cup V_s).$$

Thus, if we take $f_1 \in W_0$ such that (4.65) and (4.66) hold, then the set of fixed points of ϕ_1 in $R_{\Lambda,\pi,\mu}$ coincides with $\{w^1, \ldots, w^s\}$.

let μ_0 be the least positive root of the polynomials $p_N^1, \ldots, p_N^s$. Since the set of coordinates of $w^1, \ldots, w^s$ is finite, we can find $\mu \in (0, \mu_0)$ and a corresponding f_1 that satisfies (4.65) and (4.66) and belongs to a given neighborhood of f in $\mathcal{F}^q$. For this f_1, all fixed points of ϕ_1 in $R_{\Lambda,\pi,\mu}$ are simple. Hence, the set $\mathcal{H}_{\Lambda,\pi,\mu}$ is dense.

The given compact set K is a subset of a countable union of sets $R_{\Lambda,\pi,\mu}$. Note that if

$$\Lambda = \{\Lambda_1\},$$

i.e., if

$$\mathcal{R}_\Lambda = \{v : v_1 = \ldots = v_N\},$$

then $\mathcal{R}_\Lambda \cap K$ belongs to one set $R_{\Lambda,\pi,\mu}$.

Hence, the set $\mathcal{H} = \cap \mathcal{H}_{\Lambda,\pi,\mu}$ is residual in $\mathcal{F}^q$. Since both $\mathcal{F}_s^q$ and $\mathcal{F}_w^q$ are Baire spaces [Hirs2], $\mathcal{H}$ is dense in $\mathcal{F}^q$.

Obviously, the set $\mathcal{H}^q(K)$ is open. Let us prove that $\mathcal{H}^q(K)$ is dense. Take $f \in \mathcal{F}^q$. Since $\mathcal{H}$ is dense, we may take $f \in \mathcal{H}$.

Let $u_1, \ldots, u_s$ be all the coordinates of fixed points of ϕ in K. It was shown earlier that for $f \in \mathcal{H}$ there exists a neighborhood W in $\mathcal{F}^q$ such that if $f_1 \in W$ and

$$f_1(u_i) = f(u_i), \quad i = 1, \ldots, s, \tag{4.67}$$

then the set of fixed points of ϕ_1 in K coincides with that of ϕ.

Fix $\epsilon > 0$ and take f_1 satisfying (4.67) and such that

$$f_1'(u_i) = \frac{1}{1+\epsilon}\left(f'(u_i) - \frac{\epsilon}{h}\right).$$

Then we have

$$1 + hf_1'(u_i) = \frac{1}{1+\epsilon}(1 + hf'(u_i)), \quad i = 1, \ldots, s.$$

Hence, for any fixed point $v \in K$,

$$D\phi_1(v) = \frac{1}{1+\epsilon}D\phi(v).$$

Therefore, the eigenvalues μ_j of $D\phi_1(v)$ and the eigenvalues λ_j of $D\phi(v)$ are related by

$$\mu_j = \frac{1}{1+\epsilon}\lambda_j.$$

Obviously, we can find ϵ so small that f_1 is in a given neighborhood of f in $\mathcal{F}^q$, and, for all μ_j, the inequalities

$$|\mu_j| \neq 1$$

hold. Hence, fixed points of ϕ_1 in K are hyperbolic. This completes the proof.

□

Now we describe the properties of the global attractor $\mathcal{A} = \mathcal{A}_{h,N}$ for a diffeomorphism $\phi = \phi_{h,N}$ assuming that all fixed points of ϕ are hyperbolic.

Theorem 4.4.4. *If all fixed points of ϕ are hyperbolic, then ϕ has Morse-Smale structure on $\mathcal{A}$ (see the definition in Sect. 3.4).*

Proof. Since $\mathcal{A}$ is compact and hyperbolic fixed points are isolated, $\mathcal{A}$ contains a finite number of fixed points, $w^1, \ldots, w^m$. This proves (a4) (we refer to the numbers of items in the definition of a Morse-Smale structure given in Sect. 3.4 working with $\mathbb{R}^N$ and ϕ instead of $\mathcal{H}$ and $S(t)$). As usual, we denote by $W^s(w^i)$ and $W^u(w^i)$ the stable and unstable manifolds of w^i.

To prove (a5), we first show that the nonwandering set Ω of the diffeomorphism ϕ coincides with the set $\{w^1, \ldots, w^m\}$. Obviously, fixed points are in Ω. Consider a point v such that $\phi(v) \neq v$, denote $a = \mathcal{V}(v)$ and $b = \mathcal{V}(\phi(v))$ (the function $\mathcal{V}$ is given by Theorem 4.4.1). It follows from Theorem 4.4.1 that $a > b$. Set $c = (a + b)/2$.

Since ϕ and $\mathcal{V}$ are continuous, there exists a neighborhood U of v such that

$$\mathcal{V}(v') > c \text{ and } \mathcal{V}(\phi(v')) < c \text{ for } v' \in U. \tag{4.68}$$

It follows from (4.68) and from

$$\mathcal{V}(\phi^k(v')) \leq \mathcal{V}(\phi(v')) \text{ for } k \geq 1$$

that

$$\phi^k(U) \cap U = \emptyset \text{ for } k > 0,$$

this proves that $v \notin \Omega$. Hence,

$$\Omega = \{w^1, \ldots, w^m\}. \tag{4.69}$$

Now we claim that for any point $v \in \mathbb{R}^N$ there exists a fixed point $w^i, 1 \leq i \leq m$, such that

$$\phi^k(v) \to w^i \text{ as } k \to \infty,$$

this will prove (a5). Denote by $\omega(v)$ the ω-limit set of the trajectory $O(v)$, i.e., the set

$$\{\lim_{k\to\infty} \phi^{l_k}(v) : l_k \to \infty \text{ as } k \to \infty\}.$$

It is well known that for any v we have $\omega(v) \subset \Omega$. Since $\mathcal{A}$ is the global attractor of ϕ, any positive trajectory tends to $\mathcal{A}$, hence any set $\omega(v)$ is nonempty and consists of fixed points of ϕ. To prove our claim, let us show that every set $\omega(v)$ is a single fixed point.

Assume that, for some v, the set $\omega(v)$ contains two fixed points, w^i and w^j. Since the set of fixed points is finite, we can find their neigborhoods $U^1, \ldots, U^m$ such that

$$\overline{U^k \cup \phi(U^k)} \cap U^l = \emptyset \text{ for } k \neq l.$$

It follows from our assumption that there exist two sequences,

$$l_k, m_k \to \infty \text{ as } k \to \infty,$$

such that

$$\phi^{l_k}(v) \to w^i, \quad \phi^{m_k}(v) \to w^j,$$

and we can choose these sequences so that $l_k < m_k$. For large k we have $\phi^{l_k}(v) \in U^i$ and $\phi^{m_k}(v) \in U^j$. Hence, for these k there exist numbers n_k such that $l_k < n_k < m_k$ and

$$\phi^{n_k-1}(v) \in U^i, \quad \phi^{n_k}(v) \notin U^i.$$

Set $z_k = \phi^{n_k}(v)$ and let z be a limit point of the sequence $\{z_k\}$. It follows that

$$z \in \overline{\phi(U^i)} \setminus U^i.$$

Our choice of the neighborhoods U^k implies that

$$z \notin U^1 \cup \ldots \cup U^m,$$

hence $z \notin \Omega$. On the other hand, $n_k \to \infty$, and it follows that $z \in \omega(v) \subset \Omega$. The obtained contradiction proves (a5).

Now let us prove (a6). Since $\mathcal{A}$ is invariant, for any $v \in \mathcal{A}$, the inclusions $\phi^k(v) \in \mathcal{A}$ hold for $k \leq 0$. The same arguments as above (applied to ϕ^{-1} instead of ϕ) show that for any $v \in \mathcal{A}$ there is a fixed point w^i such that $v \in W^u(w^i)$. Hence,

$$\mathcal{A} \subset \bigcup_{i=1}^{m} W^u(w^i).$$

To prove the inverse inclusion, assume that for some fixed point w^i there is a point $v \in W^u(w^i) \setminus \mathcal{A}$. Then

$$a := \text{dist}(v, \mathcal{A}) > 0. \tag{4.70}$$

Since $\mathcal{A}$ is Lyapunov stable (see (a2)), there exists a neigborhood U of $\mathcal{A}$ such that for any $v' \in U$ and $k \geq 0$ we have

$$\text{dist}(\phi^k(v'), \mathcal{A}) < a.$$

It follows from $\phi^k(v) \to w^i$ as $k \to -\infty$ that there exists $k_0 < 0$ such that $v' = \phi^{k_0}(v) \in U$. Then for $k \geq 0$ we have

$$\text{dist}(\phi^{k+k_0}(v), \mathcal{A}) < a,$$

and for $k = -k_0$ this inequality contradicts to (4.70). The obtained contradiction proves (a6). The second statement of Theorem 4.4.1 proves (a7).   □

One can repeat the proof of Theorem 3.4.2 (taking $\mathbb{R}^N$ instead of $\mathcal{M}$) to show that if all fixed points of ϕ are hyperbolic, then ϕ has the LpSP$_+$ on a neighborhood of $\mathcal{A}$.

In a computer, the time step h takes rational values, so it is reasonable to consider a fixed countable set H of h-values. Theorem 4.4.3 and the reasons above prove the following statement.

Theorem 4.4.5. *For $q \geq 1$ there exists a residual subset F^q of $\mathcal{F}^q_w$ (or $\mathcal{F}^q_s$) such that if $f \in F^q$, then $\phi = \phi_{h,N}$ has the LpSP$_+$ on a neighborhood of its global attractor $\mathcal{A}$ for any $(h, N) \in H \times \mathbb{N}$.*

Remark. Let U be a neighborhood of $\mathcal{A}$ such that ϕ has the LpSP$_+$ on U with constants L, d_0 (of course, these characteristics depend on h, N). Consider a bounded set $B \subset \mathbb{R}^N$. One can find positive numbers $d_1 = d_1(B) \leq d_0$ and $n_0 = n_0(B)$ such that if a sequence $\{w^n : n \geq 0\}$ is a d-pseudotrajectory of ϕ with $d \leq d_1$ and $w^0 \in B$, then $w^n \in U$ for $n \geq n_0$ (see [Pi3] for details).

Set $u^n = w^{n+n_0}$. There exists a point v' such that $|\phi^k(v') - u^k| \leq Ld$ for $k \geq 0$. Since ϕ is Lipschitz in $\mathbb{R}^N$ with a Lipschitz constant depending on h, N, there is a constant $L_1 = L_1(B) \geq L$ such that

$$|\phi^k(v') - u^k| \leq L_1 d \text{ for } -n_0 \leq k \leq 0.$$

It follows that, for the point $v^0 = \phi^{-n_0}(v')$, the following inequalities are satisfied:

$$|\phi^k(v^0) - w^k| \leq L_1 d \text{ for } k \geq 0.$$

4.4.2 Discretization Errors on Unbounded Time Intervals

We fix two natural numbers N and K and consider the semi-implicit discretization given by (4.47) with space step $D = 1/(N+1)$ and time step $h = 1/K$.

Denote by $\mathcal{H}_N$ the subspace of $\mathcal{H}$ consisting of continuous functions $v(x)$ linear on any segment

$$[iD, (i+1)D], \ i = 0, \dots, N.$$

Our discretization generates a dynamical system $\Phi_{K,N}$ on $\mathcal{H}_N$ in the following natural way.

Take a function $v \in \mathcal{H}_N$, it defines a vector

$$u = (v(D), \dots, v(ND)) \in \mathbb{R}^N.$$

Let ϕ be as in (4.48), consider the vector

$$w = (w_1, \ldots, w_N) = \phi(u)$$

and define a function $v' \in \mathcal{H}_N$ such that $v'(iD) = w_i, i = 1, \ldots, N$. Now we set

$$\Phi_{K,N}(v) = v'.$$

Fix a neighborhood W of the global attractor $\mathcal{A}$ for the semigroup $S(t)$ generated by (3.62). Let T be the integer number introduced for $S(t)$ in Sect. 3.4. Define

$$\tau(v) = \sigma^T(v) = S(T)v.$$

Find a bounded neighborhood W of $\mathcal{A}$ and numbers d_0, L_0 such that the statement of Theorem 3.4.1 holds with these W, d_0, L_0 for τ instead of σ (this is possible due to Lemmas 3.4.1 and 3.4.2).

Since $\mathcal{A}$ is Lyapunov stable, we can find neighborhoods W_1, W_2 of $\mathcal{A}$ and a positive number Δ such that

$$N_\Delta(W_2) \subset W_1 \text{ and } S(t)W_1 \subset W \text{ for } t \geq 0.$$

Applying the results of [Lar1], one can prove the following statement.

Lemma 4.4.3 *There exists a neigborhood W_0 of $\mathcal{A}$ in $\mathcal{H}$ and positive numbers C_0, K_0, N_0 such that if $K \geq K_0$ and $N \geq N_0$, then*
(1) $\Phi_{K,N}^n(v) \in W_2$ for $v \in W_0$ and $n \geq 0$;
(2) if $u \in W, v \in W \cap \mathcal{H}_N$, and the inclusions

$$S(t)u \in W, \ 0 \leq t \leq 2T, \ \Phi_{K,N}^n(v) \in W, \ 0 \leq n \leq \frac{2T}{h},$$

hold, then

$$|S(nh)u - \Phi_{K,N}^n(v)| \leq C_0(|u - v| + D + h) \text{ for } T \leq nh \leq 2T$$

(here $|.|$ is the norm of $\mathcal{H} = H_0^1$).

Below we assume that the numbers K_0 and N_0 satisfy the inequality

$$C_0 \left(\frac{1}{K_0} + \frac{1}{N_0 + 1} \right) < \min \left(d_0, \frac{\Delta}{L_0} \right). \tag{4.71}$$

Now we fix $K \geq K_0$, $N \geq N_0$, and write Φ instead of $\Phi_{K,N}$.

Theorem 4.4.6. *There exists $m = m(D + h)$ such that for any $v_0 \in W_0 \cap \mathcal{H}_N$ we can find u with the property*

$$|S(nh)u - \Phi^{mKT+n}(v_0)| \leq L(D + h) \text{ for } n \geq 0,$$

where $L = C_0(L_0 C_0 + 1)$.

Proof. Take $v_0 \in W_0 \cap \mathcal{H}_N$ and construct a sequence $\{v_n : n \geq 0\}$ setting $v_{n+1} = \Phi^{KT}(v_n)$. By the choice of W_0, the inclusions

$$S(t)v_n \in W, \ t \geq 0, \ \text{and} \ \Phi^k(v_n) \in W, \ k \geq 0,$$

hold. We can apply statement (2) of Lemma 4.4.3 (with $n = KT$ and $u = v = v_k$) to show that

$$|\tau(v_k) - v_{k+1}| = |S(T)v_k - \Phi^{KT}(v_k)| \leq d,$$

where $d = C_0(D + h)$. Since $hK_0 \leq 1$ and $D(N_0 + 1) \leq 1$, inequality (4.71) implies that $d < d_0$.

It follows from Lemma 3.4.4 that there exists $m_0 = m_0(D + h)$ such that

$$\text{dist}(v_k, \mathcal{M}) \leq 2d$$

for some $k \leq m_0$. Set $w_n = v_{k+n}$ for $n \geq 0$. It follows that the sequence w_n satisfies all the conditions of Theorem 3.4.1, hence there exists y such that

$$|\tau^n(y) - w_n| \leq L_0 d \text{ for } n \geq 0.$$

Set $m = m_0 + 1, u' = \tau^{m_0 - k}(y)$, and $u = \tau^{m-k}(y)$. Since

$$|u' - w_{m_0 - k}| = |\tau^{m_0 - k}(y) - w_{m_0 - k}| \leq L_0 d = L_0 C_0(D + h) < \Delta,$$

it follows from statement (1) of Lemma 4.4.3 and the choice of W_2 and Δ that $S(t)u' \in W$ for $t \geq 0$. Hence, we can apply statement (2) of Lemma 4.4.3 to show that

$$|S(T + nh)u' - \Phi^{KT+n}(v_{m_0})| \leq C_0(|u' - v_{m_0}| + D + h) = L(D + h) \quad (4.72)$$

for $0 \leq nh \leq T$. Note that

$$S(nh)u = S(T + nh)u' \text{ and } \Phi^{mKT+n}(v_0) = \Phi^{KT+n}(v_{m_0}),$$

hence inequality (4.72) establishes the statement of our theorem for $0 \leq n \leq KT$. The same reasons and the inequality

$$|u - v_m| \leq L_0 d$$

prove the statement of our theorem for $KT \leq n \leq 2KT$, and so on. $\qquad \square$

Remark. It follows from the proof of Lemma 3.4.4 that one can take

$$m_0 = \log_{(\nu+1/2)} \frac{2C(D + h)}{R} + 1,$$

where R is the radius of a ball that contains W.

References

[Af1]   Afraimovich, V.S. and Nekorkin, V.I.: Stable stationary motions in a chain of diffusively coupled maps [in Russian]. Preprint N 267. Inst. Appl. Phys. N. Novgorod (1991)

[Af2]   Afraimovich, V.S. and Nekorkin, V.I.: Chaos of travelling waves in a discrete chain of diffusively coupled maps [in Russian]. Preprint N 330. Inst. Appl. Phys. N. Novgorod (1991)

[Af3]   Afraimovich, V.S. and Pilyugin, S.Yu.: Special pseudotrajectories for lattice dynamical systems. Random Comput. Dynamics 4 (1996) 29–47

[Ak]   Akin, E.: The General Topology of Dynamical Systems. Grad. Stud. in Math. 1. Amer. Math. Soc., Providence, RI (1993)

[AlN]   Al-Nayef, A., Diamond, P., Kloeden, P., Kozyakin, V., and Pokrovskii, A.: Bi-shadowing and delay equations. Dyn. Stab. Syst. 11 (1996) 121–135

[Alo]   Alouges, F. and Debussche, A.: On the qualitative behavior of the orbits of a parabolic partial differential equation and its discretization in the neighborhood of a hyperbolic fixed point. Numer. Funct. Anal. Optim. 12 (1991) 253–269

[Ang]   Angenent, S.B.: The Morse-Smale property for a semi-linear parabolic equation. J. Diff. Equat. 62 (1986) 427–442

[Ano1] Anosov, D.V.: Geodesic Flows on Closed Riemannian Manifolds of Negative Curvature. Proc. Steklov Math. Inst. 90 (1967). Amer. Math. Soc., Providence, RI (1969)

[Ano2] Anosov, D.V.: On a class of invariant sets of smooth dynamical systems [in Russian]. In: Proc. 5th Int. Conf. on Nonl. Oscill. 2. Kiev (1970) 39–45

[Ano3] Anosov, D.V. and Bronshtein, I.U.: Topological dynamics. In: Dynamical Systems I. Ordinary Differential Equations and Smooth Dynamical Systems. EMS 1. Springer-Verlag (1988)

[Ao1]   Aoki, N.: Topological dynamics. In: Topics on General Topology. North-Holland, Amsterdam (1989) 625–740

[Ao2]   Aoki, N. and Hiraide, K.: Topological Theory of Dynamical Systems. Recent Advances. North-Holland Math. Library 52. North-Holland, Amsterdam (1994)

[Bab]   Babin, A.V. and Vishik, M.I.: Attractors of Evolution Equations. Stud. Math. Appl. 25. North-Holland, Amsterdam (1992)

[Bar]   Barge, M. and Swanson, R.: Rotation shadowing properties of circle and annulus maps. Ergod. Theory Dyn. Syst. 8 (1988) 509–521

[Beg]   Begun, E.N. and Pilyugin, S.Yu.: Uniformly Lipschitz shadowing of pseudotrajectories [in Russian]. Vestn. SPbGU, Issue 1 (1996) 3–7

[Ben]   Benaïm, M. and Hirsch, M.W.: Asymptotic pseudotrajectories and chain recurrent flows, with applications. J. Dynam. Diff. Equat. **8** (1996) 141–176

[Bey]   Beyn, W.-J.: On the numerical approximation of phase portraits near stationary points. SIAM J. Numer. Anal. **24** (1987) 1095–1113

[Bi]   Birkhoff, G.: Dynamical Systems. Amer. Math. Soc., Providence, RI (1927)

[Bl1]   Blank, M.L.: Metric properties of ϵ-trajectories of dynamical systems with stochastic behavior. Ergod. Theory Dyn. Syst. **8** (1988) 365–378

[Bl2]   Blank, M.L.: Shadowing of ϵ-trajectories of general multidimensional mappings. Wiss. Z. Tech. Univ. Dresden **40** (1991) 157–159

[Bo1]   Bowen, R.: Periodic orbits for hyperbolic flows. Amer. J. Math. **94** (1972) 1–30

[Bo2]   Bowen, R.: Equilibrium States and the Ergodic Theory of Anosov Diffeomorphisms. Lect. Notes in Math. **470**. Springer-Verlag (1975)

[Bro]   Bronshtein, I.U.: Nonautonomous Dynamical Systems [in Russian]. Kishinev (1984)

[Bru]   Brunovsky, P. and Chow, S.-N.: Generic properties of stationary state solutions of reaction-diffusion equations. J. Diff. Equat. **53** (1984) 1–23

[By]   Bylov, B.V., Vinograd, R.E., Grobman, D.M., and Nemytski, V.V.: Theory of Lyapunov Exponents [in Russian]. Moscow (1966)

[Cha]   Chaffee, N. and Infante, E.: A bifurcation problem for a nonlinear parabolic equation. J. Appl. Anal. **4** (1974) 17–37

[Che]   Chen, L. and Li, S.-H.: Shadowing properties for inverse limit spaces. Proc. Amer. Math. Soc. **115** (1992) 573–580

[Cho1]   Chow, S.-N., Lin, X.-B., and Palmer, K.J.: A shadowing lemma for maps in infinite dimensions. In: Differential Equations (Xanthi, 1987). Dekker, N.Y. (1989) 127-135

[Cho2]   Chow, S.-N. and Palmer, K.J.: On the numerical computation of orbits of dynamical systems: the one-dimensional case. J. Dynam. Diff. Equat. **3** (1991) 361–379

[Cho3]   Chow, S.-N. and Palmer, K.J.: On the numerical computation of orbits of dynamical systems: the higher-dimensional case. J. Complexity **8** (1992) 398–423

[Cho4]   Chow, S.-N. and Palmer, K.J.: The accuracy of numerically computed orbits of dynamical systems in $\mathbf{R}^k$. In: Differential Equations and Mathematical Physics (Birmingham, AL, 1990). Academic Press, Boston, MA (1992) 39–44

[Cho5]   Chow, S.-N., Lu, K. and Sell, G.R.: Smoothness of inertial manifolds. J. Math. Anal. Appl. **169** (1992) 283–312

[Cho6]   Chow, S.-N. and Van-Vleck, E.S.: A shadowing lemma for random diffeomorphisms. Random Comput. Dynamics **1** (1992/93) 197–218

[Cho7]   Chow, S.-N. and Shen, W.: Dynamics in a Discrete Nagumo Equation. Preprint (1993)

[Cho8]   Chow, S.-N. and Van-Vleck, E.S.: A shadowing lemma approach to global error analysis for initial value ODEs. SIAM J. Sci. Comput. **15** (1994) 959–976

[Cho9]   Chow, S.-N. and Van-Vleck, E.S.: Shadowing of lattice maps. Contemp. Math. **172** (1994) 97–113

[Col]   Colonius, F. and Kliemann, W.: The Lyapunov spectrum of families of time varying matrices. Rep. 504, Inst. Math., Univ. Augsburg (1994)

[Con]   Conley, R.: Isolated Invariant Sets and the Morse Index. Reg. Conf. Series in Math. **38**. Amer. Math. Soc., Providence, RI (1978)

[Coo1]   Coomes, B.A., Koçak, H., and Palmer, K.J.: Shadowing orbits of ordinary differential equations. J. Comput. Appl. Math. **52** (1994) 35–43

[Coo2]   Coomes, B.A., Koçak, H., and Palmer, K.J.: Periodic shadowing. Contemp. Math. **172** (1994) 115–130

[Coo3]   Coomes, B.A., Koçak, H., and Palmer, K.J.: A shadowing theorem for ordinary differential equations. Z. Angew. Math. Phys. **46** (1995) 85–106

[Coo4]   Coomes, B.A., Koçak, H., and Palmer, K.J.: Rigorious computational shadowing of orbits of ordinary differential equations. Numer. Math. **69** (1995) 401–421

[Coo5]   Coomes, B.A., Koçak, H., and Palmer, K.J.: Shadowing in discrete dynamical systems. In: Six Lectures on Dynamical Systems. World Scientific (1996) 163–211

[Coo6]   Coomes, B.A., Koçak, H., and Palmer, K.J.: Long periodic shadowing. Numer. Algorithms **14** (1997) 55–78

[Cor1]   Corless, R. and Pilyugin, S.Yu.: Evaluation of upper Lyapunov exponents on hyperbolic sets. J. Math. Anal. Appl. **189** (1995) 145–159

[Cor2]   Corless, R. and Pilyugin, S.Yu.: Approximate and real trajectories for generic dynamical systems. J. Math. Anal. Appl. **189** (1995) 409–423

[Cou]   Coupled Map Lattices (ed. K. Kaneko). John Wiley (1992)

[Cov]   Coven, E.M., Kan, I., and Yorke, J.A.: Pseudo-orbit shadowing in the family of tent maps. Trans. Amer. Math. Soc. **308** (1988) 227–241

[D]   Diamond, P., Kloeden, P., and Pokrovskii, A.: Cycles of spatial discretizations of shadowing dynamical systems. Math. Nachr. **171** (1995) 95–110

[DG]   De Gregorio, S.: The study of periodic orbits of dynamical systems. The use of a computer. J. Stat. Phys. **38** (1985) 947–972

[DM]   de Melo, W.: Moduli of stability of two-dimensional diffeomorphisms. Topology **19** (1980) 9–21

[Ea]   Easton, R.: Chain transitivity and the domain of influence of an invariant set. In: The Structure of Atractors in Dynamical Systems (North Dacota State Univ., June 1977). Lect. Notes in Math. **668**. Springer-Verlag (1978) 95–102

[Ei1]   Eirola, T. and Pilyugin, S.Yu.: Pseudotrajectories generated by a discretization of a parabolic equation. J. Dynam. Diff. Equat. **8** (1996) 281–297

[Ei2]   Eirola, T., Nevanlinna, O., and Pilyugin, S.Yu.: Limit shadowing property. Numer. Funct. Anal. Optim. **18** (1997) 75–92

[Ek]   Ekeland, I.: Some lemmas about dynamical systems. Math. Scand. **52** (1983) 262–268

[Fe]   Feçkan, M.: A remark on the shadowing lemma. Funkcialaj Ekvacioj **34** (1991) 391–402

[Fo]   Foias, C., Sell, G.R., and Temam, R.: Inertial manifolds for nonlinear evolutionary equations. J. Diff. Equat. **73** (1988) 309–353

[Fr1]   Franke, J.E. and Selgrade, J.F.: Hyperbolicity and chain recurrence. J. Diff. Equat. **26** (1977) 27–36

262   References

[Fr2]   Franke, J.E. and Selgrade, J.F.: A computer method for verification of asymptotically stable periodic orbits. SIAM J. Math. Anal. **10** (1979) 614–628

[Ge]   Gedeon, T. and Kuchta, M.: Shadowing property of continuous maps. Proc. Amer. Math. Soc. **115** (1992) 271–281

[Gr]   Grebogi, C., Hammel, S.M., Yorke, J.A., and Sauer T.: Shadowing of physical trajectories in chaotic dynamics: containment and refinement. Phys. Rev. Lett. **65** (1990) 1527–1530

[Gu1]   Guckenheimer, J.: A strange, strange attractor. In: The Hopf Bifurcation Theorem and its Applications. Springer-Verlag (1976) 368–381

[Gu2]   Guckenheimer, J., Moser, J., and Newhouse, S.: Dynamical Systems. Birkhäuser-Verlag (1980)

[Had]   Hadeler, K.P.: Shadowing orbits and Kantorovich's theorem. Numer. Math. **73** (1996) 65–73

[Hal]   Hale, J.K.: Asymptotic Behavior of Dissipative Systems. Math. Surv. Monogr. **25**. Amer. Math. Soc., Providence, RI (1988)

[Ham]   Hammel, S.M., Yorke, J.A., and Grebogi, C.: Numerical orbits of chaotic processes represent true orbits. Bull. Amer. Math. Soc. **19** (1988) 465–469

[Har]   Hardy, G.H., Littlewood, J.E., and Pólya, G.: Inequalities. Cambridge Univ. Press (1934)

[Hay1]   Hayashi, S.: Diffeomorphisms in $\mathcal{F}^1$ satisfy Axiom A. Ergod. Theory Dyn. Syst. **12** (1992) 233–253

[Hay2]   Hayashi, S.: On the solution of C^1 stability conjecture for flows. Preprint.

[He1]   Henry, D.: Geometric Theory of Semilinear Parabolic Equations. Lect. Notes in Math. **840**. Springer-Verlag (1981)

[He2]   Henry, D.: Some infinite-dimensional Morse-Smale systems defined by parabolic PDE. J. Diff. Equat. **59** (1985) 165–205

[He3]   Henry, D.B.: Exponential dichotomies, the shadowing lemma, and homoclinic orbits in Banach spaces. Resenhas IME-USP **1** (1994) 381–401

[Hira1]   Hiraide, K.: Expansive homeomorphisms with the pseudo-orbit tracing property on compact surfaces. J. Math. Soc. Japan **40** (1988) 123–137

[Hira2]   Hiraide, K.: Expansive homeomorphisms with the pseudo-orbit tracing property on n-tori. J. Math. Soc. Japan **41** (1989) 357–389

[Hirs1]   Hirsch, M., Palis, J., Pugh, C., and Shub, M.: Neighborhoods of hyperbolic sets. Invent. math. **9** (1970) 133–163

[Hirs2]   Hirsch, M.: Differential Topology. Springer-Verlag (1976)

[Hirs3]   Hirsch, M., Pugh, C.C., and Shub, M.: Invariant Manifolds. Lect. Notes in Math. **583**. Springer-Verlag (1977)

[Hirs4]   Hirsch, M.: Asymptotic phase, shadowing and reaction-diffusion systems. In: Differential Equations, Dynamical Systems, and Control Science. Lect. Notes in Pure and Applied Math. **152**. Marcel Dekker Inc. New York, Basel, Hong Kong (1994) 87–99

[Hu]   Hurley, M.: Consequences of topological stability. J. Diff. Equat. **54** (1984) 60–72

[Ka]   Kakubari, S.: A note on a linear automorphism of $\mathbf{R}^n$ with the pseudo-orbit tracing property. Sci. Rep. Niigata Univ. Ser. A (1987) 35–37

[Kato1]   Kato, K.: Pseudo-orbits and stabilities of flows. Mem. Fac. Sci. Kochi Univ. **5** (1984) 45–62

[Kato2]   Kato, K.: Pseudo-orbits and stabilities of flows, II. Mem. Fac. Sci. Kochi
           Univ. **6** (1985) 33–43
[Kato3]   Kato, K.: Hyperbolicity and pseudo-orbits for flows. Mem. Fac. Sci. Kochi
           Univ. **12** (1991) 43–55
[Katok1]   Katok, A.: Local properties of hyperbolic sets [in Russian]. Appendix to the
           Russian translation of [Ni1]. Moscow (1975)
[Katok2]   Katok, A. and Hasselblatt, B.: Introduction to the Modern Theory of Dy-
           namical Systems. Encyclopedia of Math. and its Appl. **54**. Cambridge Univ.
           Press (1995)
[Ki]   Kirby, R. and Siebenmann, L.C.: Foundational Essays on Topological Mani-
           folds, Smoothings, and Triangulations. Annals of Math. Stud. **88**. Princeton
           Univ. Press (1977)
[Ko1]   Komuro, M.: One-parameter flows with the pseudo orbit tracing property.
           Monatsh. Math. **98** (1984) 219–253
[Ko2]   Komuro, M.: Lorenz attractors do not have the pseudo-orbit tracing prop-
           erty. J. Math. Soc. Japan **37** (1985) 489–514
[Kr]   Kruger, T. and Troubetzkoy, S.: Markov partitions and shadowing for non-
           uniformly hyperbolic systems with singularities. Ergod. Theory Dyn. Syst.
           12 (1992) 487–508
[Lad1]   Ladyzhenskaya, O.A.: Attractors for Semi-Groups and Evolution Equations.
           Cambridge Univ. Press (1991)
[Lad2]   Ladyzhenskaya, O.A.: Globally stable difference schemes and their attrac-
           tors [in Russian]. Preprint POMI P-5-91. St.-Petersburg (1991)
[Lanf]   Lanford, O.E. III: Introduction to the mathematical theory of dynami-
           cal systems. In: Chaotic Behavior of Deterministic Systems (Les Houches,
           1981). North-Holland, Amsterdam (1983) 3–51
[Lani]   Lani-Wayda, B.: Hyperbolic Sets, Shadowing, and Persistence for Nonin-
           vertible Mappings in Banach Spaces. Pitman Res. Notes in Math. Longman
           (1995)
[Lar1]   Larsson, S.: Nonsmooth data error estimates with application to the study
           of the long-time behavior of finite element solutions of semilinear parabolic
           equations. Preprint 1992-36, Dept. Math. Chalmers Univ. Techn. Göteborg
           Univ. (1992)
[Lar2]   Larsson, S. and Sanz-Serna, J.-M.: The behavior of finite element solutions
           of semilinear parabolic problems near stationary points. SIAM J. Numer.
           Anal. **31** (1994) 1000–1018
[Lar3]   Larsson, S. and Sanz-Serna, J.-M.: A shadowing result with applications
           to finite element approximation of reaction-diffusion equations. Preprint
           1996-05, Dept. Math. Chalmers Univ. Techn. Göteborg Univ. (1996)
[Lar4]   Larsson, S. and Pilyugin, S.Yu.: Numerical shadowing near the global at-
           tractor for a semilinear parabolic equation. Preprint 1998-21, Dept. Math.
           Chalmers Univ. Techn. Göteborg Univ. (1998)
[Lo]   Lorenz, E.: Deterministic nonperiodic flow. J. Atmosph. Sci. **20** (1963) 130–
           141
[Ma]   Mañé, R.: A proof of the C^1-stability conjecture. IHES Publ. Math. **66**
           (1988) 161-210
[Me]   Meyer, K.R. and Sell, G.R.: An analytic proof of the shadowing lemma.
           Funkcialaj Eqvacioj **30** (1987) 127–133

[Mi]   Mizera, I.: Generic properties of one-dimensional dynamical systems. In: Ergodic Theory and Related Topics III (Gustrov, 1990). Lect. Notes in Math. **1514**. Springer-Verlag (1992)

[Morim1] Morimoto, A.: Stochastically stable diffeomorphisms and Takens conjecture. Surikais Kokyuruko **303** (1977) 8–24

[Morim2] Morimoto, A.: The method of pseudo-orbit tracing and stability of dynamical systems. Sem. Note **39**. Tokyo Univ. (1979)

[Morim3] Morimoto, A.: Some stabilities of group automorphisms. In: Manifolds and Lie Groups (Progress in Math., 14). Birkhäuser-Verlag (1981) 283–299

[Moriy]   Moriyasu, K.: The topological stability of diffeomorphisms. Nagoya Math. J. **123** (1991) 91–102

[Mu]   Munkres, J.: Obstructions to the smoothing of piecewise-differentiable homeomorphisms. Ann. Math. **72** (1960) 521–554

[Na]   Nadzieja, T.: Shadowing lemma for family of ϵ-trajectories. Arch. Math. **27A** (1991) 65–77

[Ni1]   Nitecki, Z.: Differentiable Dynamics. MIT Press (1971)

[Ni2]   Nitecki, Z.: On semi-stability of diffeomorphisms. Invent. math. **14** (1971) 83–122

[Ni3]   Nitecki Z. and Shub, M.: Filtrations, decompositions, and explosions. Amer. J. Math. **97** (1975) 1029–1047

[Nu]   Nusse, H.E. and Yorke, J.A.: Is every approximate trajectory of some process near an exact trajectory of a nearby process? Comm. Math. Phys. **114** (1988) 363–379

[Ol]   Oliva, W.M., Kuhl, N.M., and Magalhães, L.T.: Diffeomorphisms of $\mathbb{R}^n$ with oscillatory Jacobians. Publ. Mat. **37** (1993) 255–269

[Od]   Odani, K.: Generic homeomorphisms have the pseudo-orbit tracing property. Proc. Amer. Math. Soc. **110** (1990) 281–284

[Om1]   Ombach, J.: Equivalent conditions for hyperbolic coordinates. Topology Appl. **23** (1986) 87–90

[Om2]   Ombach, J.: The simplest shadowing. Ann. Polon. Math. **58** (1993) 253–258

[Om3]   Ombach, J.: Shadowing for linear systems of differential equations. Publ. Mat. **37** (1993) 245–253

[Os1]   Osipenko, G.S.: On a symbolic image of a dynamical system [in Russian]. In: Boundary-Value Problems. Perm' (1983) 101–105

[Os2]   Osipenko, G.S.: Periodic points and symbolic dynamics. In: Seminar on Dynamical Systems (St.-Petersburg, 1991). Progr. Nonlinear Diff. Equat. Appl. **12**. Birkhäuser-Verlag (1994) 261–267

[Os3]   Osipenko, G.S.: Morse spectrum of dynamical systems and symbolic image. Proc. 15th IMACS World Congress **1** (1997) 25–30

[Pali]   Palis, J.: On Morse-Smale dynamical systems. Topology **8** (1969) 385–404

[Palm1]   Palmer, K.J.: Exponential dichotomies and transversal homoclinic points. J. Diff. Equat. **55** (1984) 225–256

[Palm2]   Palmer, K.J.: Exponential dichotomies, the shadowing lemma and transversal homoclinic points. Dynamics Reported **1** (1988) 265–306

[Palm3]   Palmer, K.J.: Shadowing and Silnikov chaos. Nonlinear Analysis, Theory, Methods & Applications **27** (1996) 1075–1093

[Par]   Park, J.S., Lee, K.H., and Koo, K.S.: Hyperbolic homeomorphisms. Bull. Korean Math. Soc. **32** (1995) 93–102

[Pe]   Pennings, T. and Van-Eeuwen, J.: Pseudo-orbit shadowing on the unit interval. Real Anal. Exchange 16 (1990/91) 238-244

[Pi1]   Pilyugin, S.Yu.: Introduction to Structurally Stable Systems of Differential Equations. Birkhäuser-Verlag (1992)

[Pi2]   Pilyugin, S.Yu.: The Space of Dynamical Systems with the C^0-Topology. Lect. Notes in Math. 1571. Springer-Verlag (1994)

[Pi3]   Pilyugin, S.Yu.: Complete families of pseudotrajectories and shape of attractors. Random Comput. Dynamics 2 (1994) 205-226

[Pi4]   Pilyugin, S.Yu.: Shadowing in structurally stable flows. J. Diff. Equat. 140 (1997) 238-265

[Pi5]   Pilyugin, S.Yu. and Plamenevskaya, O.B.: Shadowing is generic [to appear in Topology Appl.].

[Pla1]   Plamenevskaya, O.B.: Shadowing and limit shadowing on the circle [in Russian]. Vestn. SPbGU (1997)

[Pla2]   Plamenevskaya, O.B.: Weak shadowing for two-dimensional diffeomorphisms [to appear].

[Pli1]   Pliss, V.A.: Integral Sets of Periodic Systems of Differential Equations [in Russian]. Moscow (1977)

[Pli2]   Pliss, V.A.: Uniformly bounded solutions of linear systems of differential equations [in Russian]. Differents. Uravneniya 13 (1977) 883-891

[Pli3]   Pliss, V.A.: Sets of linear systems of differential equations with uniformly bounded solutions [in Russian]. Differents. Uravneniya 16 (1980) 1599-1616

[Po]   Poon, L., Dawson, S.P., Grebogi, C., Sauer, T., and Yorke, J.A.: Shadowing in chaotic systems. In: Dynamical Systems and Chaos 2. World Scientific (1995) 13-21

[Pu]   Pugh, C. and Shub, M.: The Ω-stability theorem for flows. Invent. math. 11 (1970) 150-158

[Q]   Quinn, F.: Topological transversality holds in all dimensions. Bull. Amer. Math. Soc. 18 (1988) 145-148

[Re]   Reinfelds, A.: The reduction of discrete dynamical and semidynamical systems in metric spaces. In: Six Lectures on Dynamical Systems. World Scientific (1996) 267-312.

[Robb]   Robbin, J.: A structural stability theorem. Ann. Math. 94 (1971) 447-493

[Robi1]   Robinson, C.: Structural stability of vector fields. Ann. Math. 99 (1974) 154-175

[Robi2]   Robinson, C.: Structural stability for C^1-diffeomorphisms. J. Diff. Equat. 22 (1976) 28-73

[Robi3]   Robinson, C.: Stability theorems and hyperbolicity in dynamical systems. Rocky Mount. J. of Math. 7 (1977) 425-437

[Ru]   Ruelle, D.: Thermodynamic Formalism. Encyclopedia of Math. and its Appl. 5. Addison-Wesley, Reading, MA (1978)

[Sac]   Sacker, R.J. and Sell, G.R.: A spectral theory for linear differential systems. J. Diff. Equat. 27 (1978) 320-358

[Sak1]   Sakai, K.: The C^1 uniform pseudo-orbit tracing property. Tokyo J. Math. 15 (1992) 99-109

[Sak2]   Sakai, K.: Pseudo-orbit tracing property and strong transversality of diffeomorphisms on closed manifolds. Osaka J. Math. 31 (1994) 373-386

[Sak3]   Sakai, K.: Shadowing property and transversality condition. In: Dynamical Systems and Chaos. 1. World Scientific (1995) 233–238

[Sak4]   Sakai, K.: Hyperbolic metrics of expansive homeomorphisms. Topology Appl. 63 (1995) 263–266

[Sak5]   Sakai, K.: Diffeomorphisms with the shadowing property. J. Austral. Math. Soc. 61 (1996) 396–399

[San]   Sanz-Serna, J.M. and Larsson, S.: Shadows, chaos, and saddles. Appl. Numer. Math. 13 (1993) 181–190

[Sas]   Sasaki, K.: Some examples of stochastically stable homeomorphisms. Nagoya Math. J. 71 (1978) 97–105

[Sau1]   Sauer, T. and Yorke, J.A.: Shadowing trajectories of dynamical systems. In: Computer Aided Proofs in Analysis (Cincinnati, OH, 1989). IMA Vol. Math. Appl. 28. Springer-Verlag (1991) 229–234

[Sau2]   Sauer, T. and Yorke, J.A.: Rigorous verification of trajectories for the computer simulation of dynamical systems. Nonlinearity 4 (1991) 961–979

[Saw]   Sawada, K.: Extended f-orbits are approximated by orbits. Nagoya Math, J. 79 (1980) 33–45

[Sc]   Schwartz, I.B.: Estimating regions of existence of unstable periodic orbits using computer-based techniques. SIAM J. Numer. Anal. 20 (1983) 106–120

[Shu1]   Shub, M.: Structurally stable diffeomorphisms are dense. Bull. Amer. Math. Soc. 78 (1972) 817–818

[Shu2]   Shub, M.: Global Stability of Dynamical Systems. Springer-Verlag (1987)

[Shl]   Shlyachkov, S.V.: A theorem on ϵ-trajectories for Lorenz mappings [in Russian]. Funkts. Anal. Pril. 19 (1985) 84–85

[Si]   Sinai, Ya.G. and Vul, E.B.: Discovery of closed orbits of dynamical systems with the use of computers. J. Stat. Phys. 23 (1980) 27–47

[Sl]   Slackov, S.V.: Pseudo-orbit tracing property and structural stability of expanding maps of the interval. Ergod. Theory Dyn. Syst. 12 (1992) 573–587

[Sm1]   Smale, S.: Stable manifolds for differential equations and diffeomorphisms. Ann. Scuola Norm. Sup. Pisa 17 (1963) 97–116

[Sm2]   Smale, S.: Differentiable dynamical systems. Bull. Amer. Math. Soc. 73 (1967) 747–817

[Ste1]   Steinlein, H. and Walther, H.-O.: Hyperbolic sets and shadowing for noninvertible maps. In: Advanced Topics in the Theory of Dynamical Systems (Trento, 1987). Academic Press, Boston, MA (1989) 219–234

[Ste2]   Steinlein, H. and Walther, H.-O.: Hyperbolic sets, transversal homoclinic trajectories, and symbolic dynamics for C^1 maps in Banach spaces. J. Dynam. Diff. Equat. 2 (1990) 325–365

[Sto]   Stoffer, D.: Transversal homoclinic points and hyperbolic sets for nonautonomous maps. I. Z. Angew. Math. Phys. 39 (1988) 518–549

[T1]   Thomas, R.F.: Stability properties of one-parameter flows. Proc. London Math. Soc. 45 (1982) 479–505

[T2]   Thomas, R.F.: Topological stability: some fundamental properties. J. Diff. Equat. 59 (1985) 103–122

[V]   Van Vleck, E.S.: Numerical shadowing near hyperbolic trajectories. SIAM J. Sci. Comput. 16 (1995) 1177–1189

[Wa1]   Walters, P.: Anosov diffeomorphisms are topologically stable. Topology 9 (1970) 71–78

[Wa2]   Walters, P.: On the pseudo orbit tracing property and its relationship to stability. In: The Structure of Attractors in Dynamical Systems. Lect. Notes in Math. **668**. Springer-Verlag (1978) 231–244

[We]   Wen, L.: On the C^1 stability conjecture for flows. J. Diff. Equat. **129** (1996) 334–357

[Wh]   Whitehead, J.: Manifolds and transverse fields in Euclidean space. Ann. Math. **73** (1961) 154–212

[Y1]   Yano, K.; Topologically stable homeomorphisms of the circle. Nagoya Math. J. **79** (1980) 145–149

[Y2]   Yano, K.: Generic homeomorphisms of S^1 have the pseudo-orbit tracing property. J. Fac. Sci. Univ. Tokyo, Sect.IA Math. **34** (1987) 51–55

[Z]   Zeidler, E.: Nonlinear Functional Analysis and its Applications I. Springer-Verlag (1986)

Index

(A,2)-diffeomorphism 158
absorbing set 204
adapted norm 10
Anosov diffeomorphism 31
asymptotic pseudotrajectory 84
Axiom A 146
Axiom A′ 111
Baire space 172
basic set (for a diffeomorphism) 152
basic set (for a flow) 113
Birkhoff constant 129
C^0 transversality condition 159
C^1 Δ-disk 13
chain recurrent set 238
critical point 249
d-pseudosolution 89
d-pseudotrajectory 1
(d, T)-pseudotrajectory 89
d-homogeneous pseudotrajectory 192
d-static pseudotrajectory 192
d-travelling pseudotrajectory 192
dissipative diffeomorphism 246
(ϵ)-shadowing 2
(ϵ, ϕ)-shadowing 1
(ϵ, ϕ)-tracing 1
equivalent recurrent vertices 241
expansion constant 85
expansive system 5
expansivity constant 5
exponential dichotomy 72
finite shadowing property 3
generalized Hadamard condition 192
generic property 172

geometric STC (geometric strong transversality condition) for diffeomorphisms 146

geometric STC (geometric strong transversality condition) for flows 115

global attractor 203

Hadamard condition 192

Hausdorff distance 243

homoclinic point 20

hyperbolic matrix 182

hyperbolic set (for a diffeomorphism) 10

hyperbolic set (for a flow) 91

hyperbolic structure (for a diffeomorphism) 10

hyperbolic structure (for a flow) 92

hyperbolicity constants (for a diffeomorphism) 10

hyperbolicity constants (for a flow) 92

hyperbolicity on a segment 54

inertial manifold 206

invariantly connected set 73

isolated invariant set 22

l-interval 174

$\mathcal{L}_p$-shadowing 68

$\mathcal{L}_{\bar{r},p}$-shadowing 72

lattice dynamical system 190

lift 173

LmSP (limit shadowing property) 64

LpSP (Lipschitz shadowing property) 3

LpSP$_+$ 3

local stable and unstable manifolds 12

locally maximal invariant set 22

lower semicontinuous family 118

Lyapunov norm 10

monotonous norm 35

Morse spectrum 239

Morse-Smale structure 204

nonwandering point (for a flow) 111

nonwandering point (for a homeomorphism) 26

path 241

periodic discrete (d, h)-pseudotrajectory 224

periodic Morse spectrum 239

piecewise hyperbolic family 55

property of weakly parametrized shadowing 91

property of strongly parametrized shadowing 91

POTP (pseudoorbit tracing property) 2

POTP$_+$ 2

r-interval 174

(R, g, d)-ball 56

recurrent vertex 241
reparametrization 91
residual subset 172
rotation number 173
Sacker-Sell spectrum 72
SUP (shadowing uniqueness property) 5
simple fixed point 248
simple periodic path 241
Smale space 31
spatially-homogeneous solution 191
spherical linear transformation 185
stable and unstable manifolds (for diffeomorphisms) 146
stable and unstable manifolds (for flows) 112
steady-state solution 191
Steinlein-Walther hyperbolic set 48
stochastic stability 2
strong shadowing property 70
strong Whitney topology 205
structural stability (for diffeomorphisms) 145
structural stability (for flows) 111
symbolic image 240
topological equivalence 111
topological stability 103
topologically Anosov homeomorphism 6
topology of uniform convergence on compact sets 248
transverse homoclinic point 21
travelling wave solution 191
upper Lyapunov exponent 233
weak (ϵ)-shadowing 164
WSP (weak shadowing property) 164

Vol. 1607: G. Schwarz, Hodge Decomposition – A Method for Solving Boundary Value Problems. VII, 155 pages. 1995.

Vol. 1608: P. Biane, R. Durrett, Lectures on Probability Theory. Editor: P. Bernard. VII, 210 pages. 1995.

Vol. 1609: L. Arnold, C. Jones, K. Mischaikow, G. Raugel, Dynamical Systems. Montecatini Terme, 1994. Editor: R. Johnson. VIII, 329 pages. 1995.

Vol. 1610: A. S. Üstünel, An Introduction to Analysis on Wiener Space. X, 95 pages. 1995.

Vol. 1611: N. Knarr, Translation Planes. VI, 112 pages. 1995.

Vol. 1612: W. Kühnel, Tight Polyhedral Submanifolds and Tight Triangulations. VII, 122 pages. 1995.

Vol. 1613: J. Azéma, M. Emery, P. A. Meyer, M. Yor (Eds.), Séminaire de Probabilités XXIX. VI, 326 pages. 1995.

Vol. 1614: A. Koshelev, Regularity Problem for Quasilinear Elliptic and Parabolic Systems. XXI, 255 pages. 1995.

Vol. 1615: D. B. Massey, Le Cycles and Hypersurface Singularities. XI, 131 pages. 1995.

Vol. 1616: I. Moerdijk, Classifying Spaces and Classifying Topoi. VII, 94 pages. 1995.

Vol. 1617: V. Yurinsky, Sums and Gaussian Vectors. XI, 305 pages. 1995.

Vol. 1618: G. Pisier, Similarity Problems and Completely Bounded Maps. VII, 156 pages. 1996.

Vol. 1619: E. Landvogt, A Compactification of the Bruhat-Tits Building. VII, 152 pages. 1996.

Vol. 1620: R. Donagi, B. Dubrovin, E. Frenkel, E. Previato, Integrable Systems and Quantum Groups. Montecatini Terme, 1993. Editors:M. Francaviglia, S. Greco. VIII, 488 pages. 1996.

Vol. 1621: H. Bass, M. V. Otero-Espinar, D. N. Rockmore, C. P. L. Tresser, Cyclic Renormalization and Auto-morphism Groups of Rooted Trees. XXI, 136 pages. 1996.

Vol. 1622: E. D. Farjoun, Cellular Spaces, Null Spaces and Homotopy Localization. XIV, 199 pages. 1996.

Vol. 1623: H.P. Yap, Total Colourings of Graphs. VIII, 131 pages. 1996.

Vol. 1624: V. Brînzanescu, Holomorphic Vector Bundles over Compact Complex Surfaces. X, 170 pages. 1996.

Vol.1625: S. Lang, Topics in Cohomology of Groups. VII, 226 pages. 1996.

Vol. 1626: J. Azéma, M. Emery, M. Yor (Eds.), Séminaire de Probabilités XXX. VIII, 382 pages. 1996.

Vol. 1627: C. Graham, Th. G. Kurtz, S. Méléard, Ph. E. Protter, M. Pulvirenti, D. Talay, Probabilistic Models for Nonlinear Partial Differential Equations. Montecatini Terme, 1995. Editors: D. Talay, L. Tubaro. X, 301 pages. 1996.

Vol. 1628: P.-H. Zieschang, An Algebraic Approach to Association Schemes. XII, 189 pages. 1996.

Vol. 1629: J. D. Moore, Lectures on Seiberg-Witten Invariants. VII, 105 pages. 1996.

Vol. 1630: D. Neuenschwander, Probabilities on the Heisenberg Group: Limit Theorems and Brownian Motion. VIII, 139 pages. 1996.

Vol. 1631: K. Nishioka, Mahler Functions and Transcendence.VIII, 185 pages.1996.

Vol. 1632: A. Kushkuley, Z. Balanov, Geometric Methods in Degree Theory for Equivariant Maps. VII, 136 pages. 1996.

Vol.1633: H. Aikawa, M. Essén, Potential Theory – Selected Topics. IX, 200 pages.1996.

Vol. 1634: J. Xu, Flat Covers of Modules. IX, 161 pages. 1996.

Vol. 1635: E. Hebey, Sobolev Spaces on Riemannian Manifolds. X, 116 pages. 1996.

Vol. 1636: M. A. Marshall, Spaces of Orderings and Abstract Real Spectra. VI, 190 pages. 1996.

Vol. 1637: B. Hunt, The Geometry of some special Arithmetic Quotients. XIII, 332 pages. 1996.

Vol. 1638: P. Vanhaecke, Integrable Systems in the realm of Algebraic Geometry. VIII, 218 pages. 1996.

Vol. 1639: K. Dekimpe, Almost-Bieberbach Groups: Affine and Polynomial Structures. X, 259 pages. 1996.

Vol. 1640: G. Boillat, C. M. Dafermos, P. D. Lax, T. P. Liu, Recent Mathematical Methods in Nonlinear Wave Propagation. Montecatini Terme, 1994. Editor: T. Ruggeri. VII, 142 pages. 1996.

Vol. 1641: P. Abramenko, Twin Buildings and Applications to S-Arithmetic Groups. IX, 123 pages. 1996.

Vol. 1642: M. Puschnigg, Asymptotic Cyclic Cohomology. XXII, 138 pages. 1996.

Vol. 1643: J. Richter-Gebert, Realization Spaces of Polytopes. XI, 187 pages. 1996.

Vol. 1644: A. Adler, S. Ramanan, Moduli of Abelian Varieties. VI, 196 pages. 1996.

Vol. 1645: H. W. Broer, G. B. Huitema, M. B. Sevryuk, Quasi-Periodic Motions in Families of Dynamical Systems. XI, 195 pages. 1996.

Vol. 1646: J.-P. Demailly, T. Peternell, G. Tian, A. N. Tyurin, Transcendental Methods in Algebraic Geometry. Cetraro, 1994. Editors: F. Catanese, C. Ciliberto. VII, 257 pages. 1996.

Vol. 1647: D. Dias, P. Le Barz, Configuration Spaces over Hilbert Schemes and Applications. VII. 143 pages. 1996.

Vol. 1648: R. Dobrushin, P. Groeneboom, M. Ledoux, Lectures on Probability Theory and Statistics. Editor: P. Bernard. VIII, 300 pages. 1996.

Vol. 1649: S. Kumar, G. Laumon, U. Stuhler, Vector Bundles on Curves – New Directions. Cetraro, 1995. Editor: M. S. Narasimhan. VII, 193 pages. 1997.

Vol. 1650: J. Wildeshaus, Realizations of Polylogarithms. XI, 343 pages. 1997.

Vol. 1651: M. Drmota, R. F. Tichy, Sequences, Discrepancies and Applications. XIII, 503 pages. 1997.

Vol. 1652: S. Todorcevic, Topics in Topology. VIII, 153 pages. 1997.

Vol. 1653: R. Benedetti, C. Petronio, Branched Standard Spines of 3-manifolds. VIII, 132 pages. 1997.

Vol. 1654: R. W. Ghrist, P. J. Holmes, M. C. Sullivan, Knots and Links in Three-Dimensional Flows. X, 208 pages. 1997.

Vol. 1655: J. Azéma, M. Emery, M. Yor (Eds.), Séminaire de Probabilités XXXI. VIII, 329 pages. 1997.

Vol. 1656: B. Biais, T. Björk, J. Cvitanic, N. El Karoui, E. Jouini, J. C. Rochet, Financial Mathematics. Bressanone, 1996. Editor: W. J. Runggaldier. VII, 316 pages. 1997.

Vol. 1657: H. Reimann, The semi-simple zeta function of quaternionic Shimura varieties. IX, 143 pages. 1997.

Vol. 1658: A. Pumarino, J. A. Rodrıguez, Coexistence and Persistence of Strange Attractors. VIII, 195 pages. 1997.

Vol. 1659: V, Kozlov, V. Maz'ya, Theory of a Higher-Order Sturm-Liouville Equation. XI, 140 pages. 1997.

Vol. 1660: M. Bardi, M. G. Crandall, L. C. Evans, H. M. Soner, P. E. Souganidis, Viscosity Solutions and Applications. Montecatini Terme, 1995. Editors: I. Capuzzo Dolcetta, P. L. Lions. IX, 259 pages. 1997.

Vol. 1661: A. Tralle, J. Oprea, Symplectic Manifolds with no Kähler Structure. VIII, 207 pages. 1997.

Vol. 1662: J. W. Rutter, Spaces of Homotopy Self-Equivalences – A Survey. IX, 170 pages. 1997.

Vol. 1663: Y. E. Karpeshina; Perturbation Theory for the Schrödinger Operator with a Periodic Potential. VII, 352 pages. 1997.

Vol. 1664: M. Väth, Ideal Spaces. V, 146 pages. 1997.

Vol. 1665: E. Giné, G. R. Grimmett, L. Saloff-Coste, Lectures on Probability Theory and Statistics 1996. Editor: P. Bernard. X, 424 pages, 1997.

Vol. 1666: M. van der Put, M. F. Singer, Galois Theory of Difference Equations. VII, 179 pages. 1997.

Vol. 1667: J. M. F. Castillo, M. González, Three-space Problems in Banach Space Theory. XII, 267 pages. 1997.

Vol. 1668: D. B. Dix, Large-Time Behavior of Solutions of Linear Dispersive Equations. XIV, 203 pages. 1997.

Vol. 1669: U. Kaiser, Link Theory in Manifolds. XIV, 167 pages. 1997.

Vol. 1670: J. W. Neuberger, Sobolev Gradients and Differential Equations. VIII, 150 pages. 1997.

Vol. 1671: S. Bouc, Green Functors and G-sets. VII, 342 pages. 1997.

Vol. 1672: S. Mandal, Projective Modules and Complete Intersections. VIII, 114 pages. 1997.

Vol. 1673: F. D. Grosshans, Algebraic Homogeneous Spaces and Invariant Theory. VI, 148 pages. 1997.

Vol. 1674: G. Klaas, C. R. Leedham-Green, W. Plesken, Linear Pro-p-Groups of Finite Width. VIII, 115 pages. 1997.

Vol. 1675: J. E. Yukich, Probability Theory of Classical Euclidean Optimization Problems. X, 152 pages. 1998.

Vol. 1676: P. Cembranos, J. Mendoza, Banach Spaces of Vector-Valued Functions. VIII, 118 pages. 1997.

Vol. 1677: N. Proskurin, Cubic Metaplectic Forms and Theta Functions. VIII, 196 pages. 1998.

Vol. 1678: O. Krupková, The Geometry of Ordinary Variational Equations. X, 251 pages. 1997.

Vol. 1679: K.-G. Grosse-Erdmann, The Blocking Technique. Weighted Mean Operators and Hardy's Inequality. IX, 114 pages. 1998.

Vol. 1680: K.-Z. Li, F. Oort, Moduli of Supersingular Abelian Varieties. V, 116 pages. 1998.

Vol. 1681: G. J. Wirsching, The Dynamical System Generated by the 3n+1 Function. VII, 158 pages. 1998.

Vol. 1682: H.-D. Alber, Materials with Memory. X, 166 pages. 1998.

Vol. 1683: A. Pomp, The Boundary-Domain Integral Method for Elliptic Systems. XVI, 163 pages. 1998.

Vol. 1684: C. A. Berenstein, P. F. Ebenfelt, S. G. Gindikin, S. Helgason, A. E. Tumanov, Integral Geometry, Radon Transforms and Complex Analysis. Firenze, 1996. Editors: E. Casadio Tarabusi, M. A. Picardello, G. Zampieri. VII, 160 pages. 1998.

Vol. 1685: S. König, A. Zimmermann, Derived Equivalences for Group Rings. X, 146 pages. 1998.

Vol. 1686: J. Azéma, M. Émery, M. Ledoux, M. Yor (Eds.), Séminaire de Probabilités XXXII. VI, 440 pages. 1998.

Vol. 1687: F. Bornemann, Homogenization in Time of Singularly Perturbed Mechanical Systems. XII, 156 pages. 1998.

Vol. 1688: S. Assing, W. Schmidt, Continuous Strong Markov Processes in Dimension One. XII, 137 page. 1998.

Vol. 1689: W. Fulton, P. Pragacz, Schubert Varieties and Degeneracy Loci. XI, 148 pages. 1998.

Vol. 1690: M. T. Barlow, D. Nualart, Lectures on Probability Theory and Statistics. Editor: P. Bernard. VIII, 237 pages. 1998.

Vol. 1691: R. Bezrukavnikov, M. Finkelberg, V. Schechtman, Factorizable Sheaves and Quantum Groups. X, 282 pages. 1998.

Vol. 1692: T. M. W. Eyre, Quantum Stochastic Calculus and Representations of Lie Superalgebras. IX, 138 pages. 1998.

Vol. 1694: A. Braides, Approximation of Free-Discontinuity Problems. XI, 149 pages. 1998.

Vol. 1695: D. J. Hartfiel, Markov Set-Chains. VIII, 131 pages. 1998.

Vol. 1696: E. Bouscaren (Ed.): Model Theory and Algebraic Geometry. XV, 211 pages. 1998.

Vol. 1697: B. Cockburn, C. Johnson, C.-W. Shu, E. Tadmor, Advanced Numerical Approximation of Nonlinear Hyperbolic Equations. Cetraro, Italy, 1997. Editor: A. Quarteroni. VII, 390 pages. 1998.

Vol. 1698: M. Bhattacharjee, D. Macpherson, R. G. Möller, P. Neumann, Notes on Infinite Permutation Groups. XI, 202 pages. 1998.

Vol. 1699: A. Inoue, Tomita-Takesaki Theory in Algebras of Unbounded Operators. VIII, 241 pages. 1998.

Vol. 1700: W. A. Woyczyński, Burgers-KPZ Turbulence, XI, 318 pages. 1998.

Vol. 1701: Ti-Jun Xiao, J. Liang, The Cauchy Problem of Higher Order Abstract Differential Equations. XII, 302 pages. 1998.

Vol. 1702: J. Ma, J. Yong, Forward-Backward Stochastic Differential Equations and Their Applications. XIII, 270 pages. 1999.

Vol. 1703: R. M. Dudley, R. Norvaiša, Differentiability of Six Operators on Nonsmooth Functions and p-Variation. VIII, 272 pages. 1999.

Vol. 1704: H. Tamanoi, Elliptic Genera and Vertex Operator Super-Algebras. VI, 390 pages. 1999.

Vol. 1705: I. Nikolaev, E. Zhuzhoma, Flows in 2-dimensional Manifolds. XIX, 294 pages. 1999.

Vol. 1706: S. Yu. Pilyugin, Shadowing in Dynamical Systems. XVII, 271 pages. 1999.

Vol. 1707: R. Pytlak, Numerical Methods for Optical Control Problems with State Constraints. XV, 215 pages. 1999.